Advances in Plastination Techniques

Nicolás E. Ottone

Advances in Plastination Techniques

 Springer

Nicolás E. Ottone
Laboratory of Plastination and Anatomical Techniques, Dental School-Facultad de
Odontología (Department of Integral Adults Dentistry), Center of Excelence in
Morphological and Surgical Studies (CEMyQ)
Universidad de La Frontera
Temuco, Chile

ISBN 978-3-031-45703-6   ISBN 978-3-031-45701-2   (eBook)
https://doi.org/10.1007/978-3-031-45701-2

This Springer imprint is published by the registered company Springer Nature Switzerland AG
The registered company address is: Gewerbestrasse 11, 6330 Cham, Switzerland

Paper in this product is recyclable.

Preface

When I invented Plastination in 1977, I never could have imagined its present global success, 45 years later. I hope you will continue your hard work, innovation and dedication to democratizing anatomy.
    Prof. Gunther von Hagens, inventor of Plastination

These words were expressed by Professor Gunther von Hagens, inventor of Plastination, during the award ceremony celebrating his career on July 20, 2022, during the 20th International Conference on Plastination. As President of that Conference, the first International Conference on Plastination held in South America, I was greatly honored to present the distinction to Professor Gunther von Hagens in person at the Plastinarium in Guben, Germany. The award recognized his outstanding career and his invaluable and notable contributions to the field of anatomy and morphological sciences, including the revolutionary creation of "Plastination", a technique that broke established standards and marked a great advance and a historical milestone in the teaching, research, and experience of anatomy.

Plastination, 46 years after its invention, has become the most important anatomical preservation technique. In this sense, at present, Plastination must be considered necessary and fundamental in an era in which the donation of human bodies for science, similar to the acquisition and donation of animal bodies, presents challenges of an ethical and moral nature that must be overcome now. Medical study with real bodies is of supreme relevance due to the transcendental anatomical knowledge that is acquired and that can be applied in clinical and surgical practice, ensuring the development of more accurate diagnoses and the application of much more precise treatments. For this reason, the possibility of preserving biological material for an indefinite period, a unique characteristic of Plastination, ensures the availability of bodies for teaching anatomy, a fundamental stone in the basic and essential training of professionals in human and veterinary health sciences.

In addition, with greater relevance especially in the last 10 years, Plastination occupies an increasingly important space in the field of anatomical research and morphological sciences, with the notable development of sheet plastination techniques (or, in our case, what we also call microplastination). These techniques allow the study of microanatomy and histology from large, macroscopic samples cut below 200 μm. The visualization of anatomy in such thin slices is a notable contribution to scientific research, thus allowing access to anatomical sites never before explored or visualized directly. The sheet plastination or microplastination allows visualization of the anatomy in its real structure, intact, without collapse, perfecting a sectional anatomical visualization that could previously only be achieved with CT, MRI, or ultrasound.

The aim of this book is to provide a clear and concrete description of Plastination techniques, including variants developed in our Laboratory. Additionally, it introduces the history of anatomical techniques and plastination, discusses the applications of plastination in teaching and research, and covers relevant ethical and biosafety details when deciding to use this anatomical technique. Likewise, in each chapter we sought to include most of the scientific publications on plastination made by researchers from around the world, and available in open and massive access databases, from 1977 (the year of the invention of Plastination) to the present. In this way, we sought to review and reference each plastination technique in detail, comparing our laboratory's experience with the experiences of other anatomists, plastinators, and researchers, with the aim of giving recognition to those who contributed to the development of plastination techniques throughout the world.

Before concluding, I would like to take the opportunity to express a very special recognition to the great professors who have accompanied me from my beginning in anatomy at University of Buenos Aires, Argentina, more than 20 years ago, to the present day at Universidad de La Frontera, Temuco, Chile (in alphabetical order): Prof. Santiago Aja Guardiola (†), Prof. Rubén Daniel Algieri, Prof. Carlos Baptista, Prof. Vicente Bertone, Prof. Homero Bianchi (†), Prof. Esteban Blasi, Prof. Horacio Conesa (†), Prof. Mariano del Sol, Prof. Ramón Fuentes, Prof. Telma Masuko, Prof. Carlos Medan. In addition to them, as I will indicate in Chap. 2 of this book, other colleagues and current professors also participated in works, congresses, courses on anatomical techniques, and plastination workshops, among other activities. I would like to recognize them for their generous collaboration and express my gratitude to many of them for their friendship. To end this section, I would like to especially acknowledge Professors Mariano del Sol and Ramón Fuentes, who stand behind the possibility of developing an idea from scratch into a reality: the establishment of the Plastination Laboratory of Universidad de La Frontera, in Temuco, Chile. Our laboratory is not only where anatomical and morphological research is promoted but also where the training of undergraduate, postgraduate, academic, and technical students is promoted, becoming a true benchmark on training facility in plastination and anatomical techniques.

With this book, I hope, with great humility, to contribute to the development of Plastination and to the empowerment of a new awakening of anatomical techniques. Based on experiences obtained in our laboratory, this book will nourish this development with the experiences in Plastination developed by other researchers worldwide. I hope that this book can constitute a concrete contribution, understanding that the search for knowledge is just a beginning. We must all contribute to massifying and popularizing the free and open communication of science, allowing academic discussion to develop without any type of restriction in all fields. In particular, the renewed development of all anatomical conservation techniques follows the supreme objective, and here I paraphrase Professor Gunther von Hagens, to democratize, by all possible means, the study, research, and academic development of anatomy and morphological sciences through Plastination.

Science is one of the highest forms of spiritual activity because it is linked to the creative activity of the intellect, the supreme form of our human condition.
    Dr. René G. Favaloro (1923–2000), Father of coronary artery bypass grafting, on the hundredth anniversary of his birth

Temuco, Chile   Nicolás E. Ottone

Contents

About the Author

Nicolás E. Ottone was born in Buenos Aires, Argentina, and holds a Medical Doctor degree from the Faculty of Medicine, University of Buenos Aires (UBA), Argentina, and a Ph.D. in Morphological Sciences from the Faculty of Medicine, Universidad de La Frontera (UFRO), Temuco, Chile. He is currently an Associate Professor at Dental School, Universidad de La Frontera (UFRO). While in Medical School, he made connections with laboratories for scientific research. This drove him to join the Dissection Team of the Second Chair of Anatomy (EDSCA) and the Museum of Anatomy of the former Institute of Morphological Sciences J. J. Naón during his second year of Medical School, in 2002, until graduation. During this experience in Argentina, he improved his skills in human dissection, clinical anatomy, and anatomical techniques, with an incipient and persistent development of plastination techniques. In Chile, he obtained his Ph.D. in Morphological Sciences from Universidad de La Frontera and gained extensive knowledge and experience in anatomical techniques, histological techniques, and stereological analysis, particularly in plastination. Utilizing this knowledge, he established and founded the Plastination and Anatomical Techniques Laboratory at UFRO and made notable advancements in enhancing plastination techniques. He is the current Director of the Plastination and Anatomical Techniques Laboratory, the Director of the Diploma in Advanced Anatomical Techniques for the Conservation of Biological Material with Applications in Teaching and Morphological Research, and the Director of the International Workshop on Plastination and Anatomical Techniques. Using his expertise in anatomy, plastination, and morphological sciences, he published over 80 articles indexed in the WoS database and received multiple research grants. In addition to being a speaker at over 70 conferences on plastination, anatomical techniques, and clinical anatomy, he also organized courses on these subjects and international conferences. In this way, he served as President at the XX Congress of Anatomy of the Southern Cone, in Pucón, Chile (2018), as President of the 20th International Conference on Plastination, the first time it was held in a South American country (Chile,

2022), and President of the First and Fifth International Congress on Anatomical Techniques, among others. He currently holds several positions at Universidad de La Frontera. He is the current President and Member of the Appointment and Promotions Commission and has previously served as the Director of Postgraduate and Research Affairs for the Faculty of Dentistry (UFRO). Furthermore, he teaches as a Faculty Professor in both the Doctoral Program in Morphological Sciences and the Doctorate Program in Medical Sciences at UFRO. Regarding his teaching career, he is responsible for teaching the new chair of clinical anatomy for undergraduate dental course, as well as responsible for teaching the chair of plastination and anatomical techniques for postgraduate courses. He is also highly involved in various scientific societies, including serving as the Vice-President of the International Society for Plastination, General Secretary of the Pan-American Association of Anatomy, and founder member of the Society of Morphology of the Southern Cone. Also he is member of Chilean Society of Anatomy, the Argentinian Association of Anatomy, and foreign member of multiple societies, such as the Mexican Society of Anatomy, the Paraguayan Society of Morphological Sciences, the Paraguayan Society of Anatomy, and the Peruvian Society of Morphological Sciences. Also, he is a Member of the Anatomy Working Group of the Anatomical Terminology Committee (FIPAT) of the International Federation of Associations of Anatomists (IFAA). He holds several positions within editorial committees, including being an Associate Editor for the *International Journal of Morphology*, an International Associate Editor for the *Journal of Morphological Sciences*, and a Member of the Editorial Committee for the *International Journal of Odontostomatology* and the *Journal of Plastination*. Additionally, he serves as an ad hoc reviewer for multiple international journals.

Chapter 1
Brief Review of the Origins of Anatomical Techniques

The thought of eternal life, of the existence of an afterlife, has driven the progress of conservation techniques throughout history. Religion established that the body was necessary for the soul of the dead to enjoy the next life and was not excluded from it, and thus the need arose for the body to be preserved. The weather conditions in the regions where bodies were buried were fundamental to ensuring their preservation. Therefore, desiccation, by dry heat or cold, and freezing are the natural means of conservation [1–8].

The artificial techniques of cadaver conservation consist of preparing, maintaining, and protecting a specimen to keep it in its primitive state or a state similar to being alive indefinitely. This last one is of fundamental importance in those societies where there is little cadaver material for research, and limited to donation while still alive for the purposes of the interested parties. Artificial means of conservation include the application of simple heat or cold; powders, as in a bed of sawdust mixed with zinc sulfate; evisceration combined with immersion; drying; local incision and immersion, arterial injections; and cavity injections. In addition, simple immersion in alcohol, brine, etc. and single arterial injection, which can be combined with treatment of cavities and/or immersion, were also used [5–8].

Cadaveric dissection has a long history that dates back to ancient times [1]. The earliest known records of cadaveric dissection, the systematic study of anatomy, date back to ancient Greece. The Greek physician and philosopher, Hippocrates, is considered to be the father of modern medicine and was one of the first to advocate for the use of cadaveric dissection to understand human anatomy. Human dissection was first performed by the physician Herophilus in the third century BCE. Herophilus used dissection to study the human body and make important observations about the nervous system, blood vessels, and other anatomical structures [1, 5–10].

The practice of cadaveric dissection was later adopted by the Roman physician Galen, who made significant contributions to the understanding of human anatomy and physiology. Galen's work on the human body became the standard for medical knowledge in Europe for nearly 1500 years [1, 5–8, 10, 11].

© The Author(s), under exclusive license to Springer Nature
Switzerland AG 2023
N. E. Ottone, *Advances in Plastination Techniques*,
https://doi.org/10.1007/978-3-031-45701-2_1

During the Middle Ages, dissection was largely prohibited in Europe due to religious and cultural taboos surrounding the handling of dead bodies. It wasn't until the Renaissance, with the revival of classical learning and the emergence of humanism, that cadaveric dissection began to be accepted as a legitimate method of scientific inquiry [1, 2, 5–8].

It was not until the fourteenth century in Europe that the study of anatomy through dissection began to be widely practiced [5–8].

In the sixteenth century, Andreas Vesalius, a Belgian anatomist, published a groundbreaking work on human anatomy, *De humani corporis fabrica* (Of the structure of the human body), which was based on extensive dissections he conducted himself. Vesalius' work marked a turning point in the history of anatomy and helped establish the importance of cadaveric dissection in medical education [5–8].

Today, cadaveric dissection remains a crucial part of medical education, allowing students and healthcare professionals to gain an in-depth understanding of the human body's structure and function [12].

Anatomy is one of the fundamental pillars of medicine, from which doctors and professionals in all areas of health acquire knowledge indispensable for the proper development of diagnoses and safe clinical and surgical practices. The practice of human cadaver dissection began in ancient Greece, with Herophilus being considered the first dissector, and was restored officially in the Renaissance, first with Mondino de Luzzi and then with Andrea Vesalius. Therefore, the anatomical dissection practiced uninterrupted since the Renaissance constitutes the first contact with the cadaver that health science students experience on their first day in anatomy class. Therefore, this is the cornerstone, the fundamental start of the student's relationship with death, the first interaction to advance the beginning of the student's ethics training, and which consists of the unrestricted respect by the human body, for the human being who decided to donate their body to science, respect that will later be transferred to the patient during clinical practice. Moreover, the interaction with the human and animal body, in the case of veterinary sciences, will allow students to approach the three-dimensional reality of the body through the manipulation and observation of the anatomical structures; this 3D contact ensures the recognition of the true and variable human proportions, the relations between the different anatomical portions, as well as the volume, shape, and size of the organs and anatomical regions. This is the great difference in training in human or veterinary medicine and other health sciences, that contact with the real human and animal body, the value of which is incalculable, establishes in relation to the current new ways of teaching that in some institutions completely replace the cadaver, such as virtual dissection tables, virtual reality software, and artificial intelligence. In addition, some studies have established that the student's interaction with the cadaver in the anatomy laboratory, in every sense, reinforces the knowledge and, in the case of surgeons, increases confidence and precision during surgery [3, 5–8, 13–19].

From this point of view, the development and creation of various anatomical techniques of preservation and conservation of biological tissues enable their use for a long time, in some cases practically forever, of human and animal cadavers subjected to these anatomical techniques, thereby avoiding tissue decomposition. Many of these anatomical techniques seek to preserve tissues with characteristics similar to those of living tissues. Others aim to replicate and highlight different anatomical structures by combining the use of several anatomical techniques at the same time. This facilitates, for example, the tasks of cadaver dissection (e.g., fixation techniques with reduced formalin concentrations, along with vascular injection techniques that emphasize the vascular structures, facilitating their identification and dissection) [3, 5–8, 15].

Preservation can be defined as an action to keep something "safe from damage, destruction or decomposition," while conservation is defined as the process of "a careful preservation and protection of something," and finally embalming is defined as "the treatment (of a cadaver), with special chemical agents to protect it from decomposition." These definitions show that although the terms "preservation" and "conservation" can be interchangeable, different languages favor them differently. While German-speaking countries are based more often on the term "conservation" of a human body, in English, the preferred term is "preservation." However, "conservation" and "preservation" cover more than the mere embalming process, the use of chemical substances in a body. Suitable storage, protection during use, and final elimination must also be taken into consideration [2, 5–8].

Therefore, historically, and based on the respect of different religious rites, a set of practices were developed for the cadaver by developing and applying methods for cleaning, conserving, embalming, restoring, reconstructing, and aesthetic care of the cadaver, a process called "thanatopraxy" [2, 5–8].

The existence of these techniques has a history, even prehistory, such as the mammoth of Siberia, the Man of Tellunt. However, they were preserved, thanks to natural phenomena where the bodies remained practically undamaged through the intervention of an acid preparation or permafrost. That is to say, no preservatives have been found in mummies (beyond linen wraps and bandages), but instead, they are processes of natural mummification, by atmospheric physical agents, such as the structure of the land (porous and dry, with saline substances) or very dry and warm air, or frigid temperatures [5–8, 20, 21].

Mummified remains have been found from the Neolithic Period covered by a red "paste." These are essentially purification rituals rather than a conservation technique. These rituals were based on the belief in the existence of spirits, good and evil, present in everything that surrounded that civilization, whether human, plant, or animal, and the belief established that they should be kept happy to avoid their possible evil influence. This is the basic idea that drove humans in this period to develop conservation of the body, using rudimentary techniques, such as placing the cadaver on platforms or tables and subjecting the body to heat to achieve its desiccation. The next step was mummification by smoking [5–8, 20, 21].

It is in this way that the first mummified people are considered to be those found in the north of Chile and south of Peru, from the Chinchorro culture, dating back to 8000 BCE. Thus, they are seen as the first artificial embalmers, with the development of mummified people, before Egypt [22]. The environment helped the Chinchorro greatly, with a local climate in that South American region, where extreme dryness prevented the decomposition of cadavers, and was associated with an increase in population over the years, so that the dead were converted in a significant part of the landscape [2, 5–8, 22]. Max Uhle in 1917 [5–8, 23, 24] differentiated archaeological vestiges of people who did not know ceramics or agriculture, depended on marine resources, and subjected their dead to artificial mummification. These groups were called the "Aborigines of Arica," a name that later was suggested to be replaced by Quiani, Chinchorro culture, Chinchorro complex, Chinchorro tradition, and Chinchorro phase. The studies mentioned indicated that three classes of mummies can be distinguished: (1) simple mummies, (2) mummies of complex preparation, and (3) mummies completely covered with a layer of mud, 1 cm thick. It says that the first could be taken as the original type; the second as a development of the first, produced by contact with the Peruvian civilizations. The third could be considered a step back toward the simple procedures. The technique of complicated mummification would have been obtained according to the following procedure:

(a) Opening of the abdomen and complete evisceration.
(b) Dissection of the body cavity with fire and its filling with other less changeable substances; in addition, the introduction of poles in the trunk and extremities to straighten and stiffen the cadaver; and, finally, closure and sewing of the natural tegument.
(c) Reconstruction of the outside of the deceased, adding bulk to the trunk and limbs with reeds. The face was frequently covered with a mud mask painted in different ways. It was very common to add detachable hair to the mummies of babies [5–8, 23, 24].

Decades later, the analysis of the material contributed by the Morro 1 site, also in Arica, allowed the archaeologists to perfect the classification proposed by Uhle (Allison et al. in 1984, [25]), and, later, one of them, Arriaza in 1995 [26], involving the universe mummies mentioned in the literature, would complement and perfect the classification:

1. Mummies of simple treatment: natural desiccation of the bodies by the environment; extended, wrapped with mats of reeds and camelid skins; supported on their backs and some with slightly bent legs
2. Mummies of complex treatment: bodies with internal treatment, among which the following subtypes and varieties are distinguished:

    (a) Black mummies: painted externally with a layer of manganese. They were the most complex of all. The bodies were practically reconstructed; most soft tissues were removed, and the skeleton was reinforced. The face, neck, trunk, extremities, and genitals were modeled with white paste and then completely painted black. The treatment that produced the black mummies was not the same in every case. These mummies have been reported for five sites between Arica and Camarones.
    (b) Red mummies: the bodies were not disassembled with the same intensity as the black mummies; in most cases, the skin was not completely removed; the bones of the arms and legs were not scraped, and, apparently, the body

was not completely dismembered. The body was filled with a mixture of materials: white or black paste, camelid wool, feathers, grass, animal and bird skins, and earth. The body was modeled and covered with red paint (iron oxide). The face was left black, but other colors, such as reddish-black, can also be found. As part of the process, incisions were made in the shoulders, wrists, abdomen, and groin, which were later stitched using threads of human hair or reed fibers. This contrasts with the black mummies, where neither incisions nor sutures have been observed. Generally, the treatment of the head of the red mummies was more complex than that of the black ones. This subtype of mummies has been found in Arica, Camarones, Punta Pichalo, Bajo Molle, and Patillos. Ring-shaped cranial deformation appears among them. Chronologically, they are found between 2570 and 2090 BCE.

(c) Mummies in bandages: the procedure combines those used for the red and black mummies. The body was completely skinned, and longitudinal poles reinforced it. After this, strips of animal leather or human skin, around 2 cm wide, were used to wrap the body. The head and face were treated as the red mummies and the bodies painted red. This subtype is documented for three infants and one adult who only had the legs bandaged. It is believed that they were contemporaries of the red mummies.

(d) Mud-covered mummies: the bodies were smoked and then covered with sand, clay, and a protein ligament. Two variants are recognized: one with evisceration and the other without. In the case of the eviscerated ones, the cavities were stuffed with ash. Then, they were prepared in the same place where they were buried, adhered to the floor of the grave. All are recorded for Arica, and it is assumed that those without evisceration are later.

In New Guinea, the preservation process was carried out with resins; the pre-Columbians covered the cadaver with oils and waxes; the Incas placed the body in dry, cold caves; and the ancient Guanche, to all this, added techniques for extracting the viscera and stuffing with straw, ground pumice stone or resins [2, 5–8, 27].

The Egypt of the Pharaohs is where the empirical approach of true anatomical conservation techniques was recovered, with preservation improving mainly from the first dynasty in 3200 BCE. An example of this is the large number of human and animal mummies made by the Necropolis of the Nile. At that time, and as we established at the beginning, it was a religious matter: the individual's postmortem survival could not be ensured if the body did not last or at least the image. There were true specialists in several categories, teachers and officials with their servants, not only for conservation of the body in general but also specialized in conserving specific anatomical structures: viscera, head, thorax, etcetera [1, 2, 5–8, 28].

These specialists carried out the preservation tasks, many of them being key members of the priest caste. Embalming fell into various categories; it had to be paid in advance, and there were even "insurance policies for mummification." The operations took place in the "House of the Dead" [1, 2, 5–8].

But there were milestones that marked the transition between using natural conservation means and using sophisticated embalming protocols. In this sense, one milestone was the use of additional substances such as grass, cedar oils, natron, natural resins derived from trees, incense, pitch, and tar [1, 2, 5–8]. And another fundamental milestone established the beginning of the development of direct interventions on the cadaver through exenteration or evisceration, a milestone that characterized the preservation of human remains during the following millennia [1, 2]. In addition, Brenner [2], citing Baadsgaard et al. [29], indicates the existence of tests toward the end of the early dynastic phase (2500 BCE), in which cadavers buried in the royal cemetery of Ur were preserved using heat and mercury.

In the most expensive, first-class embalming, prayers were repeated during the work, the same that Horus, Anubis, Thoth, and other deities had pronounced during the embalming of Osiris, which gave the guarantee of finding happiness. The priests followed the general process to preserve the bodies: they submerged the cadaver in a natron bathroom (soda carbonate) and salt for 30 days to achieve lixiviation (lixium = lye) [30]. Then, they removed it, placed it with the head facing south, and extracted the brain and cerebellum with hooks after breaking the nasal roof, fracturing the ethmoid and sphenoid bones. Next, they removed the thoracoabdominal viscera using longitudinal incisions on one side of the body. They filled the cavities (trunk and head) with myrrh, cassia, various aromatic resins, and oils. The openings were sewn, and all of which was done while reciting prayers and religious formulas. Next, they again submerged the cadaver in a natron bath with salt for 60 days, in special tanks [1, 2, 5–8].

The body left the tank with an extraordinary loss of muscle mass and fat. Then, they dried the body, filled the abdomen with wood sawdust soaked in aromatic essences, and smeared the skin with resins, cedar oil, and cinnamon essence. The arms were bandaged, crossing the body and joining the legs. The bandages were submerged in rubber or bitumen (from which the wordy mummy is derived); after positioning them and consecrating each bandage to a divinity, the body was sprinkled, at the same time, with pieces of natron, aromatic herbs, and palm oil, placing rings and the sacred beetle. In addition, between the turns of bandages, they put papyrus leaves with copies of passages from the "Book of the Dead," while the priest prayed a suitable prayer and, at the end, gave instructions into the ear of the deceased, so that they had a good journey to Osiris. The viscera were placed in the "Canopi" or containers next to the cadaver, in the coffin, with the Book of the Dead as a guide. The coffin was double, and on the upper cover, it had the image of the deceased, smiling and happy, reflecting the happiness obtained in the other world with this "version of luxury" [1, 5–8, 28, 31, 32].

In the second category, cheaper than the first, after the first bath dissolved the viscera with special oils, all the natural cavities and orifices were plugged. Then, the cadaver was dried and bandaged. And in the third category, cheapest of all, multiple washings took place in brine. The body was marinated for only 30 days in natron, then dried and bandaged, and buried standing up, without a coffin [1, 5–8, 28, 31, 32].

Moreover, the development of portraits of the deceased was meant to replace their decomposition and, essentially, improve techniques to preserve the body from putrefaction for as long as possible. Mummification is not applied to humans exclusively, since there are thousands of feline mummies of Bubastis, Hierapolis, and the sacred bulls of the Theban Necropolis. Besides being privileged images of divinity, the animals benefitted from the immortality of the deceased conferred by mummification after death. The Egyptian climate, characterized by great drought, has favored this type of conservation [5–8].

Conservation by the Egyptians can be defined as "permanent," in that they developed great art and skill, with mummies like Rameses II, Seti, Sesostris, and Hatshepsut being discovered, maintaining their shape and volume, and even the eyebrows, and only certain desiccation of the factions [5–8].

The Persians used wax to cover the cadaver completely. Ethiopians did so with transparent rubber. The pre-Hellenic Greeks, to express respect for their dead, did not use preservation, but cremation, although in the case of Alexander the Great, his body was preserved in a honey-based solution, the characteristic of which was the need to transfer the body a long distance and for a long time, from Babylon to Alexandria [1, 2, 5–8, 31]. In addition, and sometimes, copying the Persians, they wrapped the body with wax. Conservation that they could carry out on the cadaver was simply to have more time to render funeral honors or transfer it. The ancient Romans, like the Greeks, washed the cadaver, perfumed it, dressed it in their clothes in full dress, and exhibited them on the road so relatives, friends, and the rest of the population could see them and say goodbye. During the Republic, cremation was ordered, and the ashes were placed in columbaria and graves. On the other hand, Brenner [2], citing Longrigg [11], indicates there is no certainty whether the Ptolemaic "anatomists" (in the first half of the third century BCE), like Herophilus of Chalcedon and Erasistratus of Ceos, considered the first "dissectors," used embalming techniques.

Herophilus of Chalcedon (Bithynia, 335–280 BCE) was a Greek doctor identified as the Father of Scientific Anatomy. He is also considered the first anatomist since, although many believe the Renaissance was when the dissection of human bodies began, it was in the School of Alexandria (third century BCE) when cadaver dissection began. Herophilus belonged to this school and was the first to perform systematic anatomical dissections of human bodies in public, and for that he is identified as the Father of Scientific Anatomy. With this systematic practice of cadaver dissection, he established the foundations of anatomy. Erasistratus of Ceos participated with him. From this follows an endless number of discoveries associated with Herophilus, such as the torcular Herophili (venous confluence in the brain). Herophilus also performed dissections and vivisections on animals to study the brain, describing the meninges, choroid plexus, the fourth ventricle, and the cerebral sinuses [1, 5–8].

With respect to the development of anatomy and anatomical techniques in India, according to Lain Entralgo [33], in the history of Hindu anatomical knowledge, three stages were recognized: the first stage, "Vedic," where the body parts were enumerated in an orderly way. This was done during rituals to expel demons. The second stage was represented by animal sacrifices, where the animal was sectioned and its organs were offered to the gods according to different rituals; they were named as they were extracted. The third stage led to the dissection of human cadavers [5–8, 33].

The development of anatomical techniques can be associated with this Ayurvedic anatomy, which used dissection: the cadaver was emptied of its contents, placed in a drawer, submerged in river water for 7 days, and surrounded by herbs or hemp so animals did not devour it. When "the meats" became soft and separable, the doctor opened the skin with a rod and examined the organs [5–8, 33].

At the time of "Sushruta," cremation was obligatory for all who died after 2 years of age. Therefore, the bodies of children were the only source of dissection Hindu doctors had; this explained why it was believed that the body had 300 bones. When arriving at the post-Vedic period, the literary texts of Sushruta, Charaka, and Bhela were based on the Vedas and consolidated the long earlier tradition of knowledge [5–8, 33].

The Middle Ages were considered a period of great obscurantism with respect to anatomical practice, with a certain paralysis and even cultural regression; however, conservations were carried out, but not usually, but rather on kings or tycoons. The study of medicine was mainly developed in monasteries, based on the transmission and interpretation of the work of Galen. From the twelfth century on, the Church did not prohibit human dissection specifically, and certain edicts were generated that defined the anatomical practices that could be undertaken [5–8, 34]. A misinterpreted policy was that of the Council of Tours (1163), when Pope Alexander III prohibited priests from participating in studies of a physical nature, identifying this policy as *Ecclesia abhorret a sanguine* ("the Church abhors blood"), but it was not meant to prohibit the study of anatomy or surgery [5–9].

During this period, everything related to "the material of things" was unimportant. Given that material things are temporary, the human body was not studied. Anatomical dissection was considered blasphemy and prohibited. Instead, bodies were eviscerated, and the cavities filled with resins, in the classical method, roughly imitating Egyptian practices [5–8].

With the edict of Theodosius in 392 CE, which vetoed all non-Christian rites, Egypt lost its institutions and, with them, the official status of embalming. In the low Middle Ages and in the heart of the Roman Catholic Church, the practice of embalming princes and monks was confirmed: the integrity of the body became the conservation of memory, the ever-living presence of the moral significance of a character. The institutionalization of Catholicism consolidated these funeral rites that anticipated the exhibition of the body and the search for methods that avoided their decomposition [1, 5–8].

The techniques did not present the dedication or care that the Egyptian masters placed on it. Nevertheless, evisceration of the body took place, with its immersion in alcohol and insertion of conservative herbs through incisions, finally wrapped the bodies with waxed or tarred fabrics, as described by Petrus Forestus (1522–1597) and Ambroise Pare (1510–1590) (Tompsett, Brenner, 2014). An example of this was the embalming of the kings of France before being buried in Saint-Denis. The destiny of the Count of Guesalin is an example of the rudimentary technique of this period: he died in Anverne, France, and was buried in Saint-Denis (Paris) after having been possibly boiled previously to avoid premature decomposition [1, 5–8].

Mondino de Luzzi (1270–1326) made the first official human dissection in the medieval period. He was an Italian medical anatomist and surgeon from Bologna. Being the first to perform anatomical dissections in the Middle Ages, he is considered the restorer of anatomy; he also developed the first modern anatomical text and reinitiated public dissections of human bodies. Human dissections had not occurred since the Alexandrian Era, i.e., 1500 years earlier. This was done at the University of Bologna, Italy, by Mondino of Luzzi, who published his results in 1315, and this publication is considered a true manual of dissections, with the development, for the first time, of a text that "showed" the anatomy, different from the existing classical descriptions until that time, "showing" the anatomy of the dissection and not the description [1, 5–8, 35].

Mondino de Luzzi (1270–1326), also known as Mundinus, was an Italian physician and anatomist who is known for his contributions to the field of human anatomy. He was born in Bologna, Italy, and received his medical education at the University of Bologna. Mondino de Luzzi is credited with being the first anatomist to perform human dissections in the Western world since the time of Galen, a Greek physician who lived in the second century. He was able to obtain permission from the authorities to perform dissections on human cadavers, which were previously only allowed on animal specimens [1, 5–8, 35].

In 1316, Mondino published his seminal work, *Anathomia Corporis Humani* (Anatomy of the Human Body), which was the first comprehensive textbook on human anatomy in the Western world. The book was written in Latin and included detailed descriptions of the bones, muscles, and organs of the human body, as well as illustrations to accompany the text [1, 5–8, 35].

Mondino's work was widely influential and was used as a standard textbook on human anatomy for over 200 years. His anatomical discoveries and methods of dissection greatly advanced the understanding of human anatomy and paved the way for further advancements in the field of medicine [1, 5–8, 35].

Despite the fact that some of Mondino's findings were later proven to be inaccurate, his work laid the foundation for the study of human anatomy and helped establish the field as a scientific discipline. He is widely regarded as one of the pioneers of modern anatomical study, and his contributions have had a lasting impact on the field of medicine [1, 5–8, 35].

Mondino died in 1326, but his legacy continued to influence the field of anatomy for centuries to come. His work laid the foundation for the study of human anatomy, and his methods of dissection and observation are still used in medical education today. Mondino's work helped to establish the importance of empirical observation and dissection in medical education, and paved the way for later anatomists such as Andreas Vesalius [1, 5–8, 35].

The achievements of antiquity began to reappear in Europe and became a model in all the fields of science. It shaped the humanist worldview and inspired later generations of scientists and artists. Humanism is manifest in the interest of humanity: its issues, daily life, and spiritual and corporal beauty. The Renaissance painters showed the beauty of the human body, which in the Middle Ages was painted wrapped in robes. Leonardo da Vinci carefully studied anatomy and proportions. The naked ones even enter into sacred paintings. Cadaver conservation lost its religious nature, preparing for life in the hereafter, adopting a social, political, cultural, and scientific character, beginning to be necessary for education and research. Impregnation with "balsams," the origin of the embalming, ceased to be used, entering the real period of conservation, looking for a prolonged or indefinite duration of the cadaver or treated anatomical structure [1, 5–8].

The use of chemical substances introduced into the circulation by an intravascular current, similar to what the body itself uses while still alive, is the system used, with several variants, using those substances with a strong fixation power of organic matter without altering the relations of the organs or their normal macroscopic appearance. This way, anatomical injection techniques were also used, but in hollow structures (Brenner, 2014). For example, Leonardo da Vinci (1452–1519) developed an injection fluid based on turpentine, lavender oil, camphor, vermilion, rosin, sodium nitrate, and potassium nitrate [2, 5–8, 36].

In addition, after Mondino de Luzzi in 1315, Leonardo da Vinci (1452–1519) seems to have been one of the first artists of the Renaissance to make systematic topographic dissections and serial sections to illustrate the structure of the body [1]. In the words of F. J. Cole [1], this "modern biologist disguised as a medieval artist" was the first to give a representation of the cerebral ventricles using the injection of a solidification substance (wax), and also prepared a wax mold of the heart of which he made a model to determine the blood flow within the organ [2, 5–8].

Five hundred and nine years after his birth, Andreas Vesalius is considered a before and after in anatomy, defining anatomy as it is known today. Vesalius studied medicine at the Universities of Paris and Louvain (1533–1536) and graduated cum laude with a doctorate from the University of Padua (1537) at 23 years of age. He was immediately named a Professor of Anatomy and Surgery at that university. He stood out for performing human anatomical dissections himself, and this way radically changed anatomy teaching, being against the habitual practice for the time in which the dissections were done by barbers, and generally on animals. Thus, December 6, 1537, can be identified as the first anatomy lesson by Vesalius, as this was the date of his first anatomical dissection, avoiding the reading of classic texts (mainly Galen). Throughout his investigations, he followed the principle of not accepting any authority but his own eyes [1], as indicated by Romero-Reverón [37], direct observation was for Vesalius the only trustworthy source for teaching and learning anatomy, breaking with medieval practice, which consisted essentially of texts. The fundamental work by Vesalius, *De Humani Corporis Fabrica*, written between 1539 and 1542 and published in 1543, is unique as a dissection manual and anatomy textbook both for its clarity of presentation and perfection in the illustrations [1, 37]. When the lack of material threatened to limit his research, he risked his life and reputation with his colleagues by removing cadavers from the gallows of Montfauçon and the tombs of the Holy Innocents' Cemetery in Paris. The result was that he was able to demonstrate that Galen had frequently based his observations on findings in mammals other than humans [1, 37]. This incunabulum was needed to refute Galen's doctrine, which had predominated for over 12 centuries. Also, Vesalius was supported in the use of anatomical lamina, in turn stimulating his students to create their own laminae from the practice of dissection. Thus, the development of human dissections and the knowledge stored in *De Humani Corporis Fabrica* made it possible to rewrite anatomy and to improve the dominant anatomical diagrams of the time, with Vesalius being considered the founder of modern anatomy [5–8, 37].

Continuing with other anatomists who developed intravascular injections of hollow structures, these include Bartolomeo Eustachi (1520–1574), who performed injections with warm ink; Reinier de Graaf (1641–1673), who added mercury to his injections; and Jan Swammerdam (1637–1680), who developed a solidification substance similar to wax with the capacity for later hardening [2, 5–8, 38].

Also, as Tompsett [1] notes, this Renaissance movement motivated Paracelsus (1490–1541) to publicly burn works by Galen and Avicenna in Basel as a protest against the blind acceptance of the authenticity of antiquity. Linked to this renaissance of knowledge, interest was aroused in natural history and the foundation of museums, with an objective not directly associated with teaching but controversial and with the impact that the preserved samples could cause the general public.

In Holland, Frederik Ruysch (1638–1731), who used paraffin and cinnabar, created a museum of anatomical pieces. To preserve his anatomical samples in bottles, he used balsamic liquor, injected vascularly [1, 2, 5–8, 38]. The museum created by Ruysch, called the "Cabinet of Curiosities," preserved many anatomical samples, which were later sold to Peter the Great. According to a study by Driessen-Van het Revé in 2006, cited by Brenner [2], the balsamic liquor or fixation liquid used by Ruysch consisted of clotted pig's blood, Berlin blue, and mercury oxide. Ruysch's museum contained more than 1300 anatomical preparations, of great preservation quality. Unlike other researchers, however, Ruysch kept his conservation techniques a secret [1, 5–8].

Frederik Ruysch (1638–1731) was a Dutch anatomist, botanist, and a leading figure in the fields of anatomical preparations and natural history collecting during the late seventeenth and early eighteenth centuries. He is particularly known for his pioneering work in preserving and displaying anatomical specimens using a technique he developed involving injecting the circulatory system with colored wax [1, 5–8].

Ruysch began his career as a botanist and was appointed as a professor of botany and anatomy at the University of Amsterdam in 1667. However, he soon became more interested in anatomy and began to focus on the study of the human body. He was known for his meticulous dissections and his innovative preservation techniques, which allowed him to create stunning and lifelike anatomical specimens that were widely admired [1, 5–8].

In addition to his work in anatomy, Ruysch was also a passionate natural historian and collector. He amassed a vast collection of botanical and zoological specimens, as well as human anatomical preparations. His collection became famous throughout Europe and attracted many prominent visitors, including Peter the Great of Russia and the philosopher Gottfried Wilhelm Leibniz [1, 5–8].

Despite his accomplishments, Ruysch was also a controversial figure. He was accused of using corpses that had been stolen from graves for his anatomical preparations, and his critics claimed that his methods were unscientific and relied too heavily on artistic embellishment. However, his contributions to the field of anatomy and his innovative preservation techniques are still recognized and celebrated today [1, 5–8].

This greatly influenced the formation of anatomy museums throughout Europe and America. Guibert, who published a treatise on techniques, used salt, vinegar, orpiment, and cinnabar. Pérez Fadrique also developed this technique. Swadermann used the bath prior to the substance injection. Debils used alcoholic mixtures and vinegar. Chaissier based his technical on evisceration, using the sublimate with quinine, vinegar, and alcohol (this procedure was used on the cadaver of Louis XVIII), spreading it afterward with a mixture of storax, copaiba, lavender oil, and thyme [1, 2, 5–8].

The medieval museum largely contained collections of nonperishable objects such as shells, bones, and fossils. The only anatomical samples that could be included were dry and varnished viscera that looked little like the real structure [1]. The best-conserved specimens were in a "wine spirit," basically alcohol, extremely expensive for the time. Hence, the museum technique developed by Robert Boyle is considered of incalculable value. This procedure, long sought to stop the decomposition of tissues, did not mean an advance in the development of museums, due evidently to the prohibitive cost of liquor and the glass containers for such specimens, which limited the number of samples included in the collections [1, 5–8].

Therefore, the need to obtain greater durability of anatomical preparations became urgent, and so less expensive alternative fixatives continued to be sought [5–8].

Johann Jacob Wepfer (1620–1695) injected saffron water into the vessels of the brain, being the first to describe the course and branching of the carotid artery correctly. For his part, Jean Pecquet (1622–1674), from studies on the physiological effects of alcohol, tested the technique but with little success, and it was Jan Swammerdam (1637–1680) who, thanks to the application of finely stretched glass tubes, managed to inject the smallest vessels and, by mounting them in turpentine, made his findings even more visible [1, 5–8]. Thus, Swammerdam enjoyed particular success with the solidifying wax injections, and in 1672 he sent his spleen, gallbladder, and uterus samples to the Royal Society of London. Once the advantages of the solidifying injections had been generalized, their anatomical application became widespread, and testing began on a large number of materials in the search for new alternatives. Daniel Duncan (1649–1735), associated with the discovery of the fifth ventricle (cavum septum pellucidum), began with the differential dye for wax injections in the vessels. On the other hand, Pierre Simon Rouhault (1740) stood out for the perfect vascular injection of the placenta [1, 5–8].

Following Tompsett [1, 5–8], it was established that Guillaume Homberg (1652–1715) was indeed the first to discover the potential value of metal injections. To obtain these injections, his technique was based on the use of equal part lead, tin, and bismuth, discovering that they would remain sufficiently liquid to do the injection at a lower temperature than required to burn paper, in addition to developing the device that enabled the development of these injections (published in Memoires de l'Academie des Sciences, 1699) [1]. Taking as an example the experiments of Homberg, Govert Bidloo (1649–1713), Michael Bernhard Valentin (1651–1729), and William Cowper (1666–1709) developed pulmonary molds using fusible metals. However, Marcello Malpighi (1628–1694) is considered the first to perform experiments with mercury injections, demonstrating the structure of the lungs in 1661 [1], in addition to being considered the discoverer of capillary circulation, and, mainly, the "Founder of Microscopic Anatomy." On the other hand, Anthony Nuck (1650–1692) was the first to use mercury to visualize the lymphatic system, but as Tompsett [1] points out, his findings were poorly interpreted due to the extravasation of the injection mass. He achieved, nevertheless, the general description of the lymphatic structures, being one of the most complete prior to the description of Paolo Mascagni (1752–1815). In the eighteenth century, British researchers, like William Hunter (1718–1783), John Hunter (1728–1793), William Hewson (1739–1774), William Cumberland Cruikshank (1746–1800), John Sheldon (1752–1806), and Matthew Baillie (1761–1823), notably perfected the technique of arterial injection remarkably, using new conservative fluids, who used an arterial injection of several oils, mainly turpentine oil, to which they added Venice turpentine, chamomile oil, and lavender oil [1, 2, 5–8].

According to Brenner [2], in France, Cuvier (1769–1832) began to use pure alcohol, whereas Chaussier (1746–1823) submerged bodies eviscerated in a solution of mercury dichloride. On the other hand, Thenard (1777–1857) injected a liquid composed of an alcoholic solution of mercury dichloride; Sucquet (1840–1870) developed a 20% zinc chloride solution. Mayer [38] indicates that Jean Nicolas Gannal (1721–1783) was the first to investigate, publish, and apply embalming procedures directed to the general population as a funeral service [5–8, 39]. He was the first to fully publish the formulas used, obtaining the patent for his embalming fluid, which contained an aluminum acetate solution [5–8, 39].

In Italy, important representatives were also developing embalming techniques by arterial injection. In this sense, Mayer [5–8, 39] provides information on Giuseppe Tranchina (1797–1837), a well-known anatomist who successfully applied arsenic solutions arterially, without the need to eviscerate. Later, Alfredo Salafia (1869–1933) immortalized an embalming technique for funerals for famous people of the era, becoming renowned for his work on an Italian girl, Rosalia Lombardo, who died at 2 years of age in 1920 in Palermo, Sicily, because of pneumonia. The characteristic of the embalming process developed on this girl is that she was placed in a coffin with a glass cover, and later her coffin was moved to the Capuchin Catacombs in Palermo and is on display to this day. Although it was suspected that the embalming fluid contained arsenic, on the contrary, a study conducted on Alfredo Salafia's technique, by Piombino-Mascali, Aufderheide, Johnson-Williams, and Zink, published as a letter to the publisher in the Virchows Archiv in 2009 [40, 41], discovered the existence of a handwritten note by Salafia himself titled "New special method for the preservation of an entire human cadaver in a permanently fresh state" and in which the exact composition of this embalming formula was described that could be recognized as the first to contain formalin. The complete formulation of the technique, revealed by Piombino-Mascali et al. [40, 41], consisted of one part glycerin, one part formalin saturated with sulfate and zinc chloride, and one part alcohol solution saturated with salicylic acid. In addition, it indicated the injection by a single vascular route, preferably the femoral artery, without recommending any other type of procedure.

Alfredo Salafia (1869–1933) was an Italian anatomist and embalmer who became famous for his work in the field of embalming. He is best known for his skill in creating lifelike and naturalistic appearances of embalmed bodies, which earned him the nickname of "the wizard of death." Salafia was born in Marsala, Italy, in 1869, and studied medicine and surgery at the University of Palermo. He began working as an embalmer in the early 1900s, and soon gained a reputation for his exceptional skill and attention to detail. He was known for using a variety of techniques, including injecting the body with embalming fluids, removing the internal organs, and preserving the body with plaster or wax [1, 5–8].

Salafia's most famous work was the embalming of the body of Rosalia Lombardo, a 2-year-old girl who died of pneumonia in 1920. Salafia's technique was so successful that Lombardo's body remained remarkably well-preserved over the years, and she became known as "the sleeping beauty of Palermo." Her body is now on display in the Capuchin Catacombs of Palermo. Salafia died in 1933, but his legacy as an embalming expert lives on. His techniques and methods continue to be studied and admired by experts in the field, and his work with Lombardo remains one of the most famous examples of successful embalming in history [40, 41].

Thus, the way is clear to the crucial discovery in relation to fixation for the preservation and conservation of biological material, which was formaldehyde, and then formalin. Formaldehyde was discovered accidentally by Aleksandr Butlerov (1828–1886) in 1859, erroneously calling it dioxymethylene. However, it was chemist Wilhelm von Hofmann (1818–1892) who, in 1868, identified and synthesized it correctly into formaldehyde [1, 5–8]. For his part, German chemist Oskar Loew (1844–1941) is recognized as the one who finally discovered the antiseptic properties of formalin in 1888 [1, 5–8]. Then, in 1892, French biologist Auguste Trillat (1861–1944) revealed he had determined its preservative power, and the following year, in 1893, this highly versatile chemical substance was used for the first time as a tissue fixative by Ferdinand Blum in 1893, who discovered the fixative properties of formalin fortuitously while studying the antiseptic effects of the substance [40, 41]. The solution of formaldehyde gas at 40%, formalin, was used for many purposes, such as a preservative in plants by botanist Ferdinand Julius Cohn (1828–1898). Zoologists and anatomists soon found a formula suitable for their use as a preservative and fixative, thanks to its qualities, like not causing excessive shrinking or hardening, and being nonflammable and low cost [1, 5–8]. Thus, 30 years after its discovery, formalin, as of 1893, was adopted in anatomy, zoology, and histology. However, it would not be until 1895 that formalin would be proposed as a fundamental ingredient for embalming whole human bodies. Later, and as indicated previously, the fixative solution used by Salafia on the girl in Palermo is considered one of the first to contain formalin in its constitution [40, 41].

Since the discovery of formaldehyde, and its subsequent introduction as a fixing liquid, it became the component par excellence for the development of the cadaveric preservation process, with careful attention to tissue putrefaction, and ensuring its preservation and disinfection, constituting as the essential element in modern embalming solutions. During the twentieth and twenty-first centuries, and especially from the indication of formaldehyde as potentially toxic to health, alternative fixation and embalming techniques were developed, with reduced concentrations of formaldehyde and also seeking to replace it with different types of alcohols and substances, all with the aim of replacing it (see Chap. 3). And the appearance of plastination contributed in this sense. Plastination was invented by Professor Gunther von Hagens in 1977 in Heidelberg, Germany [5–8, 42–46]. Chapter 2 will develop the history of plastination and the life of its inventor, whose primordial and essential objective is to "Democratize Anatomy through Plastination."

References

1. Tompsett DH. Anatomical techniques. 2nd ed. Edinburgh: E. & S. Livingstone; 1970.
2. Brenner E. Human body preservation old and new techniques. J Anat. 2014;224(3):316–44. https://doi.org/10.1111/joa.12160.
3. Johnson JH. Importance of dissection in learning anatomy: personal dissection versus peer teaching. Clin Anat. 2002;15(1):38–44. https://doi.org/10.1002/ca.1090.
4. Glob PV. The bog people: iron-age man preserved. New York: NYRB Classics; 2004.
5. Ottone NE. Lecture: "Anatomical techniques for teaching morphological sciences". In: Theoretical-practical workshop on the use of digital tools for the teaching-learning of morphofunction. Association of Ecuadorian Faculties of Medical and Health Sciences (AFEME) and Pontificia Universidad Católica de Quito, Quito, Ecuador. February 14 to 17, 2018.
6. Ottone NE. Lecture: "History of anatomical techniques". In: 1st Workshop on plastination and anatomical techniques. Laboratory of Plastination and Anatomical Techniques, Faculty of Dentistry, University of La Frontera, Temuco, Chile. October 16, 2017.
7. Ottone NE. Lecture: "History of anatomical techniques". In: Course of advanced anatomical techniques. Doctorate in morphological sciences. Faculty of Medicine, University of La Frontera, Temuco, Chile. January 23, 2017.
8. Ottone NE. Lecture: "History of anatomical techniques". In: XIV National and International Congress of Morphophysiological Sciences—I Paraguayan Congress of Anatomy. Eastern University. Paraguayan Society of Morphophysiological Sciences. Paraguayan Society of Anatomy, City Pte, Franco, Paraguayan. September 11 and 12, 2015.
9. Ghosh SK. Human cadaveric dissection: a historical account from ancient Greece to the modern era. Anat Cell Biol. 2015;48(3):153–69. https://doi.org/10.5115/acb.2015.48.3.153.
10. Papageorgopoulou C, Xirotiris NI, Iten PX, Baumgartner MR, Schmid M, Rühli F. Indications of embalming in Roman Greece by physical, chemical and histological analysis. J Archaeol Sci. 2009;36(1):35–42. https://doi.org/10.1016/j.jas.2008.07.003.
11. Longrigg J. Anatomy in Alexandria in the third century B.C. Br J Hist Sci. 1988;21:455–88.
12. Aziz MA, McKenzie JC, Wilson JS, Cowie RJ, Ayeni SA, Dunn BK. The human cadaver in the age of biomedical informatics. Anat Rec. 2002;269(1):20–32. https://doi.org/10.1002/ar.10046.
13. Eisma R, Mahendran S, Majumdar S, Smith D, Soames RW. A comparison of Thiel and formalin embalmed cadavers for thyroid surgery training. Surgeon. 2011;9(3):142–6.

14. Feigl G, Kos I, Anderhuber F, Guyot JP, Fasel J. Development of surgical skill with singular neurectomy using human cadaveric temporal bones. Ann Anat. 2008;190(4):316–23. https://doi.org/10.1016/j.aanat.2008.05.001.
15. Barton DP, Davies DC, Mahadevan V, Dennis L, Adib T, Mudan S, Sohaib A, Ellis H. Dissection of soft-preserved cadavers in the training of gynaecological oncologists: report of the first UK workshop. Gynecol Oncol. 2009;113(3):352–6. https://doi.org/10.1016/j.ygyno.2009.02.012.
16. Brenner E, Maurer H, Moriggl B. The human cadaver as an educational tool—classification and comparison with other educational tools. Ann Anat. 2003;185:229–30.
17. Coleman R, Kogan I. An improved low-formaldehyde embalming fluid to preserve cadavers for anatomy teaching. J Anat 1998;192 (Pt 3):443–6. doi: https://doi.org/10.1046/j.1469-7580.1998.19230443.x.
18. Bergman EM, van der Vleuten CP, Scherpbier AJ. Why don't they know enough about anatomy? A narrative review. Med Teach. 2011;33(5):403–9. https://doi.org/10.3109/0142159X.2010.536276.
19. Souchon E. Embalming of bodies for teaching purposes. Anat Rec. 1908;2(6):244–7. https://doi.org/10.1002/ar.1090020605.
20. Udoaka AI, Oghenemavw L, Ebenezer T. Ancient embalming techniques amongst the Ogoni tribe in Southern Nigeria. J Exp Clin Anat. 2009;8(2) https://doi.org/10.4314/jeca.v8i2.55635.
21. Seidler H, Bernhard W, Teschler-Nicola M, Platzer W, zur Nedden D, Henn R, Oberhauser A, Sjøvold T. Some anthropological aspects of the prehistoric Tyrolean ice man. Science. 1992;258(5081):455–7. https://doi.org/10.1126/science.1411539.
22. Marquet PA, Santoro CM, Latorre C, Standen VG, Abades SR, Rivadeneira MM, Arriaza B, Hochberg ME. Emergence of social complexity among coastal hunter-gatherers in the Atacama Desert of northern Chile. Proc Natl Acad Sci U S A. 2012;109(37):14754–60. https://doi.org/10.1073/pnas.1116724109.
23. Uhle M. Los aborígenes de Arica. Publicaciones del Museo de Etnología y Antropología de Chile. 1917;1(4–5):151–76.
24. Llagostera MA. Patrones de momificación chinchorro en las colecciones Uhle y Nielsen. Chungará (Arica). 2003;35(1):5–22. https://doi.org/10.4067/S0717-73562003000100002.
25. Allison MJ, Focacci G, Arriaza B, Standen V, Rivera M, Lowenstein JM. Chinchorro momias de preparación complicada: métodos de momificación. Chungará (Arica). 1984;13:155–73.
26. Arriaza BT. Beyond death: the Chinchorro mummies of ancient Chile. Washington (DC): Smithsonian Institution Press; 1995. https://archive.org/details/B-001-000-806/page/n197/mode/2up.
27. Seyfullah LJ, Beimforde C, Dal Corso J, Perrichot V, Rikkinen J, Schmidt AR. Production and preservation of resins—past and present. Biol Rev Camb Philos Soc. 2018;93(3):1684–714. https://doi.org/10.1111/brv.12414.
28. Chunhong Y. Chinese lady Dai leaves Egyptian mummies for dead. CHINAdaily. 2004. http://www.chinadaily.com.cn/english/doc/2004-08/25/content_368631.htm.
29. Baadsgaard A, Monge J, Cox S, Zettler RL. Human sacrifice and intentional corpse preservation in the Royal Cemetery of Ur. Antiquity. 2011;85(327):27–42. https://doi.org/10.1017/S0003598X00067417.
30. Brier B, Wade RS. Surgical procedures during ancient Egyptian mummification. Chungará (Arica). 2001;33(1):117–23. https://doi.org/10.4067/S0717-73562001000100021.
31. Sharquie KE, Najim RA. Embalming with honey. Saudi Med J. 2004;25(11):1755–6.
32. Kaup Y, Schmid M, Middleton A, Weser U. Borate in mummification salts and bones from Pharaonic Egypt. J Inorg Biochem. 2003;94(3):214–20. https://doi.org/10.1016/s0162-0134(03)00002-3.
33. Lain Entralgo P. La anatomía en la Antigua India. Medicamenta. 1945;4(89):249–51.
34. Aufderheide AC. The scientific study of mummies. Cambridge: Cambridge University Press; 2003.
35. Mavrodi A, Paraskevas G. Mondino de Luzzi: a luminous figure in the darkness of the middle ages. Croat Med J. 2014;55(1):50–3. https://doi.org/10.3325/cmj.2014.55.50.

36. McKone HT. Embalming—chemistry for eternity. ChemMatters. 1999;17:12–3.
37. Romero-Reverón R. Andreas Vesalius (1514-1564). Founder of the modern human anatomy. Int J Morphol. 2007;25(4):847–50. https://doi.org/10.4067/S0717-95022007000400026.
38. Mayer RG. Embalming: history, theory, and practice. New York: McGraw-Hill; 2012.
39. Trompette P, Lemonnier M. Funeral embalming: the transformation of a medical innovation. Sci Stud. 2009;22:9–30.
40. Piombino-Mascali D. Il Maestro del sonno eterno. Palermo: La Zisa; 2009.
41. Piombino-Mascali D, Aufderheide AC, Johnson-Williams M, Zink AR. The Salafia method rediscovered. Virchows Arch. 2009;454(3):355–7. https://doi.org/10.1007/s00428-009-0738-6.
42. von Hagens G. Impregnation of soft biological specimens with thermosetting resins and elastomers. Anat Rec. 1979;194(2):247–55. https://doi.org/10.1002/ar.1091940206.
43. Whalley A. Pushing the limits. 2nd printing. Heidelberg: Arts and Sciences Verlagsgesellschaft GmbH; 2007.
44. Ottone NE. Gunther von Hagens, creator of plastination. Historical review and technical development. Rev Argent Anat Online. 2013;4(2):70–6. https://www.revista-anatomia.com.ar/archivos-parciales/2013-2-revista-argentina-de-anatomia-online-f.pdf.
45. Ottone NE, Cirigliano V, Bianchi HF, Medan CD, Algieri RD, Borges Brum G, Fuentes R. New contributions to the development of a plastination technique at room temperature with silicone. Anat Sci Int. 2015;90(2):126–35. https://doi.org/10.1007/s12565-014-0258-6.
46. Ottone NE. Plastination: techniques fundamentals and implementation at Universidad de La Frontera. J Health Med Sci. 2018;4(4):293–302.

Chapter 2
Brief Review of the Origins of Plastination

The history of plastination is the story of its Inventor, Professor Gunther von Hagens. Plastination was founded in 1977 by Professor Gunther von Hagens in Heidelberg, Germany [1, 2]. Plastination was developed as a biosafe method to preserve biological material, resulting in dry, odorless and stable samples. In this process, water and lipids in tissues are replaced by plastic polymers such as silicone, epoxy, and polyester resins. This makes the sample dry, odorless, and very stable. The type of polymer used determines the optical properties (clear or opaque) and the motion (transparent or stable) it can affect the sample being immersed. When immersed, the sample will be more stable than frozen, dehydrated, or waxed samples. It also has the advantage of keeping the original pressure and cellular energy at a microscopic level [3–5].

This process usually replaces the liquid medium (water and lipids) by replacing the liquid with a polymer so that the body does not lose its texture and color [6, 7].

Decomposition of organic matter is an important process in nature, but is also a problem for morphology studies. This is especially important for biological organisms, which will shrink in size when exposed to the environment. For this reason, it has always been a target for anatomists.

Plastination is a real option for the preservation of perishable tissues (whole body, brain, liver, lung, kidney, heart, muscle, etc.). Its dry and indestructible state is achieved by using different polymers and special plastics.

Plastination has been developed by many researchers around the world after the birth of plastination technology. This is clearly visible in each chapter of the book, describing the type of plastination process, showing all scientific publications by anatomists and scientists using the technique developed by Gunther von Hagens, and introducing modifications to the system. In this way, in every part of the article, these studies are explained to all researchers who have gained experience and can be well established by developing different plastination techniques.

Similarly, since 1986, when the International Society for Plastination was created, it has been possible to organize scientists into these research.

© The Author(s), under exclusive license to Springer Nature Switzerland AG 2023

N. E. Ottone, *Advances in Plastination Techniques*, https://doi.org/10.1007/978-3-031-45701-2_2

With advances in plastination technology, Gunther von Hagens was able to create and define genuine concepts in the history of anatomy, marking a pivotal moment in science and changing the way the morphology is taught and researched, compared with the changes made by the great Andrea Vesalius of the Renaissance.

The following is a brief introduction to the life of Professor Gunther von Hagens. Some of the information to be disclosed will be shared by Dr. Angelina Whalley's 2007 book *Pushing the Limits* [8].

Gunther von Hagens (born Günther Gerhard Liebchen) was born on January 10, 1945, in Alt-Skalden, Posen, Poland, then part of Germany. His family fled Poland to escape the Russian occupation and made their way to Berlin, where they did not stay long and eventually settled in the small town of Greiz, where von Hagens lived until the age of 19 years old (year) [8].

von Hagens had an interest in science and medicine from an early age, and was sent to the intensive care unit for several months due to circumstances that he had to live with, especially when he was close to death at the age of 6. During this long stay, his contact with doctors and nurses made him interested in medicine from his childhood, and he decided to become a doctor. Then, at the age of 12, he devoted himself to the publication of Russian Sputnik and became a well-known person in his school in Sputnik-related topics and information [8].

In 1965, von Hagens enrolled at the University of Jena Medical School, south of Leipzig and the birthplace of authors Schiller and Goethe. His indecisiveness and demeanor made him so popular that he was quoted in a school report...:

> Gunther Liebchen is a personality who does not approach tasks systematically. This characteristic and his great imagination, which sometimes made him forgetful of reality, occasionally led him to develop unusual ways of working and to be obstinate—but never in a way that was detrimental to the collective of his seminary group. On the contrary, his forms often encouraged his peers to critically analyze their own work. [8]

During his time at the university, von Hagens began to question communism and socialism, and expanded his political knowledge by gathering information from cultural, Western media. He later took part in student protests against the Warsaw Pact army's occupation of Czechoslovakia. Disguised as a student on vacation, von Hagens crossed Bulgaria and Hungary in January 1969, crossing the Czechoslovak border into Austria on January 7 and attempting to regain his freedom. He failed, but tried a second time the next day at a different location on the border. This time the police arrested him. "When I was in prison, a good guard opened the door for me to escape. I was shocked and couldn't make a decision. I feel better. Very good," he said [8].

Gunther von Hagens was arrested and sent to East Germany where he was imprisoned for 2 years. von Hagens, a 23-year-old anti-congenitalist, was seen as a threat to his own way of life and in need of rehabilitation and public education. According to Günter Liebchen's prison records... by the norms and rules of our society... The prisoner must be aware of the danger of his actions, and in doing so he makes his decision on the prisoner's future behavior before the citizen of the country (state relations) [8].

Gunther von Hagens may have felt, or even redeemed, in his lost years after serving 36 years in prison. "The horrors of captivity I had to overcome through dreams and deep friendships with other inmates who were there helped me to share my heart with others, to trust my body and mind. Everything I learned before my release and in prison helped me later in my career as a scientist" [8].

In 1970, after West Germany gained independence, von Hagens entered the University of Lübeck to complete his medical studies. After graduating in 1973, he worked at a hospital on Helgoland. A year later, after completing his medical degree, he enrolled in the Department of Anesthesiology and Emergency Medicine at the University of Heidelberg, where he realized that he had no desire for the routine work he wanted. In June 1975, his classmate Dr. Cornelia von Hagens took her last name. There are three children from this marriage: Rurik, Bera, and Tona [8].

In 1977, von Hagens began an 18-year career as assistant professor at the Institute of Pathology and Anatomy of the University of Heidelberg, where he developed the plastination:

> I was looking at a series of specimens wrapped in plastic. This was the most advanced preservation technique at the time, as these specimens are kept in one of the clear plastics. I wondered why the plastic was poured and then cured around the specimens instead of being pushed into the cells, which would stabilize the specimens from the inside and literally allow us to grip them. [8]

Prof. von Hagens continues:

> A few weeks later, I went to prepare a series of human kidney sections for a research project. The traditional method of making and then cutting it into small pieces required a lot of effort, as only 50 pieces were needed per piece. One day in college, I was in a butcher's shop and I saw the butcher cut the ham, and that's when I realized he had to cut the kidney with a meat slicer. Thus, the "spinning blade" (as I used the name I used in my college application) became my first investment in plastination. I cut a kidney out of pexiglass plates, put the liquid there, and used a vacuum to remove the air bubbles created by the pressure from the polymer and curing agent. [8]

"But when I looked at the bubbles, I realized this: you have to inject plastic into a kidney saturated with acetone and put it in a vacuum that can remove the bubbles of acetone in this form." A lot of acetone gushed out when I tried it, but after an hour the kidney was black and had shrunk. At this point most people would have thought the experiment had failed, and the only reason I tried again a week later with silicone rubber was because his experience in chemistry and physics told him they saw it tarnished. Indicator. Plexiglass and its shrinkage are due to the very rapid immersion of the sample. After correcting these points in this way, he continued to try to practice. On January 10, 1977, "...when I decided to make the runway a part of my life...."

The first article of the new technique "impregnancy of large samples with polymers" was published in the German journal *Verhandlungen der Anatomischen Gesellschaft* (Archives of the Anatomical Society) (von Hagens & Knebel, 1978). A year later, an article on "Impregnation of Materials with Chemicals and Chemicals" was published in the *Anatomical Records* (von Hagens, 1979), followed by the first use of the word "plastination" for "emulsifying resins" in 1979 (von Hagens, 1979).

The process was later patented by Professor von Hagens, and he spent the next year with all his energy completing his work. The first step in plastination is to stop decomposition:

> While cadavers are embalmed by injecting formaldehyde into the vessels, small samples are placed inside the same material. After dissection, all body fluids and soluble oils are removed from the structure and then replaced with reactive resins and elastic through the vacuum impregnation body, such as silicone rubber and epoxy, he said. [8]

During this time, von Hagens founded his company, BIODUR Products, which distributes equipment, technology, and specialty polymers for coatings to hospitals around the world. Today, more than 400 schools in 40 countries around the world are using Gunther von Hagens' invention to store anatomical models for medical education.

Plastination technology has spread all over the world, and Prof. Harmon Bickley as a well-known plastinator organized the first International Plastination Congress, at the University of Texas, San Antonio, Texas, USA, in 1982, and was actually titled "Preservation of Biological Materials by Plastination." It was held in San Antonio, Texas, USA, on April 16, 1982, and lasted only 1 day. Eighty people attended, all from the USA [9]. This same year the article was published in *Anatomical Record*, "The technique of heart plastination," from the authors Klaus Tiedemann and Gunther von Hagens. Also in 1982, the first patents related to the plastination technique appear [10].

Related to the plastination in Latin America, in 1983, according to an article published in *Journal of Plastination* in 2015 [11], by the renowned plastinator Carlos Baptista, Prof. Santiago Aja Guardiola (1949–2021) was the first to carry out plastination in Latin America, at the Faculty of Veterinary Sciences and Zootechnics of the *Universidad Nacional Autónoma de México* (UNAM). Later, in 1984, Prof. Carlos Baptista himself was the first to develop plastination in Brazil, after a visit to the USA, with the installation of the first plastination laboratory in Brazil, at the University of Sao Paulo. As also indicated in this article, Prof. Baptista, together with Prof. Esem Cerqueira, produced the first plastinated specimens in Brazil [10].

In 1983, the Catholic Church, Dr. von Hagens made the bones of St. Petersburg, Hildegard of Bingen (1090–1179), a prominent figure in Germany, theologian, and a respected writer, Pope Von Hagens II. A later idea of plasticizing John Paul failed before it became a major controversy [8].

The "Second International Conference on Plastination" was held in San Francisco, USA, in April 1984. The attendance was close to 100 people and included some from outside the USA [9].

The Third International Plastination Conference was held in San Antonio, USA, from April 21 to 25. It was widely promoted in North America and Europe. Therefore, the attendance was very good, and the conference eventually started to gain an international character. With this conference, the 5-day format was adopted: 2 days of lectures about plastination fundamentals, 1 day of informal meetings, and 2 days of advanced research presentations [9]. At this meeting, the International Society for Plastination (ISP) was founded, and plans for a journal were announced.

The *Journal of the International Society for Plastination Issue 1* was published in January 1987, including many of the papers presented at this meeting [9].

Harmon Bickley PhD was the first president (Executive Director) of ISP, from 1986 to 1995. The International Society for Plastination is a multidisciplinary organization, which includes people from all fields of science interested in the plastination technique. Plastination refers to the use of polymers to infiltrate and preserve any material for teaching, research, or diagnostic purposes [9].

At its founding, the following goals were established for the ISP (which remain to this day) [9]:

- Provide and maintain an international association of individuals and institutions that perform plastination techniques or are interested in preservation methods such as plastination.
- Serve as a forum for the exchange of information on plastination.
- Define plastination as a specialty area of professional activity, encourage other institutions to adopt these preservation methods, and invite people to learn and practice plastination as a scientific career.
- Periodically publish the *Journal of the International Plastination Society.*
- Holding regular meetings, workshops, and conferences to promote and teach plastination techniques.
- Keep a register of member institutions and people who perform plastination, their particular specialty, and other people interested in plastination.

In 1987 another relevant article on plastination was published, which to this day inspires all of us who work with this technique, "The current potential of plastination," published in Anatomy and Embryology (Berlin), by von Hagens G., Tiedemann K., and Kriz W.

In 1988, the "Fourth International Conference on Plastination" was held at the Mercer University School of Medicine, Macon, Georgia, USA, from March 21 to 25 [9]. This year, during the First National Congress of Veterinary Anatomy of Mexico, held from October 12 to 15 in Mexico City, Prof. Santiago Aja Guardiola presented the work "Plastination, the modern technique for obtaining more useful macro-specimens in the teaching-learning process," together with Prof. Martínez-Galindo.

John Bohannon, in Science [12], indicated that it was not until one night in 1988 when, on a tour of his University, Prof. von Hagens found a janitor totally astonished by the presence of plastinated pieces of a human body, to which stared. There he envisioned the possibility that plastination was not only intended for scientific and academic activity but could also be a tool to bring anatomy to the general public. In this way, continues Bohannon, a modest exhibition of plastinated samples in the small German town of Pforzheim that same year turned out to be a minor success, and there the creativity of Prof. Gunther von Hagens began to awaken [12].

In 1989, at the XIII World Congress of Anatomy, of the International Federation of Associations of Anatomists (IFAA), held in the city of Rio de Janeiro, Brazil, and under the Presidency of Prof. Moscovici, Prof. von Hagens attended and presented for the first time outside of Germany cuts plastinated in epoxy resin of a complete human body, having great success [8].

The "Fifth International Conference on Plastination" was one of the most prominent. It was held in Heidelberg, the "Birthplace of Plastination" [9]. It was from July 22 to 27. At this conference, Prof. Gunther von Hagens achieved an incredible goal by presenting the first full body plastination in history [8].

In 1992 von Hagens married Dr. Angelina Whalley, a physician who works in the capacity of Commercial Director as well as being the designer of the Body Worlds exhibits [8]. The "Sixth International Conference on Plastination" was held in Kingston, Ontario, Canada, this same year [9].

A year later, in 1993, Dr. von Hagens founded the Institute for Plastination in Heidelberg, which offers plastinated specimens for teaching use and for Body Worlds exhibitions [8].

Prof. von Hagens developed the first Body Worlds exhibition in Tokyo, Japan, in 1995, invited by Tatsuo Sakai, anatomist from Juntendo University [12]. A point of criticism from visitors to that exhibition was the terrifying appearance of the bodies that presented themselves as dissections of normal anatomy. Gunther von Hagens noted that during the Renaissance, anatomical specimens and drawings were often shown in poses denoting action. Such action positions were suitable for conveying feelings [13] but at the same time could act as a means of provocation which resulted in much publicity and media coverage for the Body Worlds exhibition. More than 400,000 people attended this first expo in Japan. After this exhibition, in 1997 the first Body Worlds exhibition in Germany was held, in the city of Mannheim.

After his first exhibition in Germany, von Hagens returned to Japan twice before continuing his exhibitions in Austria, in Switzerland, and again in Germany [14]. Later it came to London, where the exhibition was open for 11 months. In 2002, Body Worlds came to Seoul, South Korea, with over two million visitors. Finally, in March 2003, the exhibition was located simultaneously in Munich, Germany [12].

To date, Body Worlds has been watched by more than 30 million people in more than 50 countries in Europe, Asia, and North America.

As if determined to push the boundaries of free living, Dr. von Hagens made an effort to travel the world and expand his interests. In 1996, he took a visiting professor position at Dalian Medical University in China and became the director of the Plastination Research Center at the Bishkek/Kyrgyz National Academy of Medicine [8].

The first communication on a plastination technique at room temperature was made in 1996, by Zheng et al. [15], in the *Journal of the International Plastination Society*. Later, in 1998, two new communications associated with this new type of plastination appear, which basically corresponded to the use of vacuum chambers without a freezer, in the forced impregnation process, that is, located in a room at room temperature (20–22 °C), they also established modifications in the combination of reagents (silicone, catalyst, and curing agent), compared to the original technique created by Gunther von Hagens (cold plastination), but the definition of room temperature refers to the aforementioned, the use of vacuum chambers directly located in rooms, without the need to place them inside freezers. This was explained in two publications that appeared in the same issue of the *Journal of Plastination* (former *Journal of the International Society for Plastination*) in 1998. The first was an original article entitled: "Plastination at Room Temperature" by authors Zheng Tianzbong, Liu Jingren, and Zhu Kerming of the Department of Anatomy, from Shanghai Medical University, Shanghai, China [16], where they describe in detail the plastination technique at room temperature without the need to use a freezer. And in the same issue of the *Journal of the International Plastination for Society* appears after the paper of Zheng, the published abstracts of the ninth International Conference on Plastination held in Quebec City, Canada, a summary of a scientific paper presented at the conference, led by Roy Glover (University of Michigan, USA), and accompanied by Robert Henry (University of Tennessee, USA) and Robert Wade (University of Maryland, USA), entitled: "Polymer preservation technology: Poly-Cur, a next generation process for biological specimen preservation" [17]. Clearly, it is this summary that indicates the following: "Specimens can be impregnated at room temperature thus eliminating the need for a low temperature freezer. This saves cost and lab floor space. In addition, processing specimens at warmer temperatures makes them less rigid and increases their flexibility." In this way, it is stated that the definition of plastination at room temperature refers to the performance of the plastination technique without the need for a freezer at low temperature. This reaffirms our research from 2015 [18], and that will be developed in the silicone plastination chapter, where we proposed a new plastination technique at room temperature, correctly defined by performing forced impregnation in a vacuum chamber outside of a freezer, and refutes what was indicated in a paper of 2019 [19], where plastination at room temperature is wrongly defined.

In 2001, Prof. von Hagens founded Von Hagens Dalian Plastination Ltd., a private company. It was located in Dalian, China, which currently employs 250 people. Other plastination laboratories began to be established in China (Shanghai, Beijing, and Canton).

In 2004, Prof. von Hagens began as a faculty member at the NYU School of Dentistry. "The human body is the only planet left in a single generation," he declared. "I want the exhibition to be a place of inspiration and thought, even thought and self-belief, knowledge and interpretation, whatever the visitor's background and the people's philosophy" [8, 20].

Also in 2004, Dalian Hoffen Bio-technique Co. Ltd., a company focused on research and development of plastination, was founded in Dalian, China, by Prof. Hong-Jin Sui [21].

In 2011, at the tenth International Interim Conference on Plastination Toledo, Ohio, USA, Prof. Gunther von Hagens, due to his diagnosis of Parkinson's, made a farewell to the instances of participation in Plastination Conferences, in the magazine of the International Society for Plastination published his farewell letter [22].

First International Conference on Plastination in South America: Recognition of the Inventor of Plastination Prof. Gunther von Hagens

The 20th International Conference on Plastination was hosted by *Universidad de La Frontera* in Temuco, Chile (www.icp2022chile.com) [22], by Dr. Nicolas E. Ottone, President of the Conference. Originally, the 20th International Conference on Plastination, approved at the ISP General Assembly in Dalian, China, during the organization of the 19th International Conference on Plastination, was planned for July 2020. However, the tragic global pandemic that imprisoned us and that began in March 2020 prevented the holding of the Conference, which was definitively held online in July 2022, 40 years after the first International Conference on Plastination held in Texas, USA.

The event was notable for being organized in South America for the first time. Held from July 18 to 21, 2022, the event featured international key participants and regional representatives who come together to discuss topics related to fundamentals, education, and research applications of plastination and topics on classical anatomical techniques and human donation. At the same time, the fourth International Congress on Anatomical Techniques was held [23].

This meeting had two important and special moments. The first moment was in the opening meeting, with the participation of the inventor of plastination, Professor Gunther von Hagens, who opened the First International Conference on Plastination in 1982, in Texas, USA, and with his participation in the 20th International Conference on Plastination, he was invited to open this conference held for the first time in South America, in Temuco, Chile. Also, on the third day of the meeting, there was a very special moment when the Professor was given a recognition (see "ISP Career Achievement Award"). Professor von Hagens was awarded with the *Maximum Distinction* in recognition of his outstanding career and his invaluable and notable contribution in the fields of anatomy and morphological sciences for the revolutionary creation of "plastination," technique that broke the established standards and marked a breakthrough and a historical milestone in the teaching, the research, and the experience of anatomy [24, 25]. Professor Nicolás Ottone traveled especially to Guben, Germany, days before the start of the 20th International Conference on Plastination in the first days of July 2022, to personally deliver the distinction to Professor Gunther von Hagens. On that occasion Prof. Ottone was received by Angelina Whaley and Rurik von Hagens.

At the Conference, speakers from the USA, China, Germany, Turkey, New Zealand, Russia, Brazil, Chile, Spain, Peru, Costa Rica, and Mexico participated. In addition, 23 presentations were made by various universities from 12 countries [24].

The 20th International Conference on Plastination was attended by more than 110 attendees from 43 countries, on all continents, reflecting the internationalization of the technique. The outstanding participation of academics and students from

South America was highlighted, as well as undergraduate and postgraduate students from *Universidad de La Frontera*, and other universities of the Region, especially at the level of oral presentations, thus allowing to verify the great work developed by International Society for Plastination, and in this specific Conference, by *Universidad de La Frontera* for the dissemination in South America of Plastination, through Plastination Workshops, the Diploma in Anatomical Techniques, and our presence in Anatomy Congresses. The aim is to continue on this path, promoting from International Society for Plastination the development and application of plastination in anatomical teaching and research in all the world's educational institutions [24].

Below is the Scientific Program of the 20th International Conference on Plastination (https://www.plastinacion.com/20icp).

1° Day: Monday, July 18, 2022

Fundamentals of Plastination

(Moderators: Dmitry Starchik, Phillip Adds)
   Conference 1. Official Welcome:

Opening of the 20th International Conference on Plastination
Nicolás Ottone (President 20th ICP, *Universidad de La Frontera*, Temuco, Chile)
Rafael Latorre (President ISP, *Universidad de Murcia*, Spain)—All Council ISP

   Conference 2. "40 Years of Reality":

Robert Henry (Lincoln Memorial University, Tennessee, USA)
   Conference 3. "Principles of Plastination":

Kees de Jong (Center for Morphology, Zhejiang University Medical School, Hangzhou, China)
   Conference 4. "Setting Up of a Plastination Laboratory":

Volker Schill (BIODUR GmbH, Germany)

   5. Oral Presentations:

1.1. "Transforming Prosected to Plastinated Specimens using the Room Temperature Technique to Facilitate Anatomy Education." Ameed Raoof, Avelin Malyango, Mange Manyama, Charles Msuya, Nasnass Nassir. Medical Education Division, Weill Cornell Medicine-Qatar, Qatar Foundation, P.O. Box 24144, Doha, Qatar
1.2. "Novel En Bloc Dissection and Plastination of Full Body Systems." Gustilo K.1, Curran S.1, Goldberg C.1, Lohman Bonfiglio C.M.1, Frank P.W.2,

Baptista C.A.C.2, Stabio M.E.1 1. Modern Human Anatomy Program, Department of Cell & Developmental Biology, University of Colorado School of Medicine, Aurora, CO, USA. 2. Department of Medical Education, College of Medicine & Life Sciences, University of Toledo, Ohio, USA

1.3. "Imidazole Effects in Color Conservation of Plastinated Domestic Animals Heads and Encephalon with S10 and P40 Methods." Ovando F.D.1, Candanosa A.E.2 1. Departamento de Morfología, Área de anatomía, Facultad de Medicina Veterinaria y Zootecnia, Universidad Nacional Autónoma de México, Coyoacán CDMX, México. 2. Centro de Enseñanza, Investigación y Extensión en Producción Animal en Altiplano (CEIEPAA), Facultad de Medicina Veterinaria y Zootecnia, Universidad Nacional Autónoma de México, Tequisquiapan, Querétaro, México

1.4. "Painting Protocol for Plastinated Specimens." Menezes F.V.1, Monteiro Y.V.2,3, Silva M.V.F.4, Miranda R.P.4, Bittencourt A.S.3 1. Graduate Program in Anatomy of Domestic and Wild Animals-University of São Paulo/SP, Brazil. 2. Graduate Program in Biotechnology, Federal University of Espírito Santo, Vitória/ES, Brazil. 3. Department of Morphology, Federal University of Espírito Santo, Vitória/ES, Brazil. 4. Federal University of Espírito Santo, Brazil

1.5. "How to Adapt a Space to Transform It Into a Plastination Laboratory? Our Experience." Popp A.I.1,2, Lodovichi M.V.2, Castillo D.1,2, Sidorkewicj N.S.1,2, Casanave E.B.1,2 1. Instituto de Ciencias Biológicas y Biomédicas del Sur (INBIOSUR-CONICET), Argentina. 2. Departamento de Biología, Bioquímica y Farmacia, Universidad Nacional del Sur, Argentina

1.6. "Anatomical Variations of Radial and Ulnar Arteries in Plastinated Upper Limbs." Guzman D.1, Bianchi H.2, del Sol M.3, Ottone N.E.4. 1. Undergraduate student, School of Dentistry, Universidad de la Frontera, Temuco, Chile. 2. Universidad de Buenos Aires, Argentina. 3. Center of Excellence in Morphological and Surgical Studies (CEMyQ), Universidad de La Frontera, Temuco, Chile. 4. Laboratory of Plastination and Anatomical Techniques, Faculty of Dentistry, CEMyQ, Universidad de la Frontera, Temuco, Chile

1.7. "Plastination at Room Temperature. Modified Technique Using Long-Conserved Reagents." Borges Brum G.1, Diaz, M.1, Ochoa O.2, Blanco C.3 1. Auxiliar de primera; 2. Auxiliar de Segunda; 3. Profesor Adjunto. Cátedra de Anatomía, Facultad de Ciencias Veterinarias, Universidad de Buenos Aires, Argentina

6. Questions and Answers

2° Day: Tuesday, July 19, 2022

Education and Clinical Application of Plastination

(Moderators: Telma Masuko, Nicolás Ottone)
Conference 1. "Anatomy Instructional Workflows: The Role of Plastination for Creating Extended Reality Online Learning Assets":

Scott Lozanoff (University of Hawaii, USA)

Conference 2. "Ethics in Anatomy and Body Donation for Plastination":

Rurik von Hagens (Gubener Plastinate GmbH, Germany)

Conference 3. "International Fascial Net Plastination Project: The Challenges and Achievements":

Vladimir Chereminskiy (Gubener Plastinate GmbH, Germany)

4. Oral Presentations:

2.1. "Plastination, An Ideal Method to Preserve Surgical Specimens in Oral Pathology." Aldape B.1, Dieguez L.1, Candanosa I.E.2 1 Faculty of Dentistry, National Autonomous University of Mexico (UNAM), Mexico City, Mexico. 2. Highlands Teaching and Research Farm, Faculty of Veterinary Medicine, UNAM, Tequisquiapan, México

2.2. "S10 Plastination Technique in Human Anatomical Models, As An Innovative Resource for the Academy." Hurtado E., Pusselt K. Departamento de Ciencias Morfológicas, Universidad Evangélica de El Salvador, San Salvador, El Salvador

2.3. "Chicken Plastination to Assist with the Teaching of Avian Anatomy." Olivier W., Roux L. Department of Anatomy and Physiology, University of Pretoria, Gauteng, South Africa

2.4. "Interactive Atlas of the Canine Brain and Kidney created from Plastinated Samples." Alvear V.E.1, Velasco B.2, Toaquiza A.B.1, Guanoluisa C.A.1, Morales C.2, Revelo-Cueva M.1 1. Laboratorio de Anatomía Animal, Facultad de Medicina Veterinaria y Zootecnia, Universidad Central del Ecuador—UCE—Quito, Ecuador. 2. Carrera de Ingeniería en Computación Gráfica, Facultad de Ingeniería y Ciencias Aplicadas, Universidad Central del Ecuador, UCE, Quito, Ecuador

2.5. "Plastination of Bovine Brain Sections with Polyester Resin." Borges Brum G., Ochoa O., Blanco C. Cátedra de Anatomía, Facultad de Ciencias Veterinarias, Universidad de Buenos Aires, Argentina

2.6. "Design of a Digital Resource Based on Plastinated Specimens for the Learning of Dilated Cardiomyopathy in Dogs." Laguna-García I., López-Albors O, Latorre R. Department of Anatomy, Faculty of Veterinary Medicine, University of Murcia, Spain

2.7. "Plastination of Crab-Eating Foxes (Cerdocyon Thous, Linnaeus, 1766) Sectionated and its Possible Applicabilities." Alcobaça M.M.O.1, Bittencourt A.S.2, Assis Neto, A.C.1 1. Department of Surgery, Faculty of Veterinary Medicine and Animal Science, University of São Paulo, São Paulo, Brazil. 2. Anatomy Laboratory, Museum of Life Sciences, Federal University of Espírito Santo, Vitória, Brazil

5. Questions and Answers
Conference 6. "Outreach Activities for Teaching Plastination":

Selcuk Tunali (TOBB Ekonomi ve Teknoloji Üniversitesi, Turkey)

3° Day: Wednesday, July 20, 2022

Research with Plastination

(Moderators: Rafael Latorre, Carlos Baptista)
Conference 1. "Mesoscopy and Epoxy Sheet Plastination":

Ming Zhang (University of Otago, New Zealand)

Conference 2. "Epoxy Plastination for Anatomical and Clinical Research":

Dmitry Starchik (North-Western State Medical University, Saint Petersburg, Russia)

Conference 3. "Plastination of Sperm Whale":

Hong-Jin Sui (Dalian Medical University, Dalian, China)

Conference 4. "The Journal of Plastination: The second decade (1997–2007)":

Phillip Adds (Editor-in-Chief, *Journal of Plastination*)

5. Oral Presentations:

3.1. "Optimal Temperature and Duration for Impregnation and Curing of Epoxy Blocks for Ultra Thin Sections." Zheng F, Xu Z., Zhang M. Department of Anatomy, University of Otago, Dunedin, New Zealand
3.2. "Plastination as a Tool for Scientific Investigation: Three Dimensional Reconstruction of Anatomical Structures by Using Plastinated Cross-Section." "La Plastinación como Herramienta para la Investigación Científica: Reconstrucción Tridimensional de Estructuras Anatómicas mediante el Uso de Secciones Transversales Plastinadas." Sora M.C. Center for Anatomy and Molecular Medicine, Sigmund Freud University of Vienna, Austria
3.3. "E12 Sheet Plastination Technique. Protocols and Applications. Bibliographical Review with Systematic Search. Preliminary Communication." Montecinos H.1, Quevedo M.F.1, Badilla N.1, Ottone N.E.2. 1. Undergraduate student, School of Dentistry, Universidad de la Frontera, Temuco, Chile. 2. Laboratory

of Plastination and Anatomical Techniques, Faculty of Dentistry, CEMyQ, Universidad de la Frontera, Temuco, Chile

3.4. "Plastination at 4150 meters above sea level in La Paz, Bolivia." Rodriguez Torrez VH.1, Ottone NE2. 1. Catedra de Anatomía Humana y Neuroanatomía, Carrera de Medicina y Odontología, Universidad Privada del Valle, sub sede La Paz, Bolivia. 2. Laboratory of Plastination and Anatomical Techniques, Faculty of Dentistry, CEMyQ, Universidad de la Frontera, Temuco, Chile

3.5. "P40 Sheet Plastination Technique. Protocols and Applications. Bibliographical Review with Systematic Search. Preliminary Communication." Badilla N.1, Quevedo M.F.1, Montecinos H.1, Ottone N.E.2. 1. Undergraduate student, School of Dentistry, Universidad de la Frontera, Temuco, Chile. 2. Laboratory of Plastination and Anatomical Techniques, Faculty of Dentistry, CEMyQ, Universidad de la Frontera, Temuco, Chile

3.6. "Staining Technique of White Matter Tracts for Cold Temperature S10 Plastination." Carr C.1, Cotner K.2, Baptista C.A.C.3, Frank P.W.3 1. Department of Chemistry and Biochemistry, University of Toledo, Toledo, Ohio, USA. 2. Department of Biological Sciences, University of Toledo, Toledo, Ohio, USA. 3. Department of Medical Education, College of Medicine & Life Sciences, University of Toledo, Toledo, Ohio, USA

6. Questions and Answers

7. Special Presentation: "Acknowledgment to Prof. Gunther von Hagens, Inventor of Plastination." Nicolás Ottone (President 20th ICP—Universidad de La Frontera, Temuco, Chile)

8. ISP Business Meeting

4° Day: Thursday, July 21, 2022

Classical Anatomical Techniques and Donation

(Moderators: Gonzalo Borges Brum, Marco Guerrero)
Conference 1. "Fixation: New Technique":

Telma Masuko (Universidade Federal de Bahia, Salvador, Brazil)

Conference 2. "Description of the Classic Taxidermy Technique. Its Origin, History and Development in Natural Science Museums":

Diego Jara Silva (Museo Nacional de Historia Natural, Santiago, Chile)

Conference 3. "Anatomical Photogrammetry and Its Use In Post-Pandemic Teaching":

Jorge Moscol (Universidad Nacional de Piura, Perú)

4. Oral Presentations:

4.1. "Use of Glycerin for Preservation of Teaching Specimens." Archana Pathak, M.M. Farooqui, Ajay Prakash. Department of Veterinary Anatomy. College of Veterinary Science and Animal Husbandry. Pt. Deen Dayal Upadhyaya Pashu Chikitsa Vigyan Vishwavidyalaya Evam Go Anusanshan Sansthan (DUVASU), Mathura, India

4.2. "A Technical Note of Improvement of a Technique for Tissue Preservation in Veterinary Anatomy." Bernal V.1, Aburto P.1, Pérez B.1, Gómez M.1, Gutierrez J.C.2. 1. Institute of Pharmacology and Morphophysiology, Austral University of Chile, Chile. 2. Department of Anatomy, Physiology and Cell Biology, University of California, Davis, USA

4.3. "Evaluation of the Effects of Different Traditional Methods of Cleaning Skeletal Material by Means of Scanning Electron Microscopy." Popp A.I.1,2, Lodovichi M.V.2, Sidorkewicj N.S.1,2, Castillo D.1,2, Casanave E.B.1,2 1. Instituto de Ciencias Biológicas y Biomédicas del Sur (INBIOSUR-CONICET), Argentina. 2. Departamento de Biología, Bioquímica y Farmacia, Universidad Nacional del Sur, Argentina

5. Questions and Answers

6. Round Table on Human Donation

Moderators: Dr. Mariano del Sol (Universidad de La Frontera, Temuco, Chile), Rubén Daniel Algieri (Universidad de Buenos Aires, Argentina):

(a) "Body Donation Program of the University of Costa Rica (PRODOCU)." Jessica González (Universidad de Costa Rica, Costa Rica)

(b) "UC Body Donation Program. Impact on Science and Teaching." Oscar Inzunza (Pontificia Universidad Católica de Chile, Chile)

(c) "The Body Donation Program and its Interconnected Activities: Results of 14 Years of Experience." Andrea Oxley (Universidade Federal das Ciências da Saúde de Porto Alegre, Brazil)

(d) "How to implement Body Donation Programs for Educational Institutions?". Diego Pineda (Universidad Nacional Autónoma de Mexico, Mexico)

7. Official Farewell:

"Oral Presentations Awards"
"Closing Ceremony of the 20th International Conference on Plastination"
Nicolás Ottone (President 20th ICP, Universidad de La Frontera, Temuco, Chile)
Rafael Latorre (President ISP, Universidad de Murcia, Spain)—All Council ISP

International Conferences on Plastination

1982   First International Conference on Plastination—University of Texas Health Science Center, San Antonio, Texas, USA. April 1982
1984   Second International Conference on Plastination—University of Texas Health Science Center, San Antonio, Texas, USA. April 1984

1986   Third International Conference on Plastination—University of Texas Health Science Center, San Antonio, Texas, USA. April 1986
1988   Fourth International Conference on Plastination—Mercer University School of Medicine Macon, Georgia, USA. March 1988
1990   Fifth International Conference on Plastination—Faculty of Medicine University of Heidelberg, Heidelberg, Germany. July 1990
1992   Sixth International Conference on Plastination—Department of Pathology, Queen's University, Kingston, Canada. July 1992
1994   Seventh International Conference on Plastination—Anatomisches Institut, Karl-Franzens-University, Graz, Austria. July 1994
1996   Eighth International Conference on Plastination—Division of Radiology, Mater Misericordiae Hospital, University of Queensland, Brisbane, Australia. July 1996
1998   Ninth International Conference on Plastination—Departement du Chimie-biologie, Universite du Quebec, Trois-Rivieres, Quebec, Canada. July 1998
2000   Tenth International Conference on Plastination-Faculte de Medecine Jacques Lisfranc, Jean Monnet University, St. Etienne, France. July 2000
2002   11th International Conference on Plastination—Medical Sciences, San Juan, Puerto Rico. July 2002
2004   12th International Conference on Plastination—Departamento de Anatomia y Embryologia, Facultad de Veterinaria, Universidad de Murcia Spain, July 2004
2006   13th International Conference on Plastination—Anatomical Institute, Medical University of Vienna, Vienna, Austria, July 2006
2008   14th International Conference on Plastination—Heidelberg and Guben, Germany, July 2008
2010   15th International Conference on Plastination—Joint Meeting AACA—University of Hawaii, College of Medicine, Hawaii, USA, July 2010
2012   16th International Conference on Plastination—Beijing and Dalian, China, July 2012
2014   17th International Conference on Plastination—Saint Petersburg, Russia, July 2014
2016   18th International Conference on Plastination—Toledo, Ohio, USA, June 2016
2018   19th International Conference on Plastination—Dalian, China, July 2018
2022   20th International Conference on Plastination—Temuco, Chile (Online), July 2022 (Full Conference: www.plastinacion.com)
2024   21th International Conference on Plastination—Istanbul, Turkey, July 2024

Available   from   [26]   https://isp.plastination.org/annual-meetings/previous-international-conferences/

Interim Meetings on Plastination

1989   First Interim Meeting on Plastination—College of Veterinary Medicine, University of Tennessee, Knoxville, TN, USA. November 1989
1991   Second Interim Meeting on Plastination—Chaffey College, Rancho Cucamonga, California, USA, August 1991
1993   Third Interim Meeting on Plastination—Department of Pathology, University of South Alabama, Mobile, Alabama, USA. August 1993
1995   Fourth Interim Meeting on Plastination—College of Veterinary Medicine, Ohio State University, Columbus, Ohio, USA. July 1995
1997   Fifth Interim Meeting on Plastination—College of Veterinary Medicine, University of Tennessee, Knoxville, TN, USA. June/July 1997
1999   Sixth Interim Meeting on Plastination—University of Rochester Medical Center, Rochester, New York, USA. July 1999
2001   Seventh Interim Meeting on Plastination—Medical Center of Fudan University, Shanghai, China and Su Yi Plastination Factory, Nanjing, China, June 2001
2005   Eighth Interim Meeting on Plastination—Department of Morphology Faculty of Veterinary Medicine, University of St "Cyril and Methodius" Macedonia. July 2005
2007   Ninth Interim Meeting on Plastination—Division of Anatomical Sciences, University of Michigan Medical School, Ann Arbor, Michigan, USA. July 2007
2011   Tenth Interim Meeting on Plastination—College of Medicine, University of Toledo, Toledo, Ohio, USA, July 2011
2015   11th Interim Meeting on Plastination—Universidade Federal do Espirito Santo, Vitoria, Brazil, July 2015
2017   12th Interim Meeting on Plastination—University of KwaZulu Natal—Durban, South Africa—July 2017
2021   First Online Interim Meeting—Universidad de Murcia—Murcia, Spain—April 2021

Available   from   [27]   https://isp.plastination.org/annual-meetings/previous-interim-meetings/

Historical Governance of the International Society for Plastination

Presidents

Harmon Bickley,[1] Macon, Georgia (USA), Period 1986–1995

---

[1] Executive director

Robert Henry, Knoxville, Tennessee (USA), Period 1996
Andreas H. Weiglein, Graz (Austria), Period 1997–2003
Mircea-Constantin Sora, Vienna (Austria), Period 2004–2008.
Carlos A. C. Baptista, Toledo, Ohio (USA), Period 2008–2016
Rafael Latorre, Murcia (Spain), Period 2016–2022
Dmitry Starchik, St. Petersburg (Russia), Period 2022–2024

Vice-Presidents

Regis Olry, Trois-Rivieres, Quebec (Canada), Period 1997–2001
   Mircea-Constantin Sora, Vienna (Austria), Period 2002–2003
   Andreas H. Weiglein, Graz (Austria), Period 2004–2008
   Rafael Latorre, Murcia (Spain), Period 2008–2016
   Dmitry Starchik, St. Petersburg (Russia), Period 2016–2022
   Nicolas E. Ottone, Temuco (Chile), Period 2022–2024

Secretaries

Peter Cook, Auckland (New Zealand), Period 1997–2001
   Tim Barnes, Columbus, Ohio (USA), Period 2002–2003
   David Hostler, Pittsburgh, Pennsylvania (USA), Period 2004–2007
   Barbara Weninger, Graz (Austria), Period 2007–2008
   Christoph von Horst, Mainburg (Germany), Period 2008–2014
   Selcuk Tunali, Ankara (Turkey), Period 2014–2018
   Nicolas E. Ottone, Temuco (Chile), Period 2018–2022
   Octavio López Albors, Murcia (Spain), Period 2022–2024

Treasurers

Ronald Wade, Baltimore, Maryland (USA), Period 1997–1998
   Wolfgang Weber, Ames, Iowa (USA), Period 1999–2000
   Robert Henry, Knoxville, Tennessee (USA), Period 2001–2008
   Ameed Raoof, Ann Arbor, Michigan (USA), Period 2008–2014
   Joshua Lopez, Tucson, Arizona (USA), Period 2014–2016
   Carlos A. C. Baptista, Toledo, Ohio (USA), Period 2016–2022
   Carlos A. C. Baptista, Toledo, Ohio (USA), Period 2022–2024
   Available from [28] https://isp.plastination.org/about-us/governance/

ISP Distinguished Member Award

Harmon Bickley, MD

Seventh International Conference on Plastination—Anatomisches Institut, Karl-Franzens-University, Graz, Austria. July 27, 1994

Gunther von Hagens, MD

Seventh International Conference on Plastination—Anatomisches Institut, Karl-Franzens-University, Graz, Austria. July. 27, 1994

Robert W. Henry, DVM, PhD

Eleventh International Conference on Plastination—Medical Sciences, San Juan, Puerto Rico. July 16, 2002

Andreas H. Weiglein, PhD

Sixteenth International Conference on Plastination—Beijing, China, July 18, 2012

Carlos A. C. Baptista, MD, MS, PhD, MPH

Eighteen International Conference on Plastination—Toledo, Ohio, USA, June 30, 2016

Available from [29] https://isp.plastination.org/awards/distinguished-members-awards/

ISP Career Achievement Award

Gunther von Hagens, MD

This award was presented by Dr. Nicolas Ottone, President of the 20th International Conference on Plastination, held by *Universidad de La Frontera*, Temuco, Chile, online, in July 2022 [30] (https://www.youtube.com/watch?v=nwgDPhNbsMI).

Available from [25] https://isp.plastination.org/awards/career-achievement-award/

Brief Details About the Beginnings of the Author in the Plastination Technique and the Present of the Plastination Laboratory at Universidad de La Frontera

Our beginnings in plastination were in 2006, in Buenos Aires, Argentina, in the Dissection Team of the Second Chair of Anatomy of the Faculty of Medicine, University of Buenos Aires, Argentina. There, with the support of Professors Vicente Hugo Bertone, Esteban Blasi, and Carlos Medan, the first attempts at the plastination technique were developed, with local development of the equipments and the implementation of silicone and other local chemicals and instruments, which allowed to start with few resources. To develop the plastination technique, although the first attempts were unsuccessful, with some accidents included, this did not prevent further progress to achieve good results. Subsequently, Prof. Dr. Homero Bianchi granted a space in the Institute of Morphology J. J. Naón (University of Buenos Aires, Argentina) for the development of the technique, dedicating a specific space for the establishment of a plastination laboratory, in addition to having the support of personnel, such as the Technical Assistant Gaspar Gonzalez, as well as the collaboration of medical students and anatomy assistants of that time, today medical doctors, such as Vanina Cirigliano, Lucero Oloriz, and Daniela Caamaño. Also, we had the collaboration of Gonzalo Borges Brum (DVM), to whom we advise and collaborate in the assembly of the Plastination laboratory in the Faculty of Veterinary Medicine of the University of Buenos Aires, Argentina, in addition to carrying out joint scientific work. A large number of presentations on the plastination technique could also be developed between 2008 and 2013 in Argentine congresses and then, from 2010 onward, in international congresses, highlighting the first international participation in the XVII Pan-American Congress of Anatomy held in the city of Temuco, Chile, under the presidency of Prof. Dr. Mariano del Sol. In this sense, from 2006 to date, the Plastination Laboratory of the *Universidad de La Frontera* has delivered 49 conferences in the area of plastination [31–79], presenting 35 scientific papers on the technique [80–114], and published 17 scientific articles [18, 20, 115–129], demonstrating the dissemination and scientific promotion of the plastination technique. In 2014, I joined the *Universidad de La Frontera*, specifically in the Faculty of Dentistry as a professor, and in the Doctorate in Morphological Sciences as a student, obtaining a doctorate in 2017. Prof. Dr. Mariano del Sol and Prof. Dr. Ramón Fuentes were crucial in providing with the necessary support so that we could set up the Laboratory of Plastination from scratch. There, at *Universidad de La Frontera*, the development of the plastination technique is notable, allowing all the knowledge developed and the experiments carried out in plastination to be transferred not only to scientific articles but also to research projects. During these years, there was also the support of Prof. Dr. Santiago Aja Guardiola (1943–2021) from the National Autonomous University of Mexico, Mexico, as well as Prof. Dr. Carlos Baptista, from the University of Toledo, Ohio, USA, who colaborated in all versions of the plastination workshops developed at *Universidad de La Frontera*, together with Prof. Dra. Telma Masuko, from the Federal University of Bahia, Brazil. Likewise, the Laboratory of Plastination at the *Universidad de La Frontera* has the collaboration of Dr. Carlos Veuthey, Veterinary Doctor, and Prof. Mg. Ruth Prieto, with whom the line of developmental biology is developed. Likewise, we are developing protocols to apply histological stains to microplastinatedized samples, in a collaborative work with Dr. Bélgica Vásquez. In addition, these microplastinated samples are visualized using the TissueFaxs tissue cytometer (TissueGnostics GmbH, Austria), for which it is also necessary to develop special protocols for scanning and quantifying cell structures in microplastinated samples, thinking in the future of the possibility of applying immunofluorescence protocols on the microplastinized sections. Also noteworthy was the participation in research activities of doctoral students, now PhDs, Jaime Correa, Marco Guerrero, Mariela Muñoz, Camila Panes, Nikol Ponce, Aurora Prado, and Claudia Vargas and dentistry student Diego Guzman. In relation to international collaboration, it is also necessary to highlight the collaborative research work and advice from our laboratory for the start-up of plastination laboratories in other South American countries, such as Bolivia (La Paz, Univalle) with Dr. Victor Hugo Rodríguez Torres, and Ecuador (Quito, Central University of Ecuador), with Dr. María Revelo Cueva, in addition to the training of academics from universities throughout the continent, attending our plastination courses (see next section). In this way, currently, with the development of all the plastination techniques for the preservation of samples for undergraduate and postgraduate teaching, as well as with the proposal of the micro-plastination technique and the search for intact DNA in plastinated samples, the Laboratory of Plastination of *Universidad de La Frontera* is positioned at the forefront of anatomical research, seeking to improve the developments of the plastination technique for its application in teaching, research, and academic extension. Next, the scope of the plastination laboratory in the field of graduate and postgraduate training activities will be developed.

Activities Developed in the Laboratory of Plastination and Anatomical Techniques of Universidad de La Frontera

From the point of view of undergraduate and postgraduate training, in the Laboratory of Plastination, with the support of the Faculty of Dentistry, the PhD Program in Morphological Sciences, and the Center of Excellence in Morphological and Surgical Studies (CEMyQ), the International Workshop on Plastination and Anatomical Techniques has been developed since 2017, the only one in its version in South America, with the institutional support of the International Society for Plastination (ISP), currently becoming a unique training instance in all plastination techniques, especially for anatomists and morphologists from all over South America. These workshops were directed by Prof. Nicolás Ottone (ISP Secretary during the years 2018–2022 and current ISP Vice-President), and had the participation as international instructors of Prof. Dr. Carlos Baptista, from the University of Toledo, Ohio, USA (president of ISP 2008–2016 and current treasurer); Prof. Dr. Telma Masuko, from the Federal University of Bahia, Bahia, Brazil (member of the ISP board of directors, 2016 2020); and Prof. Dr. Rafael Latorre, from the University of Murcia, Spain (president of ISP 2016–2022).

During the pandemic, especially in 2021 and 2022, and as a result of the forced isolation and the suspension of face-to-face activities, we had to suspend the development of the plastination workshops, and they were replaced by an online Diploma in Advanced Anatomical Techniques. It was a great success and allowed us to maintain training activities in anatomical techniques and plastination during the pandemic. In this Diploma, we have the participation of the following academics: Prof. Dr. Mariano del Sol (*Universidad de La Frontera*, Temuco, Chile), Prof. Dr. Rubén Daniel Algieri (School of Medicine, *Universidad de Buenos Aires*, Argentina), Prof. Dr. Telma Masuko (Federal University of Bahia, Brazil), Prof. Dr. Oscar Inzunza (*Pontificia Universidad Católica de Chile*, Santiago, Chile), Prof. Dra. Carolina Brofman (School of Medicine, *Universidad de Buenos Aires*, Argentina), Prof. Dr. Pablo Lizana Arce (*Pontificia Universidad Católica de Valparaíso*, Chile), Prof. Dr. Juan Pablo Fernández (School of Medicine, *Universidad de Buenos Aires*, Argentina), Prof. Dr. Marco Guerrero (*Universidad Central del Ecuador*, Ecuador), Prof. Dr. Rafael Latorre (University of Murcia, Spain), Prof. Dr. Carlos Baptista (University of Toledo, Ohio, USA), Prof. Dr. Gonzalo Borges Brum (School of Veterinary, *Universidad de Buenos Aires*, Argentina), Prof. Dr. Carlos Blanco (School of Veterinary, *Universidad de Buenos Aires*, Argentina), Prof. Dr. Diego Jara (*Museo Nacional de Historia Natural*, Chile), and Prof. Dr. Volker Schill (BIODUR GmbH, Germany).

In the workshops and diplomas, in postgraduate activities, and in plastination and anatomical techniques developed from the Laboratory of Plastination and Anatomical Techniques of *Universidad de La Frontera*, from 2017 to 2022, more than 50 academics, postgraduate, and undergraduate students from the following university institutions participated.

Argentina

Fundación Barceló (Buenos Aires)
Universidad de Buenos Aires
Universidad Nacional del Comahue (Cipoletti)
Universidad Nacional de Córdoba
Universidad Nacional de Río Cuarto
Universidad de Morón
Universidad Nacional del Sur
Universidad Adventista del Plata

Bolivia

Universidad Mayor de San Andrés
Universidad Mayor de San Simón
Univalle La Paz

Chile

Pontificia Universidad Católica de Chile
Pont. Universidad Católica de Valparaíso
Universidad de La Frontera
Universidad Austral de Chile
Universidad Mayor
Universidad Autónoma de Chile
Universidad de Santiago de Chile
Universidad San Sebastián (Puerto Montt, Concepción)
Universidad de Playa Ancha
Universidad Diego Portales
Universidad de Concepción
Universidad Católica de Temuco
Museo Nacional de Historia Natural

Colombia

Pontificia Universidad Javeriana
Universidad Libre (Bogotá)

Costa Rica

Universidad de Ciencias Médicas, Alajuela

Ecuador

Universidad Central del Ecuador

El Salvador

Universidad de El Salvador

Paraguay

Universidad Nacional de Asunción

Perú

Universidad Católica de Arequipa

Uruguay
Universidad de la República

United States of America
The University of Texas Medical Branch
In turn, during the ISP Assembly at the XIX International Congress on Plastination, held in the city of Dalian, China, the holding of the XX International Congress on Plastination was voted and approved, in the city of Pucón, Chile, from of July 20 to 24 of 2020, under the organization and direction of Prof. Nicolás Ottone. This was the first time that an International Conference on Plastination was organized in South America. However, the pandemic changed the plans again; the face-to-face organization of this conference had to be suspended, which was finally developed online, in 2022, from 18 to 21, through the organization of the *Universidad de La Frontera*, and with the presidency of Prof. Nicolás E. Ottone. It had the invaluable participation of Prof. Gunther von Hagens, who was given a recognition for his career. The world's leading figures in plastination also participated (see above the Congress Program).

Likewise, the Laboratory of Plastination and Anatomical Techniques, as part of the Doctoral Program in Morphological Sciences, collaborates in the delivery of the Advanced Anatomical Techniques Course, which must be taken as a compulsory subject for all students of the Doctoral Program in Morphological Sciences.

In the future, extension projects will also be sought to bring the importance of anatomical knowledge closer, through plastination techniques, to elementary and middle schools in the region.

Organization of Some Activities Associated with Anatomical Techniques and Plastination

I International Course on Anatomical Techniques and Plastination. XLVIII Argentine Congress of Anatomy, I International Congress of Anatomy. Barceló Foundation Santo Tomé Headquarters, Santo Tomé, Corrientes, Argentina. October 6–8, 2011. Directors: Prof. Dr. Santiago Aja-Guardiola, Prof. Dr. Ismael Concha, Prof. Dr. Nicolás E. Ottone

II International Course on Anatomical Techniques and Plastination. Pre-Congress Course XLIX Argentine Congress of Anatomy, II International Congress of Anatomy. Central Military Hospital, Buenos Aires, Argentina. September 5–8, 2012. Directors: Prof. Dr. Santiago Aja-Guardiola, Prof. Dr. Ismael Concha, Prof. Dr. Nicolás E. Ottone

I Pan-American Symposium on Plastination. Organized within the framework of the 50th Argentine Congress of Anatomy. National University of Rosario, Rosario, Argentina. October 17–19, 2013. President: Prof. Dr. Nicolás E. Ottone (Universidad de La Frontera, Temuco, Chile)

II Pan-American Symposium on Plastination. Organized within the framework of the XVI Congress of Anatomy of the Southern Cone. National University of the Northeast, Corrientes, Argentina. October 16–18, 2014. President: Prof. Dr. Telma Masuko (Universidade Federal de Bahia, Brazil)

III Pan-American Symposium on Plastination. Organized within the framework of the LII Argentine Congress of Anatomy. Barceló Foundation, La Rioja, Argentina. September 16, 2015. President: Dr. Nicolás E. Ottone (Universidad de La Frontera, Temuco, Chile)

IV Pan-American Symposium on Plastination. Faculty of Veterinary Sciences, University of Buenos Aires, Buenos Aires, Argentina. August 24, 2016. President: Prof. Dr. Telma Masuko (Universidade Federal de Bahia, Brazil)

I International Congress on Anatomical Techniques. Organized within the framework of the V Regional Congress of Morphology, XVIII Ibero-Latin American Symposium of Anatomical, Histological and Embryological Terminology, XVI SILAT. I International Congress on Anatomical Techniques. Pucón Campus, Universidad de La Frontera, Temuco, Chile. November 11, 12 and 13, 2019. President: Prof. Dr. Nicolás E. Ottone (Universidad de La Frontera, Temuco, Chile)

II International Congress on Anatomical Techniques. Organized within the framework of the XXII Congress of Anatomy of the Southern Cone, VI Regional Congress of Morphology, II Conference of the Pan-American Association of Anatomy. On-line. Universidad de La Frontera, Temuco, Chile. November 23 to 27, 2020. President: Prof. Dr. Nicolás E. Ottone (Universidad de La Frontera, Temuco, Chile)

III International Congress on Anatomical Techniques. Organized within the framework of the XXIII Congress of Anatomy of the Southern Cone. On-line. Universidad de La Frontera, Temuco, Chile. July 18–1, 2021. President: Prof. Dr. Telma Masuko (Universidade Federal de Bahia, Brazil)

IV International Congress on Anatomical Techniques. Organized within the framework of the 20th International Conference on Plastination. On-line. Universidad de La Frontera, Temuco, Chile. July 18–21, 2022. President: Prof. Dr. Nicolás E. Ottone (Universidad de La Frontera, Temuco, Chile)

IV International Congress on Anatomical Techniques. Organized within the framework of the XXV Congress of Anatomy of the Southern Cone. Campus Pucón—Universidad de La Frontera, Pucón, Chile. October 4–6, 2023. President: Prof. Dr. Nicolás E. Ottone (Universidad de La Frontera, Temuco, Chile)

References

1. von Hagens G. Ein Leben für die Wissenschaft. Köerperwelten Das Original. http://www.koerperwelten.com/de/gunther_von_hagens/leben_wissenschaft.html.
2. von Hagens G. n Plastinarium in Guben. http://www.plastinarium.de/en/gunther_von_hagens/etappen_wege_ziele_copy.html.
3. von Hagens G. Impregnation of soft biological specimens with thermosetting resins and elastomers. Anat Rec. 1979;194(2):247–55. https://doi.org/10.1002/ar.1091940206.

4. von Hagens G. Heidelberg plastination folder. Collection of all technical leaflets for plastination. 2nd ed. Heidelberg: Anatomische Institut 1, Universitat Heidelberg; 1986.

5. von Hagens G, Tiedemann K, Kriz W. The current potential of plastination. Anat Embryol (Berl). 1987;175(4):411–21. https://doi.org/10.1007/BF00309677.

6. Bickley HC, von Hagens G, Townsend FM. An improved method for the preservation of teaching specimens. Arch Pathol Lab Med. 1981;105(12):674–6.

7. Bravo H. Plastination an additional tool to teach anatomy. Int J Morphol. 2006;24(3):475–80. https://doi.org/10.4067/S0717-95022006000400029.

8. Whalley A. Pushing the limits. 2nd printing. Heidelberg: Arts and Sciences Verlagsgesellschaft GmbH; 2007.

9. Bickley HC. A brief chronology of international happenings in plastination. J Int Soc Plastination. 1995;9(1):11–2.

10. Pashaei S. A brief review on the history, methods and applications of plastination. Int J Morphol. 2010;28(4):1075–9. https://doi.org/10.4067/S0717-95022010000400014.

11. Baptista CAC. Letter from the president. J Plastination. 2015;27(1):2.

12. Aja Guardiola S, Martínez Galindo J. Plastination, the modern technique for obtaining more useful macro-specimens in the teaching-learning process. In: Mexico City, First National Congress of Veterinary Anatomy of Mexico, held from October 12 to 15, 1988.

13. Bohannon J. Gunther von Hagens. Plastination: putting a stopper in death. Science. 2003;301(5637):1173. https://doi.org/10.1126/science.301.5637.1173.

14. von Hagens G. Guest at Friday night with Jonathan Ross. Broadcasting date 21.3.2008 [Video file]. 2008. https://www.youtube.com/watch?v=I1lihuIkmQs.

15. Zheng TZ, Weatherhead BL, Gosling J. Plastination at room temperature. J Int Soc Plastination. 1996;11(2):33.

16. Zheng T, Liu J, Zhu K. Plastination at room temperature. J Int Soc Plastination. 1998;13(2):21–5.

17. Glover R, Henry R, Wade R. Polymer preservation technology: Poly-Cur, a next generation process for biological specimen preservation. In: Abstracts of the 9th International Conference on Plastination held in Quebec City, Canada. J Int Soc Plastination. 1998;13(2).

18. Ottone NE, Cirigliano V, Bianchi HF, Medan CD, Algieri RD, Borges Brum G, Fuentes R. New contributions to the development of a plastination technique at room temperature with silicone. Anat Sci Int. 2015;90(2):126–35. https://doi.org/10.1007/s12565-014-0258-6.

19. Starchik D, Henry RW. Room temperature/Corcoran/Dow Corning™-Silicone plastination process. Anat Histol Embryol. 2019;48(6):539–46. https://doi.org/10.1111/ahe.12505.

20. Ottone NE. Gunther von Hagens, creator of plastination. Historical review and technical development. Rev Argent Anat Online. 2013;4(2):70–6. https://www.revista-anatomia.com.ar/archivos-parciales/2013-2-revista-argentina-de-anatomia-online-f.pdf.

21. Dalian Hoffen Bio-Technique Co Ltd. 2016. Web Site. Chinadaily.com.cn.

22. von Hagens G. Letter from Gunther von Hagens. In: 10th International Interim Conference on Plastination Toledo, Ohio, USA, July 9–12, 2011. J Plastination. 2009–2012;24:21.

23. 20th International Conference on Plastination. 2022. www.icp2022chile.com.

24. Ottone NE. The 20th International Conference on Plastination, Temuco, Chile, July 18–July 21, 2022. J Plastination. 2023;34(2):JP-22-09. https://journal.plastination.org/articles/the-20th-international-conference-on-plastination-temuco-chile-july-18-july-21-2022/.

25. Recognition to Gunther von Hagens. 2022. https://www.youtube.com/watch?v=nwgDPhNbsMI.

26. International Society for Plastination (ISP). Previous International Conferences. International Society for Plastination. 2023. https://isp.plastination.org/annual-meetings/previous-international-conferences/.

27. International Society for Plastination (ISP). Previous Interim Meetings. International Society for Plastination. 2023. https://isp.plastination.org/annual-meetings/previous-interim-meetings/.

28. International Society for Plastination (ISP). Governance. Web Site. International Society for Plastination. 2023. https://isp.plastination.org/about-us/governance/.

29. International Society for Plastination (ISP). Distinguished members awards. Web Site. International Society for Plastination. 2023. https://isp.plastination.org/awards/distinguished-members-awards/.
30. International Society for Plastination (ISP). Career achievement award. Web Site. International Society for Plastination. 2023. https://isp.plastination.org/awards/career-achievement-award/.
31. Ottone NE. Lecture: "Advances in plastination techniques. Applications in anatomical teaching and research". In: XX Pan American Congress of Anatomy, Quito, Ecuador. December 2, 2022.
32. Ottone NE. Lecture: "Fundamentals and applications of the plastination technique". In: "Round table on anatomical techniques. Present and future in cadaveric fixation and conservation"—VII Clinical Anatomy Teaching and Research Conference, I Congress of the International Forum of Biomedical Sciences—Organized by the Argentine Association of Clinical Anatomy. Online. November 10, 2022.
33. Ottone NE. Lecture: "Rediscovering sectional anatomy through plastination". In: "Anatomy master classes", organized by Elsevier and the Pan-American Association of Anatomy. Online. September 1, 2022.
34. Ottone NE. Lecture: "Development of plastination as a scientific tool for its implementation in research in the field of health sciences". In: "International Scientific Conference on Medicine UNIVALLE 2022"—Organized by the Universidad del Valle, Bolivia. Online. August 26, 2022.
35. Ottone NE. Lecture: "Plastination techniques". In: "Anatomical Techniques Seminar", Organized by the Post Graduation Program in Domestic and Wild Animals Anatomy, Faculty of Veterinary Medicine and Zootechnics, Department of Surgery, University of São Paulo, Sao Paulo, Brazil. Online. March 7, 2022.
36. Ottone NE. Lecture: "Advances in plastination and micro-plastination techniques oriented to teaching, research and extension in anatomy"—XXII National and I International Conference of the Inter-American Open University, Inter-American Open University, City of Rosario, Argentine Republic. Online. October 14, 2021.
37. Ottone NE. Lecture: "Anatomical techniques for the laboratory"—II Ecuadorian Congress of Morphological Sciences and I Student Scientific Conference on Morphology, of the Ecuadorian Society of Morfunctional Sciences, sponsored by the Association of Ecuadorian Faculties of Medical and Health Sciences AFEME, the Scientific Association of Medical Students ASOCEM and the Pan-American Association of Anatomy, Quito, Ecuador. Online. March 24 to 28, 2021.
38. Ottone NE. Lecture: "New postgraduate training program in morphological sciences: diploma in advanced anatomical techniques". In: XXII Congress of Anatomy of the Southern Cone, VI Regional Congress of Morphology, II International Congress on Anatomical Techniques, II Conference of the Pan-American Association of Anatomy. Universidad de La Frontera, Temuco, Chile. Online. November 23 to 27, 2020.
39. Ottone NE. Lecture: "Use of plastinated pieces for teaching and research in head and neck anatomy". In: III COBRANCAPE—III Congress of Head and Neck Anatomy. Organized by the Brazilian Society of Anatomy. Online. October 24, 2020.
40. Ottone NE. Lecture: "Microplastination. A new concept in the development of the plastination technique for morphological research". In: National Conference on Anatomy—"Anatomy in Times of Pandemic". Organized by the Argentine Association of Anatomy and the Department of Normal Anatomy of the University of Mendoza, Argentina. Online. October 23, 2020.
41. Ottone NE. Lecture: "Plastination. Revolution in anatomical conservation for teaching and research". In: Continuing Education Program in Morphological Sciences of the Pan-American Association of Anatomy. 3rd APA WebINAR. Panamerican Association of Anatomy. Online. July 21, 2020.
42. Ottone NE. Lecture: "New advances in research with plastination". In: Board Meeting and Work of the Chilean Society of Anatomy. Doctorate in Morphological Sciences, Center of

Excellence in Morphological and Surgical Studies (CEMyQ), Universidad de La Frontera, Temuco, Chile. March 14, 2020.

43. Ottone NE. Lecture: "Plastination: application in teaching, extension and research of anatomy". In: XIX Anatomical Techniques Course "Prof. Hildegardo Rodrigues". Organized by Prof. Telma Masuko Institute of Health Sciences, Federal University of Bahia, Salvador—Bahia, Brazil, and the Brazilian Society of Anatomy. February 3 to 8, 2020.

44. Ottone NE. Lecture: "Plastination, a modern conservation method". In: XIX Pan American Congress of Anatomy—21st Congress of Anatomy of the Southern Cone—7th International Congress of Anatomy—56th Argentine Congress of Anatomy—40th Chilean Congress of Anatomy—1st Ecuadorian Congress of Morphological Sciences—17th Ibero-Latin American Symposium of Anatomical, Histological, Embryological Terminology—7th Argentine Congress of Anatomical Techniques—11th Argentine Conference on Anatomy for Health Sciences Students—I South American Meeting of the International Society for Plastination (ISP). Faculty of Medicine, University of Buenos Aires (UBA)—Buenos Aires, Argentina. May 27 to 31, 2019.

45. Ottone NE. Lecture: "Fundamentals and applications of plastination". In: XIX Pan American Congress of Anatomy—21st Congress of Anatomy of the Southern Cone—7th International Congress of Anatomy—56th Argentine Congress of Anatomy—40th Chilean Congress of Anatomy—1st Ecuadorian Congress of Morphological Sciences—17th Ibero-Latin American Symposium of Anatomical, Histological, Embryological Terminology—7th Argentine Congress of Anatomical Techniques—11th Argentine Conference on Anatomy for Health Sciences Students—I South American Meeting of the International Society for Plastination (ISP). Faculty of Medicine, University of Buenos Aires (UBA)—Buenos Aires, Argentina. May 27 to 31, 2019.

46. Ottone NE. Lecture: "ISP and 20th ICP, first time in Central and South America". In: XIX Pan American Congress of Anatomy—21st Congress of Anatomy of the Southern Cone—7th International Congress of Anatomy—56th Argentine Congress of Anatomy—40th Chilean Congress of Anatomy—1st Ecuadorian Congress of Morphological Sciences—17th Ibero-Latin American Symposium of Anatomical, Histological, Embryological Terminology—7th Argentine Congress of Anatomical Techniques—11th Argentine Conference on Anatomy for Health Sciences Students—I South American Meeting of the International Society for Plastination (ISP). Faculty of Medicine, University of Buenos Aires (UBA)—Buenos Aires, Argentina. May 27 to 31, 2019.

47. Ottone NE. Lecture: "Plastination, new paradigms in the teaching of morphological sciences". In: IV Conference on Teaching Morphological Sciences. 40 hours long. Department of Biology, University of Tarapacá, Arica, Chile. December 1, 2018.

48. Ottone NE. Lecture: "Development of plastination techniques in teaching, research and its application in clinic and surgery". In: Master's Program in Dentistry, Cohort 2018. Faculty of Dentistry, Universidad de La Frontera, Temuco, Chile. September 9, 2018.

49. Ottone NE. Lecture: "Research in morphological sciences through plastination". In: 55th Argentine Congress of Anatomy. Argentine Association of Anatomy. University of Mendoza, Mendoza, Argentina. June 21 and 22, 2018.

50. Ottone NE. Lecture: "Fundamentals of plastination". In: Course of advanced anatomical techniques. Doctorate in Morphological Sciences. Faculty of Medicine, Universidad de La Frontera, Temuco, Chile. May 2, 2018.

51. Ottone NE. Lecture: "History of plastination". In: Board Meeting and Work of the Chilean Society of Anatomy. Dr. Mario Cantin Auditorium. Doctorate in Morphological Sciences. Center of Excellence in Morphological and Surgical Studies. School of Medicine. University of the Border, Temuco, Chile. March 24, 2018.

52. Ottone NE. Lecture: "Plastination in cadavers and anatomical pieces as a support tool in the teaching of morphological sciences". In: Theoretical-practical workshop on the use of digital tools for the teaching-learning of morphofunction. Association of Ecuadorian Faculties of

Medical and Health Sciences (AFEME) and Pontificia Universidad Católica de Quito. Quito, Ecuador. February 14 to 17, 2018.

53. Ottone NE. Lecture: "Teaching and research in morphology through plastination techniques with silicone, epoxy resin and polyester". In: XIX Southern Cone Congress of Anatomy, XXXVIII Chilean Congress of Anatomy, III Anatomy Olympics for Undergraduate Students. Chilean Society of Anatomy, San Sebastian University, Concepción, Chile. November 8 to 10, 2017.

54. Ottone NE. Lecture: "Modified technical protocol for plastination of cuts with epoxy resin". In: 1st Workshop on Plastination and Anatomical Techniques. Laboratory of Plastination and Anatomical Techniques. Faculty of Dentistry, Universidad de La Frontera, Temuco, Chile. October 19, 2017.

55. Ottone NE. Lecture: "Modified technical protocol for plastination at room temperature". In: 1st Workshop on Plastination and Anatomical Techniques. Laboratory of Plastination and Anatomical Techniques. Faculty of Dentistry, Universidad de La Frontera, Temuco, Chile. October 18, 2017.

56. Ottone NE. Lecture: "Standard protocol for plastination of cuts with epoxy". In: 1st Workshop on Plastination and Anatomical Techniques. Laboratory of Plastination and Anatomical Techniques. Faculty of Dentistry, Universidad de La Frontera, Temuco, Chile. October 17, 2017.

57. Ottone NE. Lecture: "Plastination techniques and their importance in teaching and research in Anatomy". In: 54th Argentine Congress of Anatomy—6th International Congress of Anatomy. Argentine Association of Anatomy. Faculty of Medical Sciences of the National University of La Plata, City of La Plata, Province of Buenos Aires, Argentina. September 28 to 30, 2017.

58. Ottone NE. Lecture: "Development of plastination techniques at the Universidad de La Frontera". In: Board Meeting and Work of the Chilean Society of Anatomy. Dr. Mario Cantin Auditorium. Doctorate in Morphological Sciences. Center of Excellence in Morphological and Surgical Studies. School of Medicine, University of the Border, Temuco, Chile. March 18, 2017.

59. Ottone NE. Lecture: "Plastination: silicone, epoxy, polyester". In: Course of Advanced Anatomical Techniques. Doctorate in Morphological Sciences. Faculty of Medicine, Universidad de La Frontera, Temuco, Chile. January 25, 2017.

60. Ottone NE. Lecture: "Development of an alternative plastination technique at room temperature with silicone and epoxy resin". In: XV Conference of Veterinary & Human Anatomy Assistants. University Santo Tomas, Temuco, Chile. November 4 and 5, 2016.

61. Ottone NE. Lecture: "Research in morphological sciences through the plastination technique". In: LIII Argentine Congress of Anatomy, XXXVII Chilean Congress of Anatomy, VI International Congress of Anatomy. Argentine Association of Anatomy, Buenos Aires, Argentina. August 25 to 27, 2016.

62. Ottone NE. Lecture: "Plastination technique with silicone, epoxy, and polyester". In: Plastination Conference (Pre-Congress activity LIII Argentine Congress of Anatomy, XXXVII Chilean Congress of Anatomy, VI International Congress of Anatomy). Argentine Association of Anatomy. Faculty of Veterinary Medicine, University of Buenos Aires, Buenos Aires, Argentina. August 24, 2016.

63. Ottone NE. Lecture: "Introduction to plastination—plastination at room temperature—presentation of the ISP". In: Plastination Conference (Pre-Congress activity LIII Argentine Congress of Anatomy, XXXVII Chilean Congress of Anatomy, VI International Congress of Anatomy). Argentine Association of Anatomy. Faculty of Veterinary Medicine, University of Buenos Aires, Buenos Aires, Argentina. August 23, 2016.

64. Ottone NE. Lecture: "Plastination techniques". In: IV Pan American Symposium on Plastination. Faculty of Veterinary Sciences—Resolution (CD) No. 2013/16. University of Buenos Aires, Buenos Aires, Argentina. August 24, 2016.

65. Ottone NE. Lecture: "Plastination, its application in research in morphology and dentistry". In: Master's Program in Dentistry, Cohort 2016-2017. Faculty of Dentistry, Universidad de La Frontera, Temuco, Chile. March 11, 2016.
66. Ottone NE. Lecture: "Impregnation, draining, assembly, drying and curing". In: I International Course on Techniques for the Preparation and Conservation of Biological Material for the Teaching of Anatomy. Peruvian Society of Morphological Sciences, Federico Villarreal National University, "Hipólito Unanue" School of Medicine, Lima, Peru. November 27, 2015.
67. Ottone NE. Lecture: "Showing of inputs, dehydration and impregnation". In: I International Course on Techniques for the Preparation and Conservation of Biological Material for the Teaching of Anatomy. Peruvian Society of Morphological Sciences, Federico Villarreal National University, "Hipólito Unanue" School of Medicine, Lima, Peru. November 27, 2015.
68. Ottone NE. Lecture: "Plastination in morphological research". In: I International Course on Techniques for the Preparation and Conservation of Biological Material for the Teaching of Anatomy. Peruvian Society of Morphological Sciences, Federico Villarreal National University, "Hipólito Unanue" School of Medicine, Lima, Peru. November 27, 2015.
69. Ottone NE. Lecture: "Plastination of cuts at room temperature with resins". In: XIV National and International Congress of Morphophysiological Sciences—I Paraguayan Congress of Anatomy. Eastern University. Paraguayan Society of Morphophysiological Sciences. Paraguayan Society of Anatomy, City Pte, Franco, Paraguayan. September 11 and 12, 2015.
70. Ottone NE. Lecture: "Plastination at room temperature with silicone". In: XIV National and International Congress of Morphophysiological Sciences—I Paraguayan Congress of Anatomy. Eastern University. Paraguayan Society of Morphophysiological Sciences. Paraguayan Society of Anatomy, City Pte, Franco, Paraguayan. September 11 and 12, 2015.
71. Ottone NE. Lecture: "Application of anatomical techniques and plastination to research in anatomy and morphological sciences". In: XIV National and International Congress of Morphophysiological Sciences—I Paraguayan Congress of Anatomy. Eastern University. Paraguayan Society of Morphophysiological Sciences. Paraguayan Society of Anatomy, City Pte, Franco, Paraguayan. September 11 and 12, 2015.
72. Ottone NE. Lecture: "Application of the plastination technique to the study of morphology". In: Master's Program in Dentistry, Cohort 2015-2016. Faculty of Dentistry, Universidad de La Frontera, Temuco, Chile. July 10, 2015.
73. Ottone NE. Lecture: "Notions of plastination (silicones and resins)". In: Pre-seminar Course on Anatomical Techniques. XI Chilean Conference on Anatomy, XI Ibero-Latin American Symposium on Anatomical, Histological and Embryological Terminology, III Regional Meeting on Morphology. Universidad de La Frontera, Temuco, Chile. November 19 to 21, 2014.
74. Ottone NE. Lecture: "Videoconference: plastination". In: I Interdisciplinary Conference on Applied Anatomy for Health Sciences Students. I International Symposium of Anatomical Techniques. National University of Córdoba, Faculty of Medical Sciences, Cordoba Argentina. September 13, 2014.
75. Ottone NE. Lecture: "Plastination at room temperature with conventional silicones". In: 50th Argentine Congress of Anatomy. 3rd International Congress of Anatomy, 1st Argentine Congress of Anatomical Techniques, 5th Argentine Conference on Anatomy for Health Sciences Students. Central Amphitheater of the Faculty of Medical Sciences of the National University of Rosario, Argentina. October 17, 2013.
76. Ottone NE. Lecture: "Plastination of human and animal corpses at room temperature with silicones". In: 50th Argentine Congress of Anatomy. 3rd International Congress of Anatomy, 1st Argentine Congress of Anatomical Techniques, 5th Argentine Conference on Anatomy for Health Sciences Students. Central Amphitheater of the Faculty of Medical Sciences of the National University of Rosario, Argentina. October 15, 2013.
77. Ottone NE. Lecture: "Plastination with minimum requirements". In: 50th Argentine Congress of Anatomy. 3rd International Congress of Anatomy, 1st Argentine Congress of Anatomical Techniques, 5th Argentine Conference on Anatomy for Health Sciences Students. Central

Amphitheater of the Faculty of Medical Sciences of the National University of Rosario, Argentina. October 15, 2013.

78. Ottone NE. Lecture: "Plastination at room temperature: vacuum chamber evolution, low cost". In: XLIX Argentine Congress of Anatomy and II International Congress of Anatomy and IV Conference on Anatomy for Students of Health Sciences, of the Argentine Association of Anatomy, Central Military Hospital Cirujano Mayor Dr. Cosme Argerich and National Academy of Medicine. Autonomous City of Buenos Aires, Argentina. September 5 to 8, 2012.

79. Ottone NE. Lecture: "Plastination at room temperature". In: XVIII Argentine Congress of the Association of Morphological Sciences of Corrientes—XIII International Congress. Association of Morphological Sciences of Corrientes. R Campus of the Faculty of Medicine of the National University of the Northeast, City of Corrientes, Province of Corrientes, Argentina. August 24 and 25, 2012.

80. Guzman D, Bianchi H, del Sol M, Ottone NE. "Anatomical variations of radial and ulnar arteries in plastinated upper limbs". In: 20th International Conference on Plastination. 4th International Congress on Anatomical Techniques. International Society for Plastination & Universidad de La Frontera, Temuco, Chile. On-line. July 18 to 21, 2022.

81. Badilla N, Fernanda Quevedo M, Montecinos H, Ottone NE. "P40 sheet plastination technique. Protocols and applications. Bibliographical review with systematic search. Preliminary communication". In: 20th International Conference on Plastination. 4th International Congress on Anatomical Techniques. International Society for Plastination & Universidad de La Frontera, Temuco, Chile. Online. July 18 to 21, 2022.

82. Montecinos H, Fernanda Quevedo M, Badilla N, Ottone NE. "E12 sheet plastination technique. Protocols and applications. Bibliographical review with systematic search. Preliminary communication". In: 20th International Conference on Plastination. 4th International Congress on Anatomical Techniques. International Society for Plastination & Universidad de La Frontera, Temuco, Chile. Online. July 18 to 21, 2022.

83. Alvarez Guisbert OJ, Alvares Duran GO, Orozco Gonzales JJ, Ottone NE, Rodriguez Torrez VH. "First experience of plastination at height at Universidad Privada del Valle, La Paz, Bolivia". In: XXIII Congress Of Anatomy Of The Southern Cone—VII Regional Congress of Morphology—XLI Chilean Congress of Anatomy—III International Congress on Anatomical Techniques—III Conference of the Pan-American Association of Anatomy— Online. November 15 to 19, 2021.

84. Ottone NE, Muñoz Ortega, M, Baptista CAC, del Sol M. "DNA extraction from plastinated tissues". In: XIX Pan American Congress of Anatomy—21st Congress of Anatomy of the Southern Cone—7th International Congress of Anatomy—56th Argentine Congress of Anatomy—40th Chilean Congress of Anatomy—1st Ecuadorian Congress of Morphological Sciences—17th Ibero-Latin American Symposium of Anatomical, Histological, Embryological Terminology—7th Argentine Congress of Anatomical Techniques—11th Argentine Conference on Anatomy for Health Sciences Students—I South American Meeting of the International Society for Plastination (ISP). Faculty of Medicine, University of Buenos Aires (UBA)—Buenos Aires, Argentina. May 27 to 31, 2019.

85. Vargas C, Baptista C, Veuthey C, del Sol M, Sandoval Vásquez C, Ottone NE. "Plastination of ultrathin sections in the humeral joint of a rat with induced osteoarthritis, for the identification of neovascularization". In: XIX Pan American Congress of Anatomy—21st Congress of Anatomy of the Southern Cone—7th International Congress of Anatomy—56th Argentine Congress of Anatomy—40th Chilean Congress of Anatomy—1st Ecuadorian Congress of Morphological Sciences—17th Ibero-Latin American Symposium of Anatomical, Histological, Embryological Terminology—7th Argentine Congress of Anatomical Techniques—11th Argentine Conference on Anatomy for Health Sciences Students—I South American Meeting of the International Society for Plastination (ISP). Faculty of Medicine, University of Buenos Aires (UBA)—Buenos Aires, Argentina. May 27 to 31, 2019.

86. Ottone NE, Vargas C, Guerrero M, Alarcón E, Veuthey C. "Preservation of human brain sections using a plastination technique of sections with polyester resin (Biodur P40)". In:

XX Congress of Anatomy of the Southern Cone, XVI Ibero-Latin American Symposium of Anatomical, Histological, Embryological Terminology—XVI SILAT, XII Chilean Conference on Anatomy, IV Regional Meeting of Morphology, Pucon, Chile. October 4, 5 and 6, 2018.

87. Prieto R, Ottone NE "Development of a room temperature silicone plastination protocol for human placenta". In: XX Congress of Anatomy of the Southern Cone, XVI Ibero-Latin American Symposium of Anatomical, Histological, Embryological Terminology—XVI SILAT, XII Chilean Conference on Anatomy, IV Regional Meeting of Morphology, Pucon, Chile. 4, 5 and 6 October 2018.

88. Villagrán F, Riveros A, Navarrete J, Veuthey C, Guerrero M, Lizana P, Ottone NE. "Development of a protocol for plastination with silicone at room temperature of coronal sections of a head of Sus scrofa domesticus, after injection with natural latex through the arterial route". In: XX Congress of Anatomy of the Southern Cone, XVI Ibero-Latin American Symposium of Anatomical, Histological, Embryological Terminology—XVI SILAT, XII Chilean Conference on Anatomy, IV Regional Meeting of Morphology, Pucon, Chile. 4, 5 and 6 October 2018.

89. Ottone NE, Vargas C, Guerrero M, Alarcón E, Veuthey C. "Development of a P40 cut plastination technique in human brain sections". In: 55th Argentine Congress of Anatomy. Argentine Association of Anatomy. University of Mendoza, Mendoza, Argentina. June 21 and 22, 2018.

90. Ottone NE, Vargas C, Veuthey C, del Sol M, Fuentes R. "Plastination of sections with Biodur E12 epoxy resin. New fast protocol". In: XIX Southern Cone Congress of Anatomy, XXXVIII Chilean Congress of Anatomy, III Anatomy Olympics for Undergraduate Students. Chilean Society of Anatomy, San Sebastian University, Concepción, Chile. November 8 to 10, 2017.

91. Ottone NE, Vargas C, Veuthey C, del Sol M, Fuentes R. "New rapid protocol for plastination of sections with Biodur E12 epoxy resin". In: 54th Argentine Congress of Anatomy, 6th International Congress of Anatomy. Argentine Association of Anatomy, National University of La Plata, La Plata, Argentina. September 28 to 30, 2017.

92. Ottone NE, Bianchi HF, Latorre L, Borges Brum G, Blanco CJ, Fuentes R. "Three-dimensional and two-dimensional reconstruction of plastinated specimens with silicone at room temperature". In: 18th International Conference on Plastination, University of Toledo, Ohio, United States of America. June 26 to July 1, 2016.

93. Borges Brum G, Vidal Figueredo R, Ottone NE, Blanco CJ. "Set up of a plastination laboratory at the Faculty of Veterinary Science at the University of Buenos Aires". In: 18th International Conference on Plastination, University of Toledo, Ohio, United States of America. June 26 to July 1, 2016.

94. Ottone NE, Borges Brum G, Vidal Figueredo R, Consejero E, Aja-Guardiola S, Masuko T, Blanco C. "Plastination of dog brain sections stained with the Mulligan technique". In: XXXVI Chilean Congress of Anatomy—XIII Latin American Symposium of Anatomical, Histological, Embryological Terminology—XVII Congress of the Southern Cone, Valdivia, Chile. November 11, 12 and 13, 2015.

95. Ottone NE, Cirigliano V, Bianchi HF, Fuentes R. "New contributions in the development of a plastination technique at room temperature with silicone". In: XI Chilean Conference on Anatomy, XI Ibero-Latin American Symposium on Anatomical, Histological and Embryological Terminology, III Regional Meeting on Morphology. Universidad de La Frontera, Temuco, Chile. November 19 to 21, 2014.

96. Ottone NE, Cirigliano V, Oloriz L, Caamaño D, Lo Tártaro M, Algieri RD, Medan C, Bianchi HF, Fuentes R. "Plastination at room temperature with silicone. Fast technique and low cost". In: 16th Congress of Anatomy of the Southern Cone, LI Argentine Congress of Anatomy, XXXV Chilean Congress of Anatomy, II Uruguayan Congress of Anatomy. National University of the Northeast, Corrientes, Argentina. October 16 to 18, 2014.

97. Aja-Guardiola S, Dominguez Calderón G, Cajiao MN, González-Contreras IP, Tomero Ramírez M, Jimenez Mejía R, Ottone NE. "Plastination as an educational tool in favor of

animal welfare". In: XXIV Pan American Congress of Veterinary Sciences. Pan American Association of Veterinary Sciences. Havana Convention Center, Cuba. October 6 to 9, 2014.

98. Aja-Guardiola S, Barreto Oble D, García Jarquín J, Reyna Covarrubias D, Ottone NE. "Emergency curing in an exhibit of plastinated specimens after hydrometeorological disasters". In: XXIV Pan American Congress of Veterinary Sciences. Pan American Association of Veterinary Sciences. Havana Convention Center, Cuba. October 6 to 9, 2014.

99. Aja-Guardiola S, Ottone NE, Olmedo Pérez G, González IP, Domínguez Calderón RG. "Application of plastination towards parasitology". In: XXV National Congress of Anatomy "Mtra. Conception Rugerio and Vargas". 1st National Meeting of Neuromorphology. Mexican Society of Anatomy, Durango, Mexico. October 1 to 4, 2014.

100. Ottone NE, Fuentes R, Bianchi HF, Algieri RD, Cirigliano V, Caamaño D, Oloriz, L, Lo Tártaro M. "Plastination with silicon and epoxy resin: new technique and its importance in the study of anatomy". In: 18th Congress of the International Federation of Associations of Anatomists—IFAA. Beijing, China. August 8 to 10, 2014.

101. Ottone NE, Bianchi HF, Fuentes R, Cirigliano V, Oloriz L, Caamaño D, Lo Tártaro M, Medan C. "Contributions to the development of plastination technique at room temperature with silicone". In: 17th International conference on Plastination—International Society for Plastination, Saint Petersburg, Russia. July 14 to 18, 2014.

102. Ottone NE, Bianchi HF, Fuentes R, Aja-Guardiola S, Cirigliano V, Oloriz L, Borges Brum G, Blasi E, Algieri RD, Bertone VH. "Development and installation of a plastination laboratory using low cost equipment and materials". In: 17th International conference on Plastination—International Society for Plastination, Saint Petersburg, Russia. July 14 to 18, 2014.

103. Aja-Guardiola S, Barreto-Oble D, García Jarquín J, Reyna Covarrubias DA, Ottone NE. "Intra-museum emergency curing due to lack of silicone hardening in the plastination process". In: XVII International Congress of Medicine, Surgery and Zootechnics in Dogs, Cats and Other Pets. National Autonomous University of Mexico, Faculty of Veterinary Medicine and Zootechnics. Ministry of Continuing Education and Technology. Department of Medicine, Surgery and Zootechnics in Dogs and Cats. Mexican Canophile Federation, Mexico City, Mexico. March 30, 2014.

104. Ottone NE, Caamaño D, Cirigliano V, Oloriz L, Borges Brum G, Bianchi HF, Blasi E, Medan C, Bertone HV, Aja-Guardiola S. "Plastination of human and animal corpses with conventional silicones". In: 50th Argentine Congress of Anatomy. 3rd International Congress of Anatomy, 1st Argentine Congress of Anatomical Techniques, 5th Argentine Conference on Anatomy for Health Sciences Students. Faculty of Medical Sciences of the National University of Rosario, Rosario, Argentina. October 15 to 19, 2013.

105. Aja-Guardiola S, Domínguez Calderón RG, Guajardo Garza R, Olmedo Pérez G, Borges Brum G, García Jarquín J, Ottone NE. "Recycling and recovery of acetone within the plastination laboratory itself". In: 50th Argentine Congress of Anatomy. 3rd International Congress of Anatomy, 1st Argentine Congress of Anatomical Techniques, 5th Argentine Conference on Anatomy for Health Sciences Students. Faculty of Medical Sciences of the National University of Rosario, Rosario, Argentina. October 15 to 19, 2013.

106. Ottone NE, Aja-Guardiola S. "Plastination at room temperature of human and animal cadavers with conventional silicones". In: XVIII Pan American Congress of Anatomy—XX National Meeting of Morphology "Dr. Fernando Quiroz Pavía"—IX Ibero-Latin American Symposium of Terminology, Huatulco, Oaxaca, Mexico. September 29 to October 4, 2013.

107. Ottone NE, Aja Guardiola S, Borges Brum G, Blasi ED, Medan CD, Algieri RD, Bertone VH, Bianchi HF. "Room temperature plastination of human and animal brains with conventional silicones". In: XVIII Pan American Congress of Anatomy—XX National Meeting of Morphology "Dr. Fernando Quiroz Pavía"—IX Ibero-Latin American Symposium of Terminology. Huatulco, Oaxaca, Mexico. September 29 to October 4, 2013.

108. Ottone NE, Blasi ED, Medan CD, Algieri RD, Cirigliano V, Oloriz L, Frojan D, Bertone VH, Aja Guardiola S. "Plastination at room temperature: vacuum chamber and forced impregnation". In: I National and International Virtual Conference on Education and Research in

Morphological Sciences—Normal Anatomy Chairs of the School of Medical Technology, School of Kinesiology and Physiotherapy, Faculty of Medical Sciences, National University of Córdoba (FCM-UNC) and Association of Anatomists of Cordoba (ADAC)—Cordoba, Argentina. November 10 to 30, 2012.

109. Ottone NE, Blasi ED, Medan CD, Cirigliano V, Oloriz L, Frojan D, Bertone VH, Bianchi HF, Aja-Guardiola S. "Evolution of the plastination technique at room temperature". In: XLIX Argentine Congress of Anatomy and II International Congress of Anatomy and IV Conference on Anatomy for Students of Health Sciences, of the Argentine Association of Anatomy, Central Military Hospital Cirujano Mayor Dr. Cosme Argerich and National Academy of Medicine, Buenos Aires, Argentina. September 5 to 8, 2012.

110. Aja-Guardiola S, Ottone NE, Conesa HA, Bianchi HF, Medan CD. "Central nervous system response to dissection after freezing at -50 °C, -30 °C, -10 °C and -5 °C for the purpose of plastination". In: XXII International Symposium on Morphological Sciences—Sao Paulo—Brazil. February 12 to 16, 2012.

111. Bianchi HF, Bertone VH, Algieri RD, Ottone NE, Mitidieri VC, Aja-Guardiola S. "Minimum requirements to install a plastination laboratory economically". In: XXII International Symposium on Morphological Sciences—Sao Paulo—Brazil. February 12 to 16, 2012.

112. Bertone VH, Mitidieri VC, Ottone NE, Cirigliano V, Oloriz L, Aja-Guardiola S. "Replacing the rigid chamber by a flexible plastic chamber for post-impregnation curing step of the plastination technique". In: XXII International Symposium on Morphological Sciences—Sao Paulo—Brazil. February 12 to 16, 2012.

113. Ottone NE, Blasi E, Dominguez M, Lorenzo H, Medan C, Bertone VH. "Construction of a plastination laboratory at room temperature". XVII Pan American Congress of Anatomy—XII Congress of Anatomy of the Southern Cone—XXXI Chilean Congress of Anatomy, Temuco, Chile. October 25 to 30, 2010.

114. Ottone NE, Blasi E, Bertone VH, Domínguez M, Lorenzo H, Medan C. "Plastination at room temperature in the Dissection Unit of the Second Chair of Anatomy". In: XLVII Argentine Congress of Anatomy and 2nd Conference on Anatomy for Students of Health Sciences, of the Argentine Association of Anatomy, Faculty of Health Sciences of the National University of Comahue, Cipolletti, Province of Río Negro, Argentina. October 11, 12 and 13, 2010.

115. Rodriguez-Torrez VH, Ottone NE. First plastination experience at 4,150 meters above sea level, in the height of La Paz, Bolivia. Int J Morphol. 2023. In Press.

116. Toaquiza AB, Gómez C, Ottone NE, Revelo-Cueva M. Conservation of organs (heart, brain and kidney) of canine by cold-temperature silicone plastination done at an animal anatomy laboratory in Ecuador. Int J Morphol. 2023;41(4):1004–8.

117. Ottone NE, Guzmán D, Bianchi HF, del Sol M. Anatomical variations of radial and ulnar arteries in plastinated upper limbs. Int J Morphol. 2023;41(2):548–54. https://doi.org/10.4067/S0717-95022023000200548.

118. Ottone NE. Unified plastination protocol with silicone at cold and room temperature. Int J Morphol. 2021;39(2):630–4. https://doi.org/10.4067/S0717-95022021000200630.

119. Ottone NE, Baptista CAC, del Sol M, Muñoz Ortega M. Extraction of DNA from plastinated tissues. Forensic Sci Int. 2020;309:110199. https://doi.org/10.1016/j.forsciint.2020.110199.

120. Vargas CA, Baptista CAC, Del Sol M, Sandoval C, Vásquez B, Veuthey C, Ottone NE. Development of an ultrathin sheet plastination technique in rat humeral joints with osteoarthritis induced by monosodium iodoacetate for neovascularization study. Anat Sci Int. 2020;95(2):297–303. https://doi.org/10.1007/s12565-019-00500-7.

121. Ottone NE. Micro-plastination. Technique for obtaining slices below 250 μm for the visualization of microanatomy in morphological and pathological experimental protocols. Int J Morphol. 2020;38(2):389–91. https://doi.org/10.4067/S0717-95022020000200389.

122. Ottone NE, Guerrero M, Alarcón E, Navarro P. Statistical analysis of shrinkage levels of human brain slices preserved by sheet plastination technique with polyester resin. Int J Morphol. 2020;38(1):13–6. https://doi.org/10.4067/S0717-95022020000100013.

123. Guerrero M, Vargas C, Alarcón E, del Sol M, Ottone NE. Development of a sheet plastination protocol with polyester resin applied to human brain slices. Int J Morphol. 2019;37(4):1557–63. https://doi.org/10.4067/S0717-95022019000401557.
124. Prieto R, Vargas CA, Veuthey C, Aja-Guardiola S, Ottone NE. Fundamental concepts of the modified room temperature plastination protocol with silicone, with subsequent pigmentation, and its application for the conservation of human placenta. Int J Morphol. 2019;37(1):369–74. https://doi.org/10.4067/S0717-95022019000100369.
125. Ottone NE, Baptista C, Latorre R, Bianchi HF, del Sol M, Fuentes R. E12 sheet plastination—techniques and applications. Clin Anat. 2018;31(5):742–56. https://doi.org/10.1002/ca.23008.
126. Ottone NE, Vargas CA, Veuthey C, del Sol M, Fuentes F. Epoxy sheet plastination on a rabbit head–new faster protocol with Biodur® E12/E1. Int J Morphol. 2018;36(2):441–6. https://doi.org/10.4067/S0717-95022018000200441.
127. Ottone NE. Plastinación: plastination: techniques fundamentals and implementation at Universidad de La Frontera. J Health Med Sci. 2018;4(4):293–302.
128. Ottone NE, del Sol M, Fuentes R. Report on a sheet plastination technique using commercial epoxy resin. Int J Morphol. 2016;34(3):1039–43. https://doi.org/10.4067/S0717-95022016000300036.
129. Ottone NE, Cirigliano V, Lewicki M, Bianchi HF, Aja Guardiola S, Algieri RD, Cantin M, Fuentes R. Plastination technique in laboratory rats: an alternative resource for teaching, surgical training and research development. Int J Morphol. 2014;32(4):1430–5. https://doi.org/10.4067/S0717-95022014000400048.

Chapter 3
Fundamentals on Plastination

Introduction

Plastination is an anatomical technique for the microscopic preservation of biological material, both human and animal, developed by Prof. Gunther von Hagens in Heidelberg, Germany, in 1977 [1–19]. Plastination is a method of cadaveric conservation by means of which biological and especially soft specimens such as the brain, heart, kidney, lung, liver, and muscles can be preserved, as well as specimens and sections of bodies in the fields of anatomy and pathology, humans and animals [15, 16]. In this process, water and lipids in biological tissues are replaced by plastic polymers such as silicone, epoxy, or polyester resins, which are subsequently hardened, resulting in dry, odorless, and highly durable specimens. The kind of polymer used determines the optical property (transparent or opaque) and the flexibility that it could confer to the impregnated specimen. Once impregnated, the specimen is much more stable than one that has been frozen, dehydrated, or waxed. Plastination also has a great advantage, and this is that the plastinated specimens retain their original surface relief and cell identity down to the microscopic level [1–4, 6, 12, 18, 19].

The decomposition of organic matter is a vital process in nature, but it is also an impediment for morphological studies and research [6, 12, 13]. This is particularly important in biological specimens which reduce in size considerably when exposed to normal atmospheric conditions. For this reason it has always been a constantly pursued objective for anatomists. Plastination, in this sense, is a true alternative in the preservation of perishable biological tissues (whole bodies; whole organs such as brains, livers, lungs, kidneys, hearts, muscles; joint preparations; sections cut from whole corpses or isolated regions; etc.) reaching a dry and imperishable state through the use of different polymers and special plastics, ensuring that the organs, limbs, and entire bodies do not lose their apparently normal texture and disposition [5–17].

© The Author(s), under exclusive license to Springer Nature
Switzerland AG 2023
N. E. Ottone, *Advances in Plastination Techniques*,
https://doi.org/10.1007/978-3-031-45701-2_3

The aim of this chapter is to describe the fundamentals of plastination techniques created by Gunther von Hagens, and how they were implemented in the Laboratory of Plastination and Anatomical Techniques, of the Faculty of Dentistry and CEMyQ, of *Universidad de La Frontera*, in the city of Temuco, Chile, through the creation of a laboratory, scientific publications, and academic training activities.

Plastination Techniques Developed by Gunther von Hagens

Basically, von Hagens developed three plastination techniques: the cold plastination in silicone (S10, Biodur), for the preservation of organs, limbs, portions, and whole bodies; the sheet plastination with epoxy resin (E12, Biodur), for the conservation of millimeter sections of any body region, ensuring minimal retraction of the tissues and maximum transparency of the cuts; and the sheet plastination technique of sections with polyester resin (P40, Biodur), originally created for the conservation of millimeter sections of the brain, allowing a great differentiation between the gray and white matter, but later this technique also began to be used for the preservation of sections of any region body, with the disadvantage of retracting the tissues, more than with epoxy.

All plastination techniques are based on the following steps [1–4]: "dehydration/defatting," with acetone (100%) in a freezer at −25 °C, seeking replace biological fluids and/or sample fixatives with acetone; the acetone is replaced until the water content is less than 1%. Acetone is an excellent dehydrating agent in plastination due to its low boiling point and high vapor pressure, which allows for substitution by various polymer types during forced impregnation. It is important to measure the purity of acetone weekly, with an acetonometer, as the water and acetone sample reach equilibrium in about 7 days. Before performing the purity measurement of the acetone sample, it is important to mix the contents thoroughly. Due to the difference in density, water may settle at the bottom of the container, leading to stratification. The acetone aliquot collected for measurement may not be a true representation of the entire volume's average concentration if the sample is not well-mixed. Therefore, it is recommended to mix the sample well before collecting an aliquot for purity measurement. In order to renew the acetone, it is essential to maintain a stable concentration of it for at least 48 hours. Once a steady level is achieved, the acetone must be changed. This involves transferring the samples to a higher purity acetone than the one previously used and the final bath should always consist of 100% acetone. If the purity of the acetone is 99.5% or higher, it indicates that the dehydration step is complete. Then, defatting is done at room temperature, also with acetone or methylene chloride. Degreasing is a crucial step in the plastination process. It involves removing the fat from the tissue before impregnation. This is important because the lipids in the adipose tissue and cells do not impregnate well, leading to an excess of fat in the final plastinated sample. This excess fat can cause the sample to become rancid and slippery, ruining the final results. Defatting with acetone is slow at low temperatures. Therefore, the samples should be placed in acetone at room temperature after dehydration to maximize its degreasing effect. The degreasing process with acetone can take anywhere from 15 days to several months depending on sample size and fat content. Degreasing is monitored by the yellowing of the acetone as the lipids are extracted. When the acetone turns intense yellow, replace it with another pure acetone to extract more fat. On the other hand, dichloromethane degreasing is fast (1-3 days) at room temperature, but requires at least a 99% dehydration for sample use. The next stage is "forced impregnation" it is the central and most important step of plastination, in which the specimen is immersed in a mixture of polymer with a catalyst, inside a vacuum chamber (depending on the plastination technique), and in this way, a vacuum begins to be generated inside the chamber, through the operation of a vacuum pump connected to the chamber, to achieve, by a difference of pressures (vacuum), the removal of acetone and the entry of the polymer into the sample. In this way, acetone can be removed through vacuum, creating a vacuum that forces polymer entry into cells and interstice. Acetone's low boiling point (56°C) enables forced impregnation of polymers with high boiling points (silicone, epoxy, polyester) for plastination. For its part, It is not feasible to achieve an adequate substitution with polymers due to the high boiling point of ethyl alcohol at 78.4°C. Forced impregnation, and the way to carry it out and control it, will be explained in detail in Chap. 5. Finally, the last stage is "curing," in which the final drying of the sample will be sought (depending on the plastination technique, the components, and equipment used for this stage will vary).

It is important to consider that prior to the development of any plastination technique, adequate and precise planning of anatomical dissection to be carried out in the sample is necessary, which must be meticulous and well developed, with complete removal of the subcutaneous cell tissue, muscle fascia, and display of special structures, previously planned, since after plastination dry preparations will be obtained and with a relative hardness that makes subsequent dissection difficult. In addition, adequate fixation will also be necessary, the plastination technique usually requires the use of 10% formaldehyde, but this will depend on the specimen considered (central nervous system, e.g., will require special processing, for a longer fixation time with formaldehyde, to maintain its size and to avoid excessive shrinkage during the dehydration and forced impregnation steps). Also, is important to know that during the process of plastination, shrinkage is a common occurrence, especially during dehydration. The amount of shrinkage is dependent on various factors such as the duration of fixation in formalin and the amount of fat present in the sample. But when dehydration is carried out using acetone at -25°C, the shrinkage is reduced as the water and sample freeze when immersed in pure acetone, which stabilizes the shape, structure, and size of the sample. On the other hand, fresh tissue samples and those with a high percentage of water and fat are more susceptible to shrinkage.

Cold Plastination with Silicone (S10)

This technique is known by the acronym S10, since Gunther von Hagens, in 1980, created the company Biodur, for the commercialization of all the products and equipment necessary for the development of plastination techniques, and among them is silicone, which he called "S10." For this reason, the plastination technique that requires the use of silicone is defined as the S10 technique.

The S10 silicone plastination technique [1–3] allows to create rigid and flexible, resistant, and opaque samples. After dehydration with acetone, the samples are impregnated at −25°C, inside a vacuum chamber, immersed in a mixture of silicone (S10) and catalyst (S3) (100:1, respectively), which have a pressure low steam (high boiling point). The volatile intermediate (acetone) found within the specimen is constantly removed by a vacuum pump. As the acetone is removed, a pressure difference will cause the polymer to enter the specimen. The forced impregnation must be carried out slowly as the polymer enters the specimen where the acetone changes from a liquid to a gas state and is removed. The impregnation rate is carefully adjusted by a controlled addition of air into the vacuum pump via a bypass valve. The duration of the forced impregnation step will depend mainly on the size of the specimen (and the quantity), the density of the tissue, and the viscosity of the polymer used. During this period the vacuum should be intensified from a pressure of 760 mmHg (at sea level), according to the desired formation of bubbles (acetone vapor), to a pressure of approximately 10 mmHg, where the small bubbles will go to the surface (the visualization of bubbles is indicative of the release of acetone from the interior of the specimen).

Once this final level of pressure is reached, and with no bubbles (indicator of replacement of acetone by silicone), then the forced impregnation stage has come to an end, and the specimen is removed from the polymer solution, then 24 h after completion of this stage. After impregnation, the samples are subjected to the curing stage (polymerization), which consists of exposing the samples, inside a hermetic chamber, to a liquid that contains silicate and which is vaporized (S6). Catalyst (S3), contained in the impregnating mixture, initiates the curing of the silicone molecules through end-to-end polymerization. Due to cross-linking during the final gas/vapor cure, the silicone within the sample will solidify and dry out. The sample surface cures quickly, but gas/vapor diffusion into the sample is slower. To ensure proper curing throughout the process, the sample must be stored in hermetic plastic bags for as long as it allows for final drying. An issue to consider is the positioning of the specimen, that is, the adequate display of the anatomical elements in the dissected regions, through the placement of separator needles, support threads, and other elements for the adequate composition of the plastinated preparation. This can be done prior to dehydration, or after forced impregnation, but always before curing, since, after this stage, the hardening that the sample reaches prevents the mobilization of the anatomical structures.

Sheet Plastination with Epoxy Resin (E12)

Sheet plastination technique with epoxy resin (E12) allows obtaining plastinated slices of any body region with a thickness of between 2 and 5 mm [2, 3]. Obtaining slices of less than 2 mm defines the "ultrathin sheet" technique, and for this a diamond blade saw is required to carry out the sections [6, 10–12].

In the traditional technique [2, 3], the fundamental steps of plastination are respected, but it requires the previous cutting of the samples, since it will be the slices that are going to be impregnated. For its part, in the ultrathin sheet plastination technique [20–22] or also defined from 2020 by Ottone as micro-plastination [6], in this technique is generated a block of epoxy resin with the sample inside, and it is this block that will be cut with the diamond blade saw. In this way, going back to the traditional version, the sample to be cut should be placed inside a container, in turn filled with polyurethane foam, which will allow the most appropriate handling of the sample at the time of cutting with the circular saw. The block must be previously cooled to ensure adequate performance of the slices. Once cold, the block is cut, and the sections obtained must be immediately placed in 100% acetone (at −25 °C) for dehydration.

Once dehydration is complete, the sample must be subjected to defatting that will allow the removal of fat, ensuring greater transparency. This can be done with acetone or dichloromethane (most toxic component, and if used, it must be done under a fume hood, with facial protection). After degreasing, slices must be placed in an epoxy resin and catalyst mixture (E12/E1), with an infinite number of protocols in relation to the mixture percentages and components used [13]. This stage is carried out in a vacuum chamber, at room temperature (20 °C), reducing the pressure from 760 to 10 mmHg, for 24 h. Finally, the curing stage is reached, which will consist of assembling a curing chamber ("sandwich"), made up of two glass plates and acetate sheets between which the slices are placed with a new mixture of E12/ E1, and this "sandwich" is placed in an oven at 50 °C, ensuring the hardening and drying of the sections in 48–72 h. In this way, micro-plastination is defined with the aim of bringing together under this denomination the techniques of plastination in ultrathin sheets with epoxy resin applied to experimental protocols of morphology and pathology, and by means of which we seek to obtain slices of a thickness less than those of 0.25 mm/250 μm.

Sheet Plastination with Polyester Resin (P40)

Sheet plastination with polyester resin (P40) was originally invented for the preservation of brain slices, ensuring a differentiation of the gray matter and white matter; however, later it was also implemented for the rest of the body, both humans and animals [2–4, 11, 14].

Table 3.1 Comparison between classical plastination techniques created by Gunther von Hagens [1–4]

Steps	Silicone[a]	Epoxy[b]	Polyester[c]
Dehydration	Cold temperature (−25/−20 °C) Room temperature (+22/+20 °C)[d] Acetone 100%		
Forced impregnation	Cold temperature: −25/−20/−15 °C	Room temperature: +20/+22 °C	Room temperature: +20/+22 °C
	Mixture: S10 + S3	Mixture: E12 + E1[e]	Mixture: P40
	4–6 weeks of forced impregnation	18–24 h of forced impregnation	18–24 h of forced impregnation
Curing	Vaporization of S6	Stove at 50 °C	UV light

[a] The samples may correspond to whole bodies, isolated body regions, isolated organs, body sections from 1 to 2 cm thick, organ sections from 1 to 2 cm thick
[b] The samples correspond only to sections of body regions, from 2 to 5 mm thick
[c] The samples correspond only to brain slices 2–3 mm thick
[d] Brains or brain slices should not be dehydrated at room temperature or subjected to the defatting process, due to the risk of excessive shrinkage. They should only be cold dried
[e] This is the standard impregnating mixture, there being a great diversity of combinations that are presented in Chap. 6

The steps of this technique are similar to those of E12 technique, with variations fundamentally in the impregnation stage, in which the polyester resin can be used without a catalyst, and also in the curing stage. At this stage, it is necessary to create "flat chambers," also between two glass plates but separated by a silicone tube and joined both glass plates by paper clips. In this way, a curing chamber is made inside which the slices obtained are located, already impregnated, and the chamber is filled with more polyester resin. Subsequently, the chamber is closed, and the specimens are subjected to ultraviolet (UV) light, either from light tubes, or they could also be subjected to sunlight, but in the shade, not directly to sunlight, thus ensuring the proper temperature (which should not exceed 30°) and thus allowing the final hardening of the samples. In Table 3.1 a comparison is made between Gunther von Hagens' plastination techniques.

Plastination with Silicone at Room Temperature

Zheng et al. in 1998 [23] published in the *Journal of the International Society for Plastination* the fundamentals for the development of the plastination technique at room temperature, that is, using vacuum chambers for forced impregnation without the use of a freezer, at 20 °C. This was also implemented by Roy Glover, in the USA, looking for an alternative method to the plastination technique created by Gunther von Hagens. Also in 1998, in the *Journal of the International Society for Plastination*, the abstracts of the ninth International Conference on Plastination were published, among which is the work of Glover et al. [24]. In both cases [23,

24], the technique basically consists of combining the components in a different way: in the forced impregnation, they mixed the silicone with the curing agent, and as a curing agent, they used the catalyst, but they do not vaporize it on the preparation; rather, it is sprayed or brushed.

Contributions from Our Laboratory of Plastination and Anatomical Techniques

The Laboratory of Plastination and Anatomical Techniques began its construction in 2014, with the acquisition of equipment, and the final setup, in 2016, the year in which it finally began its activities. In our lab, all the plastination techniques created by Gunther von Hagens are carried out, as well as those that we have developed and implemented, with modifications to the original techniques, which allow a faster and lower-cost development, maintaining the quality in the final result of the preparations.

In 2014, we implemented a plastination technique, with silicone, but at room temperature, with substantial modifications in the forced impregnation process, and in the way the polymer, catalyst, and curing agent are used. This technique was applied in laboratory rats, indicating the need for the plastination technique for its use in this type of specimens, dedicated to the practice of surgical approach techniques, and the possibility of respecting the 3 Rs, with the reduction in the use of animals to carry out this type of training, ensuring ethical handling of animals [8].

Subsequently, in 2015, the details of these new contributions made to the plastination technique at room temperature were published in the journal of the Japanese Society of Anatomy *Anatomical Sciences International* [9]. These contributions consisted, basically, in the modification of the forced impregnation, combining active and passive periods, related to the switching on and off of the vacuum pump, respectively, during the entire process of forced impregnation. Likewise, the silicone, catalyst, and curing agent were combined in the same way as in cold plastination technique [1] but carrying out the forced impregnation process at room temperature. All this made it possible to obtain plastinated specimens of the same quality as the original technique, but with a notable reduction in times and costs, since the use of a freezer in forced impregnation was avoided, which requires modification by safety measures, removing the compressor from the freezer compartment and moving it to an adjoining room, thus ensuring that the compressor does not come into contact with possible acetone vapors [25].

In 2016, we made a contribution to the modification in sheet plastination technique with epoxy resin, using locally components, in addition to not using the catalyst during the forced impregnation stage, which ensured the fluidity of epoxy resin whose hardening was then ensured during the curing stage [10]. In this way, a reduction in costs was ensured, since all the products were local, in addition to facilitating the development of the technique, in a fundamental and central step such as forced impregnation.

We continue to investigate the technique of plastination of sections with epoxy resin, but this time a protocol with Biodur E12/E1 resins was presented, but in which the forced impregnation stage was significantly reduced, from 24 h (original) to only 45 min [12]. The technique was implemented in rabbit head sections only 2 mm thick, achieving a very adequate technique and ensuring its correct conservation.

Finally, it was presented and published in the journal of the American Association of Clinical Anatomists *Clinical Anatomy*, the first review carried out on a plastination technique [13], since its creation in 1977 (and without considering the article review carried out by Gunther von Hagens himself in 1987 "The current potential of plastination"). This article brought together all the articles made with sheet plastination technique with epoxy resin, from 1977 to 2017, and allowed the identification of all the advantages that this technique presents, fundamentally for research, due to the possibility of performing morphometry, in addition to the advantage of visualizing a microscopic anatomy, due to the possibility of maintaining the structures in their anatomical position, without retraction, with conservation of morphology, and also existing the possibility of performing histological stains, and the use of confocal microscopy for the study of the sections [13].

In addition, in 2020, from our laboratory we also made a contribution to the plastination technique of ultrafine cuts with epoxy resin, through the description of the micro-plastination technique [6] (see Chap. 6), as well as that same year we developed and described a protocol for extracting DNA from plastinated samples (subjected to a deplastination process), with the possibility of identifying fully intact DNA and with multiple possibilities of application in various areas of science [7] (see Chap. 8).

Conclusion

The plastination techniques, created by Gunther von Hagens in 1977 at the University of Heidelberg, Germany, constitutes a true revolution in the preservation of human and animal bodies for study and research. Our experience developed in the Laboratory of Plastination and Anatomical Techniques of *Universidad de La Frontera* has allowed us to implement a complete laboratory, where all the plastination techniques are developed, as well as allowing us to investigate them, publishing modifications to the original techniques, making them more feasible in their implementation in any institution. In addition, we develop improvement courses for the training of academics in plastination and other anatomical techniques. In the future, we will continue implementing plastination technique, and deepening its use in research projects with the aim of applying anatomical knowledge through plastination and the importance of its use for the study of morphological sciences and its application in the clinical, surgery, and imaging, both undergraduate and postgraduate, research, and university extension in the community.

References

1. von Hagens G. Impregnation of soft biological specimens with thermosetting resins and elastomers. Anat Rec. 1979;194(2):247–55. https://doi.org/10.1002/ar.1091940206.
2. von Hagens G, editor. Heidelberg plastination folder. Collection of technical leaflets of plastination. Heidelberg: Biodur Products GmbH; 1986.
3. von Hagens G, Tiedemann K, Kriz W. The current potential of plastination. Anat Embryol (Berl). 1987;175(4):411–21. https://doi.org/10.1007/BF00309677.
4. von Hagens G. Plastination of brain slices according to P40 procedure. A step-by-step description. In: Heidelberg plastination folder. Collection of technical leaflets of plastination. Heidelberg: Biodur Products GmbH; 1994. p. 1–23.
5. Ottone NE. Gunther von Hagens, creator of plastination. Historical review and technical development. Rev Argent Anat Online. 2013;4(2):70–6. https://www.revista-anatomia.com.ar/archivos-parciales/2013-2-revista-argentina-de-anatomia-online-f.pdf.
6. Ottone NE. Micro-plastination. Technique for obtaining slices below 250 μm for the visualization of microanatomy in morphological and pathological experimental protocols. Int J Morphol. 2020;38(2):389–91. https://doi.org/10.4067/S0717-95022020000200389.
7. Ottone NE, Baptista CAC, Del Sol M, Muñoz Ortega M. Extraction of DNA from plastinated tissues. Forensic Sci Int. 2020;309:110199. https://doi.org/10.1016/j.forsciint.2020.110199.
8. Ottone NE, Cirigliano V, Lewicki M, Bianchi HF, Aja Guardiola S, Algieri RD, Cantin M, Fuentes R. Plastination technique in laboratory rats: an alternative resource for teaching surgical training and research development. Int J Morphol. 2014;32(4):1430–5. https://doi.org/10.4067/S0717-95022014000400048.
9. Ottone NE, Cirigliano V, Bianchi HF, Medan CD, Algieri RD, Borges Brum G, Fuentes R. New contributions to the development of a plastination technique at room temperature with silicone. Anat Sci Int. 2015;90(2):126–35. https://doi.org/10.1007/s12565-014-0258-6.
10. Ottone NE, del Sol M, Fuentes R. Report on a sheet plastination technique using commercial epoxy resin. Int J Morphol. 2016;34(3):1039–43. https://doi.org/10.4067/S0717-95022016000300036.
11. Ottone NE, Guerrero M, Alarcón E, Navarro P. Statistical analysis of shrinkage levels of human brain slices preserved by sheet plastination technique with polyester resin. Int J Morphol. 2020;38(1):13–6. https://doi.org/10.4067/S0717-95022020000100013.
12. Ottone NE, Vargas CA, Veuthey C, del Sol M, Fuentes F. Epoxy sheet plastination on a rabbit head–new faster protocol with Biodur® E12/E1. Int J Morphol. 2018;36(2):441–6. https://doi.org/10.4067/S0717-95022018000200441.
13. Ottone NE, Baptista CAC, Latorre R, Bianchi HF, Del Sol M, Fuentes R. E12 sheet plastination: techniques and applications. Clin Anat. 2018;31(5):742–56. https://doi.org/10.1002/ca.23008.
14. Guerrero M, Vargas C, Alarcón E, del Sol M, Ottone NE. Development of a sheet plastination protocol with polyester resin applied to human brain slices. Int J Morphol. 2019;37(4):1557–63. https://doi.org/10.4067/S0717-95022019000401557.
15. Ottone NE, Guzmán D, Bianchi HF, del Sol M. Anatomical variations of radial and ulnar arteries in plastinated upper limbs. Int J Morphol. 2023;41(2):548–54. https://doi.org/10.4067/S0717-95022023000200548.
16. Toaquiza AB, Gómez C, Ottone NE, Revelo-Cueva M. Conservation of organs (heart, brain and kidney) of canine by cold-temperature silicone plastination done at an animal anatomy laboratory in Ecuador. Int J Morphol. 2023;41(4):1004–8. https://doi.org/10.4067/S0717-95022023000401004.
17. Rodriguez-Torrez VH, Ottone NE. First plastination experience at 4,150 meters above sea level, in the height of La Paz, Bolivia. Int J Morphol. 2023; In Press
18. Bickley HC, von Hagens G, Townsend FM. An improved method for the preservation of teaching specimens. Arch Pathol Lab Med. 1981;105(12):674–6.

19. Baptista CAC, Cerqueira EP, Conran PB. Impregnation of biological specimens with resins and elastomers: plastination with Biodur S10 resin. Rev Bras Cienc Morfol. 1988;5(1):60–2.
20. Sora MC. Epoxy plastination of biological tissue: E12 ultra-thin technique. J Int Soc Plastination. 2007;22:40–5.
21. Sora MC, Strobl B, Radu J. High temperature E12 plastination to produce ultra-thin sheets. J Int Soc Plastination. 2004;19:22–5.
22. Soal S, Pollard M, Burland G, Lissaman R, Wafer M, Stringer MD. Rapid ultrathin slice plastination of embalmed specimens with minimal tissue loss. Clin Anat. 2010;23(5):539–44. https://doi.org/10.1002/ca.20972.
23. Zheng T, Liu J, Zhu K. Plastination at room temperature. J Int Soc Plastination. 1998;13(2):21–5.
24. Glover RA, Henry RW, Wade RS. Polymer preservation technology: POLY-CUR. A next generation process for biological specimen preservation. Abstract. J Int Soc Plastination. 1998;13(2):39.
25. Baptista CAC, Bellm P, Plagge MS, Valigosky M. The use of explosion proof freezers in plastination: are they really necessary? J Int Soc Plastination. 1992;6:34–7.

Chapter 4
Cadaveric Fixation and Conservation Techniques Prior to Plastination

General Description

In the origins of anatomical techniques, very diverse fixation and conservation solutions were used, starting in antiquity with substances such as oils, resins, and wine that prevented the decomposition of tissues by slowing down this process, allowing, as alcohol does, a coagulation of the proteins, but without presenting the excellent characteristics of preservation that are associated with the fixative liquid par excellence, which is formalin [1–4]. From this point of view, tissue fixation is a physico-chemical process through which the composition of proteins is altered, as indicated above, thus preventing the decomposition of biological tissues. However, this can cause excessive hardening of the tissues, which is why it is defined that an excellent fixation liquid, such as formalin since its discovery, must stop decomposition, without distorting the morphological characteristics of the samples, in order to consequent relative hardening of tissues. Formalin was discovered by Butlerov in 1859, but adequately described, in its final conformation, by Wilhelm von Hofmann in 1868, developing the method for obtaining it from methanol [1, 2]. In 1893 the fixing properties of formalin were discovered, corresponding to Ferdinand Blum, who was the first to use formalin as a tissue fixative [1, 2].

Formaldehyde is a gas soluble in water, constituting its commercial form as formalin or formol. The chemical composition of the liquid presentation of formaldehyde, formalin, is 37–40% formaldehyde and 10% methanol diluted in water. Formaldehyde in the presence of oxygen tends to oxidize to formic acid and loses its fixative effect. This transformation process occurs more frequently when it is diluted to 10%. For this reason, once prepared, it should not be stored for long periods of time [5, 6].

An alternative to reduce its oxidation is to add phosphate salts to the preparation, in order to maintain an adequate pH. When fixing tissues with formaldehyde diluted in water, these must remain at room temperature for at least 24 h to allow the

N. E. Ottone, *Advances in Plastination Techniques*,
https://doi.org/10.1007/978-3-031-45701-2_4

reaction between the chemical and the components of the piece to occur until equilibrium is reached [7, 8]. Currently, the search for fixation techniques for the practice of surgical techniques on the cadaver implies the possibility of replicating the characteristics of fresh cadavers, simulating the characteristics in vivo. It is very common to develop surgical technique practice courses on real human cadavers using fresh cadavers; however, there are several disadvantages when using fresh cadavers: there are risks of contamination; decomposition of the corpse during the practice process; short-term use of carcasses and limited possibility of reusing carcasses; and adding to all this the reduced profitability in the development of the courses. In this way, for the development of plastination techniques, for the preservation of bodies for teaching practice, and for research or the practice of surgical techniques, different alternatives of fixative and preservative solutions are developed, with and without formaldehyde.

Considerations on the Adverse Effects of Formaldehyde

It is necessary when developing anatomical fixation and conservation techniques prior to plastination, to mainly ensure the brake on tissue decomposition, to take into account a crucial problem that must be considered in all work practices, and that is associated with toxic effects, on the health of people who not only perform the anatomical technique but also those who manipulate the biological material preserved for learning (students), teaching (teachers), research (professors/students), as well as academic extension (general population). And in this sense, formaldehyde has harmful characteristics that have led to seek its total or partial replacement, to eliminate (ideally) or significantly reduce the toxic effects of anatomical samples preserved with it. Most of the countries have begun to regulate the use of formaldehyde in anatomy laboratories due to the high toxicity that its use presents, and there are many clinical studies that support this decision. The International Agency for Research on Cancer (IARC), in 2006, classified formaldehyde as carcinogenic to humans [9], making it even more necessary to develop new protocols to limit and control the use and human exposure to it. In 2011, the National Toxicology Program, made up of various research groups from the US Department of Health and Human Services, in its 12th Report on Carcinogens, identified formalin as a human carcinogen [10]. A test carried out in rats to study the effect of exposure to formaldehyde on the kidney revealed that it causes kidney damage, specifically when finding morphological changes at the level of the nephron [11], in addition to measuring different markers, n-acetyl-ß-d-glucosaminidase, which defines the damage in the proximal convoluted tubule; antidesmin antibodies, which increase when there is damage to the podocytes; nephrin and podocin, whose distribution and expression are altered when there is injury to podocytes and the basement membrane; and terminal dUTP deoxynucleotidyl transferase, which determines the presence of cell apoptosis. Likewise, the mucous membranes of the respiratory and ocular tracts are affected by formaldehyde, causing everything from rhinitis and ocular irritation to nasopharyngeal cancer [9–11]. Finally, it has been possible to identify that chronic exposure to formaldehyde produces genotoxicity and skin sensitization [9–11].

General Characteristics of Fixing Solutions for Preservation and Conservation

In the first place, Coleman and Kogan in 1998 [4] clearly identify the important properties that fixation solutions must possess in order to achieve successful preservation of cadavers:

- Adequate long-term structural preservation of organs and tissues, obtained with minimal retraction or distortion thereof.
- Maintain adequate cervical flexibility, as well as limbs, as well as internal organs, avoiding excessive hardening.
- Fundamental prevention of desiccation of tissues.
- Avoid the appearance of fungi and/or bacteria, as well as, if they do appear, their spread within the corpse.
- Prevent the spread of infection and/or other biohazards to staff, faculty, and students.
- Comply with health and biosafety regulations in the workplace, by minimizing the use, specifically, of formaldehyde and phenol, due to the environmental chemical hazards posed by their use.
- Retention of the color of biological tissues in addition to minimizing browning, the result of oxidation of biological tissues.

These final characteristics of the preserved anatomical samples can also be achieved if the fixing solution mixture ensures compliance with the characteristics of the compounds used for their preparation, such as fixation, moisture conservation, and antibacterial and fungicidal properties:

- Fixation: formaldehyde, ethyl alcohol, isopropyl alcohol, and glutaraldehyde
- Moisture preservation: glycerin, polyethylene glycol
- Antibacterial: formaldehyde, ethyl alcohol, isopropyl alcohol, glutaraldehyde, and hydrogen peroxide
- Fungicide: phenol
- Alternatives: common salt and nitrate salts, as partial or total replacement of formaldehyde in the function of fixation and conservation

Formalin is incorporated into the cell membrane and into the molecular structure of the sample, making the sample firm and minimizing shrinkage. Its use is intended to disinfect the sample and get rid of dangerous germs; denature any tissue enzymes that may remain active, even after dehydration and plastination, and that may cause tissue breakdown; finally, they can interfere with curing after impregnation.

Continuing with the application of formalin to plastination, to prevent hardening of the superficial layer of biological tissue, which will cause a delay in the diffusion of the fixative, it is recommended to start fixation with low concentrations of formalin (from 1% to 5%), to then increase until reaching a higher concentration (10–20%), depending on the sample to be fixed. In case of wanting to generate more flexible samples, it is necessary to prepare solutions with very low concentrations of formalin (from 1% to 3%).

General Protocol for the Fixation of Cadaveric Material [12–17]

Reception of Cadaveric Material

The cadaveric material (whole body or specific anatomical regions) must be received in the anatomy laboratory, and if it will not be immediately subjected to the fixation process for its preservation, which is ideal, it should be refrigerated (+4 °C–0 °C). This refrigeration ensures adequate planning of the final destination of the cadaveric material, in relation to whether it will be necessary to section the material for its subsequent anatomical dissection, or whether it will be completely processed. Likewise, in general, the entire body is fixed, and then it is dissected totally and/or partially, depending on whether the body is subjected to a sectioning process for its division into anatomical regions.

Anatomical Dissection Planning

Before carrying out the fixation, it is necessary to carry out adequate planning associated with the final destination of the cadaveric material. It is always necessary to take into account the anatomical techniques that will later be applied to that material, be it injection/corrosion techniques, nerve staining, plastination, etc. In this sense, there will be variations in the fixation modality, since, for example, when deciding to perform a dissection technique of white nerve fibers (Klinger technique), prior to fixation, the extraction of the brain must be carried out to proceed with the particular conservation of this with the steps required by the Klinger technique. That is why this step is essential. The same happens if you decide to perform the dilation of certain organs (small and large intestines, stomach, heart, etc.) and also if at the same time it is decided to make cuts in the organs, in the manner of "windows," to visualize, from the surface of the organ, deep structures. Thus, planning prior to fixation to then carry out anatomical techniques that allow highlighting certain anatomical structures that are decided to be highlighted is essential to achieve adequate preservation of the anatomical material which can later be subjected to these anatomical techniques.

Steps for Performing Vascular Fixation

In the first place, you must remove, with razors or scissors, the hair from the face, armpits, perineal region, and other anatomical regions where it is present. Subsequently, you must proceed to wash the corpse with commercial detergent, using a brush and sponge, also considering with special care the washing of the perineal, axillary, and neck regions, as well as the nose, ears, eyes, mouth, and other orifices. Finally, the entire carcass should be rinsed under running water.

Once this is done, the channeling process can begin. This is started by a long incision following the longitudinal direction of the vessel, to avoid its destruction and allow correct placement of the cannula. The common carotid arteries and the external jugular veins on both sides are selected in priority. To do this, you must access the level of the carotid trigone or the lesser supraclavicular fossa, between the sternal and clavicular heads of the sternocleidomastoid muscle. Ideally, one plastic tube is placed in the cephalic direction and another in the caudal direction, in both arteries and veins. However, the arterial route is usually selected on an individual basis. This also occurs when channeling the femoral artery (it being possible to incise the femoral vein in the same way) at the level of the femoral triangle, below the inguinal ligament. This last route is used especially for the correct injection of the pelvic organs and lower limbs. To prevent the loss of the probe, it is necessary to fix it with a thread ligature, which can be cotton, linen, or another material. In the case of a fetus, the umbilical artery is of choice, since it allows avoiding the opening of the corpse.

Once the vessels have been channeled, to destroy possible clots present, two solutions defined as formula 1 and formula 2 must be instilled successively.

Gravitationally, formula 1 is instilled, consisting of a dilution of hot water and enzymatic detergent: 8 mL of detergent for each liter of water. The detergent will be composed of E2 (amylase and lipase) and E3 (amylase, lipase, and protease). A total of 4 L of formula 1 should be administered. Also formula 1 could be composed of saline (4 L) and heparin (1 mL). Once formula 1 is administered, you will need to wait about 20–30 min for it to take effect. Subsequently, make a small longitudinal incision in the internal jugular vein to end the drainage of blood and possible clots.

Subsequently, detergent water is instilled according to formula 2, which consists of a 10-volume hydrogen peroxide solution in tap water, also adding 5% formalin. The volume of the solution to be used is approximately 5% of the body weight (if the corpse weighs 100 kg, 5 L of hydrogen peroxide solution will be used).

Once the passage of formulas 1 and 2 for the destruction of possible clots present has finished, hot water must be instilled gravitationally, verifying the return through the canalized vessels and that were left open to verify this situation, visualizing that the drained liquid is clear, until drainage of water with open venous system is visualized.

Once most of the washing liquid has been let out, the fixation itself is carried out. In other words, it is at this moment that you must decide which fixing method to use. As will be seen in the respective chapters, 10% formalin fixation is defined as the ideal method to fix cadavers that will later be plastinated, but plastination techniques can also be used or applied to anatomical samples that have been fixed with another type of material solutions.

In this sense, the chosen fixing solution must be placed in a container, or peristaltic pump, and connecting the probes placed in the carotids (first in the cephalad direction) and then passing the fixing solution, observing that the liquid passes through the jugular veins. When the fixative solution begins to flow, the vessel must be closed, using a hemostatic clamp in each one, closing the venous system. It is very useful to detect the state of the eyeballs, which become turgid and protrude from the orbits when a sufficient amount of fixative solution is reached (between 5 and 10 L approximately).

Likewise, it is important to carry out complementary fixation of the nervous system, made up of the brainstem, cerebellum, and cerebrum, through cannulation in the cephalad direction from the carotid artery (changing the direction of the cannulation) and administering a formalin solution to the 20%. Injections may also be made through the superior orbital fissure, at the level of the lower margin of the orbits.

Complementary and direct fixations could also be made at the level of the upper and lower limbs, as well as through the back, thorax, and abdomen, making injections at the muscular level.

The classic fixative solution, based on formalin, can be classified as weak, medium, and strong.

Weak solution: 5% formalin
Medium solution: 10% formalin
Strong solution: 15% formalin
Extra-strong solution: 20% formalin

Once the turgidity of the eyeballs has been verified, the orientation of the cannulas is changed, placing them in the caudal direction. The fixative is made to flow in the direction of the lower limbs. In this case, the observation of the joints of the lower limbs, especially the interphalangeal joints, makes it possible to decide whether to continue or end the process. Its rigidity is an indicator that the fixative solution spread correctly. The nostrils should also be observed, as when foaming fixative solution begins to come out, this means that the alveoli have been destroyed and the solution is flowing out (at this point it can be allowed to flow for a few minutes and then there is no point, because the liquid will not continue penetrating the corpse). The clamps are placed on the probes that carried the fixative, and the corpse is left to rest, which is sprayed with the same fixative liquid and covered with canvas and plastic bags. There are authors who recommend the beginning of the processing of small corpses (up to 25–30 kg) at 24 h. In heavier carcasses, they recommend waiting for 48 or 72 h.

The abdomen and thorax can also be cannulated and instilled specifically, which will be fixed with the medium fixative solution (10% and 15% in 5 L of water, respectively).

Finally, prolonged conservation is carried out in pools specially built for this task, in which the corpse is completely submerged in a strong fixative solution, based on 20% formalin (in 1000 cm³ pools). Prior to this, superficial skin incisions must be made to increase the passage of the strong fixative solution and contribute to the preservation of cadaveric material.

The material must remain submerged for at least a week in order to subsequently begin its preparation (dissection, surgical techniques, etc.).

It is recommended to keep the carcass in a dry place with a controlled temperature (20–22 °C), but this technique is equally effective in places without air conditioning, with high temperatures.

It is recommended to neutralize the formaldehyde with 5% monoethanolamine in 95% water, for 48–72 h after embalming or 2 weeks before the start of the dissection.

If the presence of fungus is found, you can apply on the affected area, with spray, the fixative solution, or alcohol vinegar.

During the preparation of the fixing solution, first mix ethyl alcohol with phenol. The amount of fixing solution to be infused should be at least twice the amount of body blood (7–8% of body weight), that is, for a 70-kg corpse, you should infuse 10–12 L of solution.

If there is not a good distribution of the fixative solution through a single arterial access, another vascular access can be made in the femoral and/or axillary artery. This perfusion is discernible at the "arterial pulse palpation" sites of the common femoral, axillary, and/or carotid arteries and through flow in the superficial venous vessels.

Fixation of Brains/Encephalons [12]

In relation to brain fixation, mainly for plastination, it is ideal to achieve a final fixation with a high concentration of formalin (at least 20%). That is, in the nervous tissue, the fixation must be started, as indicated above, in an increasing way, from 5%, 10% until reaching the final fixation of 20%, on a weekly basis, leaving the samples submerged in formalin at 20% for a minimum of 4 months. Subsequently, specimens can be stored in 1–2% formalin solution.

Biological Tissue Preservative-Fixing Solutions

Various formulas of solutions with variable concentrations of formalin are transcribed below, which can be applied for the preservation of human and animal biological material, also depending on the final destination that is given to the samples subjected to these processes.

Trinity Fluids

Formaldehyde (37%)—23%
Glutaraldehyde—5%
Ethylene glycol—14%
Ethanol—56%
EDTA—1.6%
Surfactant—0.4%
Dilute the mixture with water (1:1)

Trisco Formula

Formaldehyde (37%)—2.50%
Ethanol—13.53%
Nonionic detergent—0.12%
EDTA—0.4%
Ethylene glycol—3.38%
Glutaraldehyde—1.08%
Methanol—0.47%
Deionized water—77%

Laskowski Solution [18]

Glycerin—20 parts
Ethanol—4 parts
Phenol—1 part
Boric acid—1 part

Tompsett (1970) [1]

1. Methylated industrial spirit, which contains about 96% ethyl alcohol
2. Formalin, 40% saturated formaldehyde solution
3. Liquid phenol, which is a cruder and cheaper form of phenol than the crystallized form
4. Glycerin
5. Water

Laskowski's Solution Modified (Silva et al. 2007) [19]

Glycerin—800 mL
Ethanol—200 mL
Carbolic acid—50 g
Boric acid—50 mg

Larssen Solution, Modified by Sampaio (1989) [20]

Sodium chloride—500 g
Baking soda—900 g
Chloral hydrate—1000 g
Sodium sulfate—1100 g
Formalin 10%—500 mL
Distilled water—1000 mL

Walther Thiel's Solution (1992, 2002) [21–24]

Solution A
Boric acid—3 g
Ethylene glycol—30 mL
Ammonium nitrate—20 g
Potassium nitrate—5 g
Hot tap water—100 mL

Solution B
Ethylene glycol—10 mL
4-Chloro-3-methylphenol—1 mL

Injection Solution
Solution A—14,300 mL
Solution B—500 mL
Formaldehyde—300 mL
Sodium sulfate—700 g

Immersion Solution
Ethylene glycol—10 mL
Formaldehyde—2 mL
Solution B—2 mL
Boric acid—3 g

Ammonium nitrate—10 g
Potassium nitrate—10 g
Sodium sulfate—7 g
Hot tap water—100 mL

Larssen's solution modified by Guimaraes da Silva et al. (2004) [25]

Formaldehyde 10%—100 mL
Glycerol—400 mL
Chloral hydrate—200 g
Sodium sulfate—200 g
Baking soda—200 g
Sodium chloride—180 g
Distilled water—2000 mL

Constantinescu et al.'s Solution (2007) [26]

Formaldehyde 37%—1200 mL
Propylene or ethylene glycol—400 mL
Phenol—1000 mL
Water—20 L

Salt-Saturated Solution (Hayashi et al. 2014) [27]

Sodium chloride—20 kg
Formalin 20%—1 L
Phenol—200 mL
Glycerin—500 mL
Isopropyl alcohol—4 L
Tap water—19.3 L

Telma Masuko's Solution (2022) [28]

Alcohol 1000 mL
Phenol 1000 mg
Glycerin 1000 mL

Sodium nitrate 1000 mg
Formaldehyde 2000 mL
Distilled water 14,000 mL

This technique allows the fixation and preservation of whole bodies, human and animal, at room temperature for prolonged periods, more than 2 years, with morphological and mobility characteristics similar to fresh bodies, due to the preservation of the soft and flexible characteristics of the bodies and natural colors. Likewise, Masuko et al. [28] carried out histological analyses of the tissues preserved with this technique and demonstrated the preservation of the characteristics and cellular structures of the samples analyzed, with more than 1175 days of fixation. Likewise, in addition to the advantages in tissue preservation, there are also economic advantages for the implementation of this technique, since the inputs used are low-cost, both for fixation and for the maintenance of the preserved specimens.

Chilean Preservative Fixative Solution [29–31]

1. Sodium chloride (1.5 kg) + 6 L of water.
2. Sodium nitrate (1.2 kg) + 6 L of water. Failing that, potassium nitrate or urea in concentrations >43%.
3. Glycerin (4 L).
4. Ethyl alcohol (6 L). Isopropyl alcohol can also be used.
5. Concentrated benzalkonium chloride (2 L).
6. Formaldehyde 5% (0.5 L).
7. Eucalyptus essence (0.5 L).

The Chilean Conservative Fixing Solution is fixing because it stops the natural putrefaction process, and it is conservative since it maintains the fixing conditions over time. It can be used by intravascular injection and/or immersion. The components used to create the Chilean Conservative Fixing Solution (CCFS) are the following: sodium chloride, with preservative properties; sodium nitrate, which allows the preservation of the color of the tissues; glycerin, which acts in preserving by inhibiting enzymatic changes, giving flexibility to tissues; ethyl alcohol, which ensures dehydration and elimination of fatty tissue; benzalkonium chloride (for ophthalmological use), being in high concentrations sporicidal (antifungal); formalin, which acts as a disinfectant and preservative; and eucalyptus essence (or other aroma), which serves to neutralize the irritating smell of formalin. Some details can be highlighted in relation to the application of the CCFS: as in other formulations, it is possible to eliminate the formalin by oxidative degradation, using 0.3% sodium hypochlorite (NaClO), with a subsequent washing with running water, and repeating the process until the smell of formalin is removed. To keep the samples hydrated, it is possible to immerse them in the CCFS or externally moisten the samples with an atomizer, by sprinkling or placing cloths moistened with the formulation every 15 days, without the need, in the latter case, to immerse the samples in tanks with

CCFS; once fixed with CCFS, samples should not be washed with running water, as this would alter the concentrations of the solution. This demonstrates some of the advantages of CCFS: softening of tissues, with preservation of color and degradation of fatty tissue. It is not necessary to renew the solution or use storage tanks, since the samples can be kept moist by applying the CCFS by spraying, as previously indicated.

Walther Thiel's Technique [12–17, 21–24, 32]

Walther Thiel's method of fixation, conservation, and preservation of corpses "in natural colors" is based on three processes: fixation, disinfection, and conservation/preservation, with solutions containing 4-chloro-3-methyl-phenol, together with various salts, boric acid and ethylene glycol, as basic components.

It allows prolonged conservation, maintaining the color, texture, plasticity, and flexibility of the fresh specimen. Its storage is simple and does not require the use of sinks. The vessels and canaliculi can be injected up to their finest branches.

All this enables its use with great advantages over formalized material, and even over fresh material due to its ease of handling and long duration. Its multiple uses are applicable in anatomical research, teaching purposes in the undergraduate, and training in surgical techniques in the postgraduate.

Profuse washing with warm running water of the entire vascular tree and its cavities is first carried out.

Thiel's solution is administered intravascularly (through the carotid and femoral arteries) incorporating this same mixture through the air, esophagogastric, and colonic tracts. Solutions A and B are prepared and then combined to obtain the injection solution (see above).

Once the injection solution has been completely incorporated into the corpse, the body is placed in a basin containing the third mixture proposed by the aforementioned author, corresponding to the immersion solution (see previously). The immersion time in this solution is 30 days.

After the period established for immersion, the corpse is removed from the pool and simply stored in a plastic bag with a zip closure. Periodic immersion of the corpse should be carried out only to maintain the humidity of the tissues, when signs of drying appear. Periodic immersions last approximately 7 days.

Walther Thiel's technique offers the following advantages:

(a) The duration of conservation, which can be years, with little maintenance
(b) Storage, which does not require pools of liquids, but rather is carried out in plastic bags, with the whole corpse, thus saving space
(c) The conservation of the original physical properties (color, flexibility, plasticity)
(d) Little or no emission of noxious or irritating vapors
(e) The possibility of carrying out arterial and canalicular fillings for research and anatomical demonstration

These advantages offer teaching, training, and improvement possibilities for undergraduate and postgraduate courses.

The use of this technique on cadavers for undergraduate students can help visualize joint mobility and the appearance of these preparations, being very similar to that of the living, eliminating joint and tissue rigidity, and the uniform color of the formolized material.

For postgraduate students, they can perform explorations and surgical procedures, especially arthroscopic, laparoscopic, and thoracoscopic, and endoscopic or vascular suture, without the requirement of a freezer or the inevitable decomposition of the fresh material depending on the exposure time at room temperature.

Saturated Salt Solution (SSS) [27]

The saturated salt solution, proposed by Hayashi et al. [27], is a formulation that presents salt as an essential component. Initially this was proposed in 1998 by Coleman and Kogan [4], who based themselves on the historical processes developed by the Egyptians, thought about the development of a fixation technique in which salt predominated, but maintaining both formulations, both that of Coleman and Kogan [4] and Hayashi et al. [27], the use of formalin and phenol. Both formulations will be described below.

Saturated Salt Solution Proposed by Coleman and Kogan (1998) [4]
Sodium chloride (20 kg)
Formaldehyde 37–40% (0.5 L)
Phenol (0.2 L)
Glycerin (0.5 L)
Isopropyl alcohol (4 L)
Water (20–30 L)
Total 25–35 L (depending on the size of the carcass)

Saturated Salt Solution Proposed by Hayashi et al. (2014) [27]
Sodium chloride (20 kg)
Formaldehyde 20% (1.0 L)
Phenol (0.2 L)
Glycerin (0.5 L)
Isopropyl alcohol (4 L)
Water (19.3 L)
Total (25.0 L)

The big difference between both formulas is the reduction in the percentage of formaldehyde concentration in the formula proposed by Hayashi et al. [27] compared to the original proposed by Coleman and Kogan [4]. Similarly, in this latter formulation, the final concentration of formaldehyde in the embalming solution mixture was only 0.5–0.75%.

Discussion

Since several years ago, and because it was initially commented on the toxicity and carcinogenic potential of formaldehyde, there is a tendency to reduce the amount of formaldehyde used in fixative preservatives [5, 9–17, 19–41]. Embalming and fixative fluids should provide good long-term structural preservation of organs and tissues along with prevention of excessive hardening and color retention of tissues and organs [28]. Tests of various substance mixtures that provide ideal tissue preservation, including small amounts of formaldehyde, along with substances that contribute to or replace fixation functions, have been reported. Several of these techniques have been discussed in this chapter, highlighting the widespread use at an international level of Walther Thiel's technique, as well as the techniques that incorporate salt and alcohols into their formulations in replacement of formaldehyde, or in combination with the latter, but in a final concentration that is almost insignificant compared to that originally used in the fixing and conservation formulas [12–17, 23, 24]. Among the formulas that do not incorporate formaldehyde in their composition, there is one that is based on white vinegar, glycerin, ethanol, sodium citrate, and malachite green [30]. Animal tissue prepared with this formaldehyde-free fixing formula retains similar properties to living tissue and does not differ during dissection from samples prepared with solutions containing formaldehyde. There are also other formaldehyde-free formulas, such as the one described by Janczyk et al. [42], in which the compounds used are sodium nitrite, ethanol, polyethylene glycol, oregano oil, and distilled water, in which it has been found that preparing tissues is easier than using formaldehyde and aqueous solutions. Among the proposed formulations with high salt concentrations is the solution of Coleman and Kogan [4], which presented a formulation containing low concentrations of formaldehyde (1.43%), phenol, glycerol, isopropanol, large amounts of salt, and distilled water.

Coleman and Kogan [4] in their letter to the editor indicate that they introduced novelties to their fixation process, characterized, on the one hand, by a new way of storing the corpses, in a "dissection bed," made up of a thick bag of polyethylene, inside which part of the embalming formula is placed and which are also adapted with an internal motor that allows the generation of a very low flow of vapors rich in formaldehyde, which is absorbed by a replaceable active carbon filtration system (also described in Coleman [43]). Added to this is the new proposed fixing liquid, which is characterized by a notable reduction in the percentage of final formaldehyde present in the formulation, and therefore in the fixed samples. But to this they add a very high concentration of salt, combined with the low content of formaldehyde, preventing desiccation from being significant, thanks to this high salt content that is retained in the fixed biological tissues. Also, before arranging the bodies for dissection, excess embalming fluid is removed and the corpses are left fairly dry. And also, before opening the bags that contain the corpses in the "bed," they activate the internal motor, thus removing the excess embalming fluid, leaving the corpses quite dry. This is how Coleman and Kogan [4] state that their new embalming mixture results in a dissection room practically free of toxic odors, with compliance with biosafety measures for people and environmental safety, with the generation of corpses with notable dissection properties. They were able to confirm this by taking autopsy samples of a large number of tissues from the embalmed bodies and organs, which were subjected to histological examinations [12–17].

Also Coleman and Kogan [4] established that the antiseptic and fixative properties of the SSS have proven to be excellent. The whole tissue and organ body showed minimal deformation patterns, and the tissues remained flexible and were easily dissected. There is no need to dry less and add more liquid. The textures retain most of their natural color and show no evidence of "browning" oxidation effects, even over long periods of time. Tissues that are often susceptible to poor embalming, such as the brain, retain their color. There are no signs of fungal growth. Sometimes there are small salt deposits on the skin or organs.

At present, not only the search for fixation formulas that are more suitable for the health of the people who handle the samples but also the need for improvement in surgical techniques, and above all, in models that are most similar to the living patient, have determined the development and search for alternative fixation formulas, which ensure that the tissues of the cadavers of these training courses, both human and animal, are as close as possible to the living state [12–17]. There are a lot of trainings for doctors, and there is a growing need for models with the same characteristics as humans. This is confirmed by comparing the use of human cadavers for surgical skills training with the use of live animals and virtual simulators, and by demonstrating the benefits of anatomical fidelity and the ability to assess surgical outcome when developing human material courses. In this way, when using human bodies in this type of surgical training, it is essential to choose the appropriate type of fixation technique for the preservation of the bodies and its correlation with the type of training to be developed [12–17]. Tissue flexibility and staining are among the most important requirements that pose a challenge to professor or technical in charge for the preparation of these specimens, as the main effect of formaldehyde is tissue hardening and discoloration. In addition, there is also a comparison with the use of fresh frozen human bodies, which have a color, softness, and mobility similar to the living body; however, there are great disadvantages, such as economics, having to have refrigerated storage for the samples, and perhaps, even more important, the rapid putrefaction and deterioration of the samples, since they are constantly thawed for use, and this determines that the bodies have a very limited durability in time (a few weeks), adding to this the possibility of infection, in case you do not work respecting all biosecurity measures [12–17, 27].

The characteristic of the techniques proposed by Coleman and Kogan [4] and Hayashi et al. [27] is not only the use of salt in the formulation but also the incorporation of low levels of formaldehyde in the final formula of the SSS solutions. In this sense, Coleman and Kogan [4] highlight the ability to significantly reduce formaldehyde levels in anatomy laboratories, thus reducing the exposure of students and teachers to toxic formaldehyde levels, thus avoiding classic conditions such as alterations in the respiratory tract or tearing of the eyes.

Kalanjati et al. [44] developed a low formaldehyde-containing embalming solution that showed lighter coloration and thus morphology and structure details (muscles, neurovascular and internal organs) compared to high formaldehyde embalming solutions. Also, the consistency of muscles and internal organs is more elastic and drier while they stay moist compared to high formaldehyde solutions that make darker organs and muscles and wet, and often structures are hardly differentiated one to another [44]. Related to that, the appearance of low formaldehyde formula cadavers is better for anatomy teaching and learning when compared to classic formaldehyde solutions.

Among some techniques that are currently used, Walther Thiel's formula [12–17, 21, 24] is often used in human medicine to prepare sites for surgical training, and its advantages include intact cadavers, subcutaneous tissue, fascia, and internal organs, and the muscles retain their natural color, consistency, and flexibility, similar to those in the real body, to living ones. But the Thiel technique has the disadvantages of the high cost of the inputs that the preparation of the formulation requires, the complication in the implementation of the formula, as well as the muscle disintegration suffered by these samples and the limited time for dissection.

Conservative fixation solutions suitable for this type of intervention include modified Larssen's solution [20, 25] and Laskowski's solution [18, 19]. Both solutions can preserve cadavers with the same characteristics as live humans and animals, which is especially important when teaching surgery. Silva et al. [19] applied a modified Larssen solution, and identified very satisfactory results in samples of embalmed dogs for practical anatomy classes, in which they were also able to perform surgical approach procedures. But they also describe the need to improve the embalming solution at the level of achieving better conservation of the abdominal viscera of the same samples.

It is also crucial, when deciding on the fixation and embalming techniques to be developed, to evaluate and analyze the destination that will be given to the cadaveric samples, whether it be teaching anatomy at undergraduate or postgraduate level, surgical training in specialization courses, or scientific research [12–17]. But also, not only should the application of fixation and embalming techniques be analyzed from the point of view of the final destination of the corpses, but also, and importantly, analyze the weather conditions of the place where the techniques will be implemented, as well of the exposure times to the samples, since the storage and conservation conditions will be determined by the conditions of the environment in which the samples are prepared, which could also be artificially adapted if the necessary economic resources are available [12–17].

In order to achieve this, experiments should be carried out using different solutions and methods to control the anatomical material, to measure the properties such as degree of fixation, tissue changes, resistance to desiccation, flexibility, and emission of irritating vapors, and to prevent the reproduction of bacteria that affect worker's health and spoil the results [45].

In this way, there are many fixation and embalming techniques of preserving corpses for educational purposes. The current trend is to reduce or eliminate the use of formaldehyde in its formula, due to the harmful effects reported for this substance [12–17, 45]. In this way, plastination is a process that preserves biological tissues by replacing the water and lipids in the tissue with polymers, such as silicone rubber or epoxy resin, thereby creating a dry, odorless, and durable specimen that can be used for educational or scientific purposes. The technique was invented in 1977 by Dr. Gunther von Hagens, who was interested in finding a way to create and preserve more durable and lifelike specimens for medical education and research purposes and in exhibits of anatomy and pathology [46–60].

Prof. von Hagens began developing the technique while working at the University of Heidelberg. He started experimenting with various preservation techniques, such as freeze-drying and embedding specimens in plastic, to better showcase the human body's intricate details. After several years of experimentation, von Hagens developed a technique that involved immersing the tissue in a series of solvents, such as acetone, and then impregnating it with a polymer under vacuum pressure. The result was a specimen that retained the shape and structure of the original tissue without damaging its structure but was dry, odorless, and durable, and could be manipulated directly without the need for gloves or protective clothing [46–60].

The use of plastination techniques to obtain long-term anatomical material can reduce the number of cadavers that need to be prepared in the physical examination area to meet the requirements of the method.

von Hagens' plastination technique gained popularity in the scientific and medical communities. The technique also gained mainstream attention with the "Body Worlds" exhibitions, which showcased plastinated specimens of real human bodies and animals in various poses.

References

1. Tompsett DH. Anatomical techniques. 2nd ed. Edinburgh: E. & S. Livingstone; 1970.
2. Brenner E. Human body preservation—old and new techniques. J Anat. 2014;224(3):316–44. https://doi.org/10.1111/joa.12160.
3. Tolhurst DE, Hart J. Cadaver preservation and dissection. Eur J Plast Surg. 1990;13:75–8. https://doi.org/10.1007/BF00177811.
4. Coleman R, Kogan I. An improved low-formaldehyde embalming fluid to preserve cadavers for anatomy teaching. J Anat. 1998;192(Pt. 3):443–6. https://doi.org/10.1046/j.1469-7580.1998.19230443.x.
5. Whitehead MC, Savoia MC. Evaluation of methods to reduce formaldehyde levels of cadavers in the dissection laboratory. Clin Anat. 2008;21(1):75–81. https://doi.org/10.1002/ca.20567.
6. Thavarajah R, Mudimbaimannar VK, Elizabeth J, Rao UK, Ranganathan K. Chemical and physical basics of routine formaldehyde fixation. J Oral Maxillofac Pathol. 2012;16(3):400–5. https://doi.org/10.4103/0973-029X.102496.
7. Kiernan JA. Formaldehyde, formalin, paraformaldehyde and glutaraldehyde: what they are and what they do. Microsc Today. 2000;8(1):8–12.
8. Fox CH, Johnson FB, Whiting J, Roller PP. Formaldehyde fixation. J Histochem Cytochem. 1985;33(8):845–53. https://doi.org/10.1177/33.8.3894502.
9. International Agency for Research on Cancer (IARC). Formaldehyde, 2-butoxyethanol and 1-tert-butoxypropanol-2-ol. IARC Monogr Eval Carcinog Risks Hum. 2006;88:39–325.
10. National Toxicology Program. NTP 12th report on carcinogens. Rep Carcinog. 2011;12:iii–499.

11. Njoya HK, Ofusori DA, Nwangwu SC, Amegor OF, Akinyeye AJ, Abayomi TA. Histopathological effect of exposure of formaldehyde vapour on the trachea and lung of adult Wistar rats. IJIB. 2009;7(3):160–5.
12. Ottone NE. Lecture: "Neurotechnics". In: 1st Workshop on plastination and anatomical techniques. Laboratory of Plastination and Anatomical Techniques, Faculty of Dentistry, University of La Frontera, Temuco, Chile. October 16, 2017.
13. Ottone NE. Lecture: "Alternative techniques to formaldehyde for fixation and conservation". In: I International course on techniques for the preparation and conservation of biological material for the teaching of anatomy. Peruvian Society of Morphological Sciences, Federico Villarreal National University, "Hipólito Unanue" School of Medicine, Lima Peru. November 27, 2015.
14. Ottone NE. Lecture: "Traditional fixation and conservation techniques alternatives to formalization". In: XIV National and International Congress of Morphophysiological Sciences – I Paraguayan Congress of Anatomy. Eastern University. Paraguayan Society of Morphophysiological Sciences, Paraguayan Society of Anatomy, City Pte, Franco, Paraguayan. September 11 and 12, 2015.
15. Ottone NE. Lecture: "Cadaveric conservation techniques with reduced levels of formaldehyde". In: XLIX Argentine Congress of Anatomy and II International Congress of Anatomy and IV Conference on Anatomy for Students of Health Sciences, of the Argentine Association of Anatomy, Central Military Hospital Cirujano Mayor Dr. Cosme Argerich and National Academy of Medicine, Autonomous City of Buenos Aires, Argentina. September 5 to 8, 2012.
16. Ottone NE. Lecture: "New technique of intravascular injection with reduced levels of formaldehyde". In: XVIII Argentine Congress of the Association of Morphological Sciences of Corrientes—XIII International Congress. Association of Morphological Sciences of Corrientes. Campus of the Faculty of Medicine of the National University of the Northeast, City of Corrientes, Province of Corrientes, Argentina. August 24 and 25, 2012.
17. Ottone NE. Lecture: "Conservation of the body in natural colors". In: XVIII Argentine Congress of the Association of Morphological Sciences of Corrientes—XIII International Congress. Association of Morphological Sciences of Corrientes, Campus of the Faculty of Medicine of the National University of the Northeast, City of Corrientes, Province of Corrientes, Argentina. August 24 and 25, 2012.
18. Laskowski S. L'embaumement, la conservation des sujets et les préparations anatomiques. Genève-Bâle-Lyon: H. Georg; 1886.
19. Silva RM, Matera JM, Ribeiro AA. New alternative methods to teach surgical techniques for veterinary medicine students despite the absence of living animals. Is that an academic paradox? Anat Histol Embryol. 2007;36(3):220–4. https://doi.org/10.1111/j.1439-0264.2007.00759.x.
20. Sampaio F. Study of the growth of the human kidney during the fetal period. Doctoral thesis in morphology. Sao Paulo: Escola Paulista de Medicina; 1989.
21. Thiel W. Die Konservierung ganzer Leichen in natürlichen Farben. Ann Anat. 1992;174:185–95. https://doi.org/10.1016/S0940-9602(11)80346-8.
22. Thiel W. Ergànzung für die Konservierung ganzer Leichen nach W. Thiel. Ann Anat. 2002;184:267–9. https://doi.org/10.1016/S0940-9602(02)80121-2.
23. Ottone NE, Vargas CA, Fuentes R, del Sol M. Walter Thiel's embalming method: review of solutions and applications in different fields of biomedical research. Int J Morphol. 2016;34(4):1442–54. https://doi.org/10.4067/S0717-95022016000400044.
24. Bertone VH, Blasi E, Ottone NE, Dominguez ML. Walther Thiel Method for the Preservation of Corpses with Maintenance of the main physical properties of Vivo. Rev Argent Anat Online. 2011;2(3):89–92.
25. Guimarães da Silva RM, Matera JM, Ribeiro AA. Preservation of cadavers for surgical technique training. Vet Surg. 2004;33(6):606–8. https://doi.org/10.1111/j.1532-950x.2004.04083.x.

26. Constantinescu GM, Beittenmiller MR, Mann F, et al. Clinical anatomy of the prepubic tendon in the dog and a comparison with the cat. J Exper Med Surg Res. 2007;14: 79–83.
27. Hayashi S, Homma H, Naito M, Oda J, Nishiyama T, Kawamoto A, Kawata S, Sato N, Fukuhara T, Taguchi H, Mashiko K, Azuhata T, Ito M, Kawai K, Suzuki T, Nishizawa Y, Araki J, Matsuno N, Shirai T, Qu N, Hatayama N, Hirai S, Fukui H, Ohseto K, Yukioka T, Itoh M. Saturated salt solution method: a useful cadaver embalming for surgical skills training. Medicine (Baltimore). 2014;93(27):e196. https://doi.org/10.1097/MD.0000000000000196.
28. Masuko TS, dos Santos Andrade L, Ottone NE, Baptista CAC, Barros MD. Histological study of a low-cost embalming technique with preservation for nearly three years at room temperature. FASEB J. 2022;36(S1) https://doi.org/10.1096/fasebj.2022.36.S1.R6364.
29. Rojas Oviedo JD, Ruiz Diaz SS. Empleo de solución fijadora conservadora chilena como alternativa al uso del formaldehido para la preservación de tejidos. En: V Congreso Colombiano de Morfología. Cali—Colombia, Noviembre de 2009. Int J Morphol. 2010;28:337–40. https://doi.org/10.4067/S0717-95022010000100050.
30. Muñetón Gómez CA, Ortiz JA. Conservación y elaboración de piezas anatómicas con sustancias diferentes al formol en la Facultad de Ciencias Agropecuarias de la Universidad de La Salle. Rev Med Vet. 2011;22:51–5.
31. Castillo JCF. Restauración de Piezas Anatómicas Humanas. Tesis de Magister en Morfología Humana. Bogotá: Universidad Nacional de Colombia; 2014.
32. Bertone VH, Blasi E, Ottone NE, Dominguez ML. Walther Thiel method for the preservation of corpses with maintenance of the main physical properties of vivo. Rev Argent Anat Online. 2011;2(3):89–92.
33. Ottone NE. Lecture: "Corpse conservation with the Walther Thiel technique". In: XLIX Argentine Congress of Anatomy and II International Congress of Anatomy and IV Conference on Anatomy for Students of Health Sciences, of the Argentine Association of Anatomy, Central Military Hospital Cirujano Mayor Dr. Cosme Argerich and National Academy of Medicine. Autonomous City of Buenos Aires, Argentina. September 5 to 8, 2012.
34. Ottone NE. Lecture: "Application and usefulness of Walther Thiel's cadaveric preservation technique". VII Colombian Congress of Morphology—XXXII Chilean Congress of Anatomy—XIII Congress of Anatomy of the Southern Cone. Panamerican Association of Anatomy. Estelar Santamar Hotel and Convention Center, Santa Marta, Colombia. October 27, 28 and 29, 2011.
35. Ajayi IE, Shawulu JC, Ghaji A, Omeiza GK, Ode OJ. Use of formalin and modified gravity-feed embalming technique in veterinary anatomy dissection and practicals. J Vet Med Anim Health. 2011;3(6):79–81.
36. Dixit D. Role of standardized embalming fluid in reducing the toxic effects of formaldehyde. Indian J Forensic Med Toxicol. 2008;2(1):33–9.
37. Hisamitsu M, Okamoto Y, Chazono H, Yonekura S, Sakurai D, Horiguchi S, Hanazawa T, Terada N, Konno A, Matsuno Y, Todaka E, Mori C. The influence of environmental exposure to formaldehyde in nasal mucosa of medical students during cadaver dissection. Allergol Int. 2011;60(3):373–9. https://doi.org/10.2332/allergolint.10-OA-0210.
38. Viegas S, Ladeira C, Nunes C, Malta-Vacas J, Gomes M, Brito M, Mendonca P, Prista J. Genotoxic effects in occupational exposure to formaldehyde: a study in anatomy and pathology laboratories and formaldehyde-resins production. J Occup Med Toxicol. 2010;5(1):25. https://doi.org/10.1186/1745-6673-5-25.
39. Russell AD. Bacterial spores and chemical sporicidal agents. Clin Microbiol Rev. 1990;3(2):99–119. https://doi.org/10.1128/CMR.3.2.99.
40. Vardaxis NJ, Hoogeveen MM, Boon ME, Hair CG. Sporicidal activity of chemical and physical tissue fixation methods. J Clin Pathol. 1997;50(5):429–33. https://doi.org/10.1136/jcp.50.5.429.

41. Demiryürek D, Bayramoğlu A, Ustaçelebi S. Infective agents in fixed human cadavers: a brief review and suggested guidelines. Anat Rec. 2002;269(4):194–7. https://doi.org/10.1002/ar.10143.
42. Janczyk P, Weigner J, Luebke-Becker A, Kaessmeyer S, Plendl J. Nitrite pickling salt as an alternative to formaldehyde for embalming in veterinary anatomy—a study based on histo- and microbiological analyses. Ann Anat. 2011;193(1):71–5. https://doi.org/10.1016/j.aanat.2010.08.003.
43. Coleman R. Reducing the levels of formaldehyde exposure in gross anatomy laboratories. Anat Rec. 1995;243(4):531–3. https://doi.org/10.1002/ar.1092430417.
44. Kalanjati VP, Prasetiowati L, Alimsardjono H. The use of lower formalin-containing embalming solution for anatomy cadaver preparation. Med J Indones. 2012;21(4):203–7. https://doi.org/10.13181/mji.v21i4.505.
45. Matheus JF. Preservation of anatomical pieces for teaching in medical careers. Gac Cienc Vet. 2012;17:5–10.
46. von Hagens G. Impregnation of soft biological specimens with thermosetting resins and elastomers. Anat Rec. 1979;194(2):247–55. https://doi.org/10.1002/ar.1091940206.
47. von Hagens G, editor. Heidelberg plastination folder. Collection of technical leaflets of plastination. Heidelberg: Biodur Products GmbH; 1986.
48. von Hagens G, Tiedemann K, Kriz W. The current potential of plastination. Anat Embryol (Berl). 1987;175(4):411–21. https://doi.org/10.1007/BF00309677.
49. Ottone NE. Gunther von Hagens, creator of plastination. Historical review and technical development. Rev Argent Anat Online. 2013;4(2):70–6. https://www.revista-anatomia.com.ar/archivos-parciales/2013-2-revista-argentina-de-anatomia-online-f.pdf.
50. Ottone NE, Cirigliano V, Lewicki M, Bianchi H, Aja-Guardiola S, Algieri RD, Cantin M, Fuentes R. Plastination technique in laboratory rats: an alternative resource for teaching, surgical training and research development. Int J Morphol. 2014;32(4):1430–5. https://doi.org/10.4067/S0717-95022014000400048.
51. Ottone NE, Cirigliano V, Bianchi HF, Medan CD, Algieri RD, Borges Brum G, Fuentes R. New contributions to the development of a plastination technique at room temperature with silicone. Anat Sci Int. 2015;90(2):126–35. https://doi.org/10.1007/s12565-014-0258-6.
52. Ottone NE, del Sol M, Fuentes R. Report on a sheet plastination technique using commercial epoxy resin. Int J Morphol. 2016;34:1039–43. https://doi.org/10.4067/S0717-95022016000300036.
53. Ottone NE, Vargas CA, Veuthey C, del Sol M, Fuentes F. Epoxy sheet plastination on a rabbit head–new faster protocol with Biodur® E12/E1. Int J Morphol. 2018;36(2):441–6. https://doi.org/10.4067/S0717-95022018000200441.
54. Ottone NE, Baptista CAC, Latorre R, Bianchi HF, Del Sol M, Fuentes R. E12 sheet plastination: techniques and applications. Clin Anat. 2018;31(5):742–56. https://doi.org/10.1002/ca.23008.
55. Prieto R, Vargas CA, Veuthey C, Aja-Guardiola C, Ottone NE. Fundamental concepts of the modified room temperature plastination protocol with silicone, with subsequent pigmentation, and its application for the conservation of human placenta. Int J Morphol. 2019;37(1):375–6. https://doi.org/10.4067/S0717-95022019000100369.
56. Guerrero M, Vargas C, Alarcón E, del Sol M, Ottone NE. Development of a sheet plastination protocol with polyester resin applied to human brain slices. Int J Morphol. 2019;37(4):1557–63. https://doi.org/10.4067/S0717-95022019000401557.
57. Vargas CA, Baptista CAC, Del Sol M, Sandoval C, Vásquez B, Veuthey C, Ottone NE. Development of an ultrathin sheet plastination technique in rat humeral joints with osteoarthritis induced by monosodium iodoacetate for neovascularization study. Anat Sci Int. 2020;95(2):297–303. https://doi.org/10.1007/s12565-019-00500-7.
58. Ottone NE. Micro-plastination. Technique for obtaining slices below 250 μm for the visualization of microanatomy in morphological and pathological experimental protocols. Int J Morphol. 2020;38(2):389–91. https://doi.org/10.4067/S0717-95022020000200389.

59. Ottone NE. Unified plastination protocol with silicone at cold and room temperature. Int J Morphol. 2021;39(2):630–4. https://doi.org/10.4067/S0717-95022021000200630.
60. Ottone NE, Guzmán D, Bianchi HF, del Sol M. Anatomical variations of radial and ulnar arteries in plastinated upper limbs. Int J Morphol. 2023;41(2):548–54. https://doi.org/10.4067/S0717-95022023000200548.

Chapter 5
Silicone Plastination Technique

General Description

Cold-temperature plastination with silicone has been known by the abbreviation S10 since Gunther von Hagens founded the Biodur company in 1980 for the commercialization of all the products and equipment necessary for plastination techniques, including the silicone version called "S10." For that reason, the plastination technique that requires the use of silicone is defined as the S10 technique. Figure 5.1 shows the basic steps of the classical technique of cold-temperature plastination with silicone.

The plastination technique with S10 silicone produces rigid/flexible and resistant samples [1–41]. Most plastinated specimens are embalmed, using formalin concentrations from 5% to 20%, or some other type of embalming and fixation technique with reduced formalin levels and a predominance of glycerin and other types of alcohol [42–46]. Thus, the embalming fluids with long-chain alcohols (e.g., glycerol) must be eliminated prior to initiating dehydration, because they do not favor the correct final preservation of the samples. Moreover, fixation can be done by immersion and infiltration, with this type of fixation corresponding to organs and isolated specimens. In the case of the hollow organs, it is recommended that these be dilated before and during fixation, as well as during the essential steps of plastination. This is an essential step in the plastination of all types of hollow organs, such as the heart [4]. On the other hand, fixation can be omitted when epoxy resins are used (epoxy resins have fixation properties), which results in better color preservation [1].

Dehydration is fundamental, as is defatting, since the water and lipids cannot exchange directly with curable polymers [1]. This procedure also seeks to avoid shrinkage. The degree of defatting is essential in plastination. The fundamental step consists of dehydration based on freeze substitution in acetone. An alternative

N. E. Ottone, *Advances in Plastination Techniques*,
https://doi.org/10.1007/978-3-031-45701-2_5

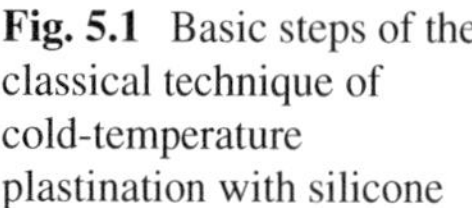

Fig. 5.1 Basic steps of the classical technique of cold-temperature plastination with silicone

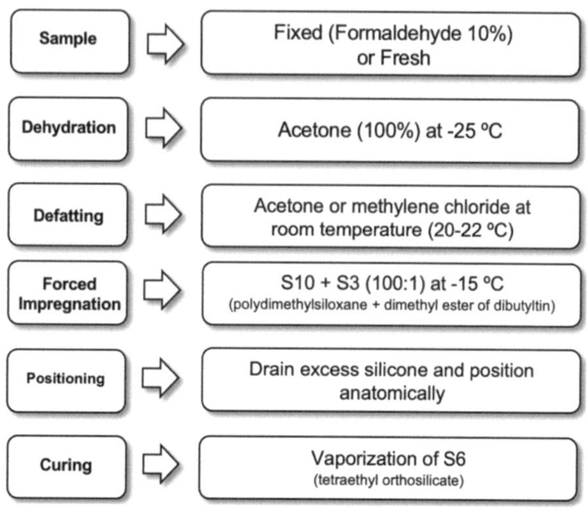

option is classic dehydration by graduated ethanol stages, known from histological techniques, and mainly in those countries where acetone is difficult to access.

After dehydration with acetone, the samples are impregnated at -25 °C in a vacuum chamber, submerged in a mixture of silicone (S10) and catalyst (called S3) (100:1, respectively]) which have a low steam pressure (high boiling point). The volatile intermediary (acetone) found in the specimen is constantly removed by a vacuum pump. Once the acetone is removed, a pressure difference will determine that the polymer enters the specimen. Forced impregnation must be carried out slowly as the polymer enters the specimen where the acetone changes from a liquid to a gaseous state and is removed. The impregnation speed is carefully adjusted by the controlled addition of air inside the vacuum pump using a bypass valve. The duration of the forced impregnation stage will depend mainly on the size (and amount) of the specimen, the density of the tissue, and the viscosity of the polymer used. During this period, the vacuum must be intensified to a pressure of 760 mmHg, according to the desired formation of bubbles (intermediate), to a pressure of approximately 5 to 7 mmHg, where the small bubbles will rise to the surface (bubbles indicate the acetone is exiting the interior of the specimen). Once this final pressure level is reached, and with no bubbles (indicating the replacement of acetone for silicone), then the forced impregnation stage is finished, and the specimen is removed from the polymer solution 24 h after this stage is complete.

After the impregnation, the samples undergo the curing stage (polymerization), which consists of exposing the samples, within a sealed chamber, to a liquid that contains silicate and which is gasified (S6). The S3 catalyst contained in the impregnation mixture begins the curing of the silicone molecules by end-to-end polymerization. Due to the cross-linking during the final curing with gas, the silicone within the sample will be solidified and dried. The surface of the sample is cured quickly, but the diffusion of the gas toward the inside of the sample is slower. To guarantee an adequate curing throughout the process, the sample must be kept in hermetically

Table 5.1 Equipment, instruments, accessories, and chemicals used in plastination techniques with silicone

Steps	Equipments and accessories	Chemicals
Fixation[a]	Cannulas Basic dissection instruments Containers Injection pump	Formalin
Sectioning (optional)	Band saw Deep freezer (−80 °C) Freezer (−25 °C)	Liquid nitrogen Gelatin Polyurethane foam
Dehydration	Freezer (−25 °C) Containers (HDPE plastic or stainless steel) Acetonometer(s) Grids (plastic or metal)	Acetone
Defatting	Containers (HDPE plastic or stainless steel) Grids (plastic or metal)	Acetone Dichloromethane Extraction hood
Forced impregnation	Vacuum chamber[b] Vacuum pump Vacuometer Valves (vacuum control) Freezer (−25 °C) Grids (plastic or metal)	Silicone: Polydimethylsiloxane (S10) Catalyst: Dimethyl ester of dibutyltin (S3)[c] Tetraethyl orthosilicate (S6)[d]
Curing	Curing chamber Plastic bag Clamps or paper clips Silica crystals Small membrane pump (small aquarium aerator) Brush/sprayer[g] Plastic wrap[g]	Curing agent: Tetraethyl orthosilicate (S6)[e] Dimethyl ester of dibutyltin (S3)[f]

[a] The sample can be fresh, without fixation (except for brains, always fixated)
[b] The vacuum chamber outside a freezer is for plastination at room temperature [6, 47–49]
[c] In cold-temperature plastination [1–3], the catalyst is S3. Also in our proposal [6], see below
[d] In classical room-temperature plastination [47–49], the catalyst is S6
[e] In cold-temperature plastination [1–3], the curing agent is S6. Also in our proposal [6], see below
[f] In the classical room-temperature plastination [47–49], the curing agent is S3
[g] These elements are used in the curing stage in the traditional plastination technique at room temperature

sealed plastic bags during this time to complete its final drying. A topic to consider is the positioning of the specimen, i.e., the suitable display of the anatomical elements in the dissected regions, by placing separating needles, support threads, and other elements for the suitable composition of the plastinated preparation. This can be done prior to dehydration, or after forced impregnation, but always before curing, since, after this stage, the hardening the sample achieves prevents the anatomical structures from moving. Table 5.1 lists the equipment, instruments, accessories, and chemicals used in plastination techniques with silicone.

Cold-Temperature Plastination Technique

The cold-temperature silicone plastination technique, invented by Prof. Gunther von Hagens [1–3], is the gold standard in plastination (Fig. 5.1), defined as the S10 plastination technique, from which the fundamental steps of the technique are defined and from which specific variations are created that allow the development of the remaining plastination and cutting techniques at room temperature. In this sense, the silicone polymers are known, and were originally used by various industries for the generation and production of products. Therefore, in plastination, although the original polymers developed by Biodur (S10) exist, generic polymers can also be used and adapted to the technique; this is why it is important to know the chemical characteristics of the products. Researchers Chaynes and Mingotand, in 2004 [50], analyzed the components used in the S10 plastination technique, revealing the three chemical components used in the technique: S10, S3, and S6. Biodur S10 is a silicone polymer identified as a polydimethylsiloxane [HO[CH3]2 Si[OSi[CH3]2]nOH], a polymer of alternating oxygen and dimethylsilane atoms. Biodur S3, the catalyst/extender of short silicone chains to form long silicone chains, is a dimethyl ester of dibutyltin, [CH3[CH2]10OCO]2Sn[[CH2]3 [CH3]2. Biodur S6, the curing agent which binds the long silicone chains to achieve the final three-dimensional structure of the silicone when combined with its corresponding catalysts and curing agent, is a tetraethyl orthosilicate [Si[OCH2CH3]4]. The knowledge of how to develop the combination of polymers is essential, and these will be described in the forced impregnation step.

Specimen Preparation

Classically, the gold standard embalming and fixation technique for plastination is defined as formalin in variable percentages according to the researcher, between 5% and 10% [1–9, 51, 52]. In this sense, "formalin-free" embalming fluids can also be used, which in fact and generally contain formalin but in minimal concentrations [up to 1%], combined with various concentrations of different types of alcohol or salts, which contribute to the disinfection and preservation process alcohol [42–46]. It is always recommended, in the latter case, to use embalming fluids, the components of which are known, and this way those chemical products which could interfere with the plastination process in some of its stages, such as dehydration, forced impregnation, and curing, and which could be precipitated either in the fluid or worse still on the surfaces of specimens subjected to the plastination process can be identified, especially after the curing stage. The samples can also be injected by vascular access with natural latex (chemically cis-1,4-polyisoprene) or colored epoxy resins before the dissection. For this reason, as previously indicated, formalin is the gold standard as the fixative for biological samples intended for plastination.

Commercial fixatives often contain strange chemical products and are not recommended. Some of these chemical products can interfere with the curing of the impregnation mixtures and/or be precipitated on the surface of the sample impregnated after curing. Formalin is the best fixative for plastination [1–9, 51, 52]. The hollow organs must be dilated and fixed with 10% formalin after being cleaned with tap water. Vascular injection is recommended to highlight the vessels or lymph vessels and must be done during this preparation time, after fixation. Epoxy- or silicone-based products seem to be the best for vascular injection of the samples to be plastinated. Latex can be used; however, the cut surfaces of the latex often remain sticky. The samples or parts of them can be stained during the preparation and even at times after the protocol, during the last acetone bath [1, 2, 52]. The color of the sample can be intensified and clarified at this point with bleaching. Generally, it is better to almost completely dissect the sample before continuing with the next step: dehydration. It is good to eliminate the excess fat. Nevertheless, some additional dissection can be successful during any of the steps.

Dehydration

According to von Hagens [1], dehydration and defatting are mandatory in the plastination technique, since the water and lipids cannot exchange directly with curable polymers. As is known from histology, an adequate dehydration procedure must avoid shrinkage, and defatting is fundamental to plastination. In this sense, the two methods used are dehydration by stages with graduated ethanol and freeze substitution in acetone.

At present, the difficulty accessing acetone in various countries makes dehydration with ethanol necessary, the main disadvantage of which is the need to replace it equally with an intermediary low-boiling solvent (like acetone or methylene chloride). In addition, dehydration with ethanol takes a long time and causes considerable shrinkage (near 50%) [53]. Dehydration with ethanol is advantageous to the embalmed samples because it easily eliminates the long-chain alcohols contained in the embalming fluid (especially after the addition of H_2O_2).

The gold standard of dehydration for plastination is freeze substitution in acetone, because it saves time, requires less work than dehydration with ethanol, and causes less shrinkage of tissues [1–3, 54], being the only way to dehydrate brain tissue with a tolerable shrinkage (less than 10%) [53]. The only disadvantage of freeze substitution is the formation of ice crystals in samples that will also be used for histology. This can be overcome by processing samples that contain 20–100% formalin, which acts as a protective agent against the freezing [3].

Therefore, the samples to be plastinated must be completely dehydrated. After preparing the sample, the fixative and/or any other chemical product used in the preservation solution must be rinsed off the sample with running tap water for several days, even weeks, depending on the size and number of the samples. After rinsing the samples completely, it is necessary to cool them to 2–4 °C for 24 h before

Fig. 5.2 Dehydration by freeze substitution, with control of acetone levels (being at 87.5%, corresponding in this case, to the control of a third acetone bath). Plastic container for samples subjected to the dehydration step

beginning the dehydration. On the following day, the samples must be submerged in acetone (100% laboratory grade) at a temperature of -25 °C, as this dehydration consists of "freeze substitution" (Fig. 5.2). The recommended proportion of acetone in relation to the samples is 1:10 [1–3, 54]. However, it is possible to adapt the amount of acetone to a size associated with the sample, which can then be placed inside containers of a size similar to the sample (Fig. 5.3). Consequently, the acetone use can be reduced to practically half what is recommended (1:5 or less) [6–9]. The general protocol indicates that between three and four changes of 100% acetone must be done weekly; however, this will depend on the number and size of the samples, and above all the control of the acetone percentage (purity) in the dehydration container, which must be checked daily with an acetonometer/alcoholmeter. Dehydration will be complete when the final acetone bath has an acetone percentage >99.5% (Fig. 5.4). To record and control the acetone percentage (purity), a small amount must be extracted from the dehydration container where the samples are located, for example, extracting 250 mL and placing it in a similar glass sampling bottle to view the defatting as well which, although in cold it is less, also occurs. The acetonometer/alcoholmeter must be placed inside the sampling bottle to record the percentage that the instrument marks when floating in the liquid. It is recommended, at the beginning, that acetone purity be recorded daily to adequately monitor the process. When the record is the same on two consecutive days, it means that the acetone can be changed for a new one at 100% and this way advance with the

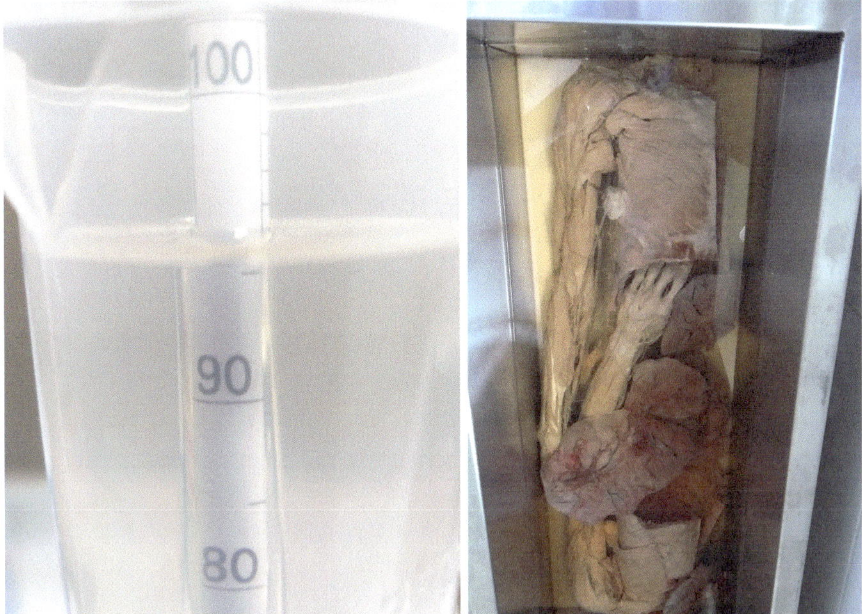

Fig. 5.3 Dehydration at room temperature, with control of acetone levels (being at 86.0%, corresponding in this case, to the control of the second acetone bath). Stainless steel container for samples subjected to the dehydration step

dehydration process. For this record to be suitable, it is also very important to know the calibration temperature of the acetonometer/alcoholmeter that will be used in the laboratory, which can be −20 °C, −10 °C, or 15 °C, 20 °C, with the last two being the most common. Cold acetone will show a much lower acetone percentage (purity) than if the acetone reaches the calibration temperature of the acetonometer/ alcoholmeter. Therefore, an incorrect purity percentage of the acetone will be indicated. Hence, to obtain a correct record of the acetone purity, it is necessary for the extracted acetone to be at the calibration temperature of the acetonometer/alcoholmeter. There are several ways to achieve this, since generally the dehydration will be done in cold (−25 °C), and one option is to let the acetone rest at room temperature, and in 2 or 3 h it will reach the required temperature. Another option is to place the acetone in a bain-marie, i.e., to warm up the acetone uniformly and slowly, submerging the container into another larger one with lukewarm water (36–40 °C), ensuring the increase in temperature in the acetone container, so that it reaches the calibration temperature of the acetonometer/alcoholmeter. Thus, the acetone purity values can be recorded during the time required for dehydration. Once 99.5% acetone purity has been reached, the dehydration process is complete. The specimens are now ready to impregnate. However, it is better to defat [eliminate the excess lipids] the dehydrated samples [1–3, 6, 9, 54, 55].

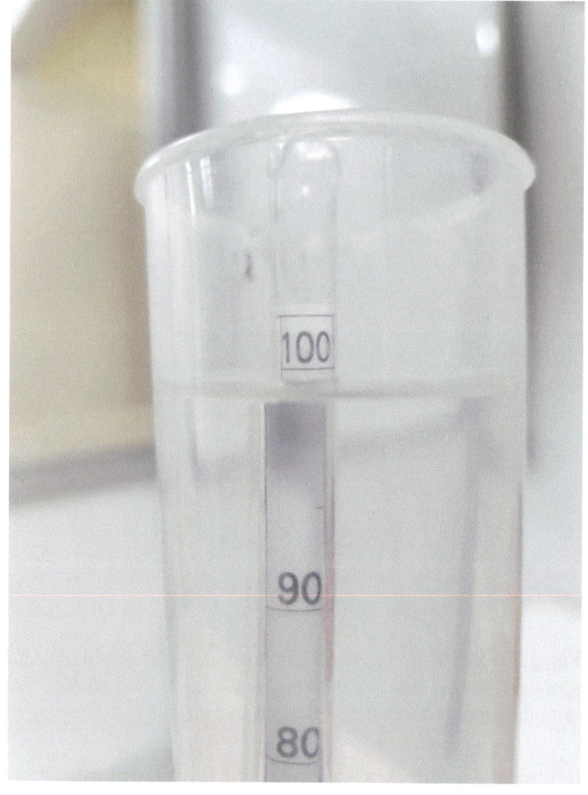

Fig. 5.4 Control of acetone levels in the dehydration process, reaching a value of 99.5%, corresponding to the value necessary to determine the completion of the dehydration step

Parsai, Frank, and Baptista in 2020 [56] proposed the use of sucrose to reduce the contraction of brain tissue during dehydration. To do this, after fixing, cutting, and pre-cooling the brain sections at 5 °C, they were placed in 10% sucrose or 10% DMSO at 5 °C overnight prior to dehydrating the brain sections. Then, they continued with the classic plastination process with silicone at cold temperatures. At the end of the plastination, they verified that the brain tissues treated with sucrose suffered the least retraction. So the authors [56] concluded that the treatment of brain tissue with 10% sucrose, before carrying out the plastination process at cold temperature with silicone, is effective in reducing tissue shrinkage.

Defatting

Defatting is fundamental and must be continued immediately once the dehydration process is finished, because fat does not impregnate well, and so it is recommended that the excess fat that can be eliminated mechanically be eliminated during the initial preparation of the sample prior to impregnation. If dehydration was done by freeze substitution in acetone, then the samples must immediately go in a container

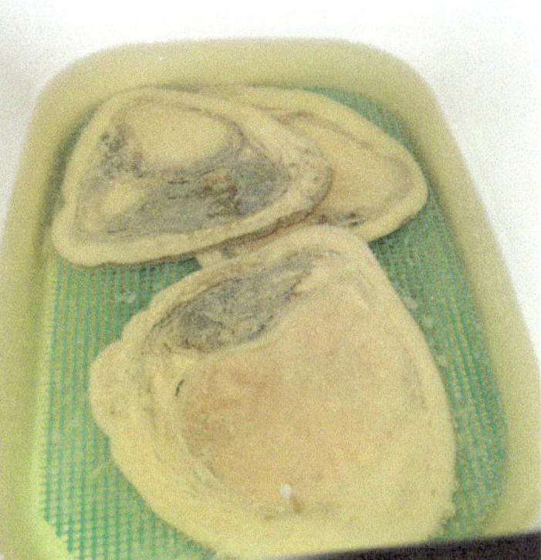

Fig. 5.5 Degreasing step of samples in acetone at room temperature. Note the yellowish coloration of acetone, due to its powerful degreasing action at room temperature

with acetone at room temperature (20–22 °C) (Fig. 5.5). The dehydrating power of acetone is at its maximum at room temperature. The time for which the samples will be left defatting will depend on the size and number of samples, as well as the anatomical region being treated (according to presence of fatty tissue to defat). This acetone is changed when the acetone turns yellow instead of staying clear. The acetone percentage must be measured and changed for new acetone (100%) and the samples left to rest for a few more weeks (could be two or three) or until the acetone becomes yellow again. In the case of brains, these should never be subjected to the defatting process at room temperature [31, 57, 58]. They must simply be dehydrated by the freeze substitution in acetone process and then submerged in the silicone-catalyst impregnation mixture (S10/S3) for the forced impregnation. On the other hand, dehydration can also be done at room temperature, mainly on samples corresponding to limbs or large regions of the body. In this case, the defatting occurs at the same time as the dehydration and extra defatting steps in acetone at room temperature would not be required. This also happens in dehydration with ethanol [3]. For samples rich in lipids (that contain bones, subcutaneous or subserous fatty tissue), a correct defatting can be obtained in a final bath of methylene chloride. If the sample has fat at the end, methylene chloride/dichloromethane (MCL) may be the solvent to use. MCL is a powerful defatter, but highly toxic to health; thus, it must be used under an exhaust hood and with face and respiratory protection. MCL is not the dehydrating agent par excellence, but it does defat; therefore, the sample must be completely dehydrated before changing it to MCL. Generally, a few days will be required to defat the samples in MCL. In addition, the steam pressure of MCL is easy to extract and, therefore, excellent for impregnation.

Finally, for plastination, acetone is ideal because it acts as a dehydrating agent, a defatting agent, and an intermediary solvent all at the same time, and it is easily mixed with all the different resins used for plastination [3].

Fig. 5.6 Mixture of the silicone (polydimethylsiloxane) with the catalyst (dimethyl ester of dibutyltin) in cold-temperature plastination

Forced Impregnation

This is the central, fundamental, and most important step in plastination. It consists of the replacement of the intermediary solvent (acetone) by curable polymers, in this case, the S10/S3 mixture (ratio 1% S3 to total volume of S10) [1–4, 6, 59–67]. It is necessary to consider that when mixing S10 and S3, the reaction between the two begins, determining an increase in the viscosity of the mixture due to the elongation of the silicone molecule (Fig. 5.6). This entire process must be done inside a vacuum chamber that will have a tempered glass cover (safety glass) (20 mm thick) which is also located inside a freezer (Fig. 5.7a, b). The impregnation temperature must be −15 °C, which slows down the elongation process of the silicone molecule (S10) combined with the catalyst (S3), and this way it remains less viscose, thanks to the low temperature. Thus, the mixture becomes hardened at high temperatures, and this means that the mixture must be kept in the cold (at least −15 °C) in case it should need to be stored. In addition, a colder temperature, below −15 °C, slows down the reaction of the impregnation mixture, so that it does not become more viscous. There are experiences where the S10/S3 mixtures can be kept for several years inside the cold vacuum chambers, replacing the silicone levels and catalyst as they shrink. After these considerations, and continuing with the process, the mixture must be placed inside the vacuum chamber, in a plastic container that is put inside the vacuum chamber, and to achieve the acetone-impregnation mixture exchange, vacuum must be applied during the forced impregnation step. The samples, totally dehydrated by the volatile intermediary solvent (acetone), must be submerged in the S10/S3 mixture (Fig. 5.8). The intermediary solvent, acetone, has a high steam pressure and a low boiling point (+56 °C), whereas the silicone-catalyst mixture (S10/S3) has low steam pressure and a high boiling point. Therefore, when vacuum is applied, only the intermediary solvent is extracted continuously from the samples

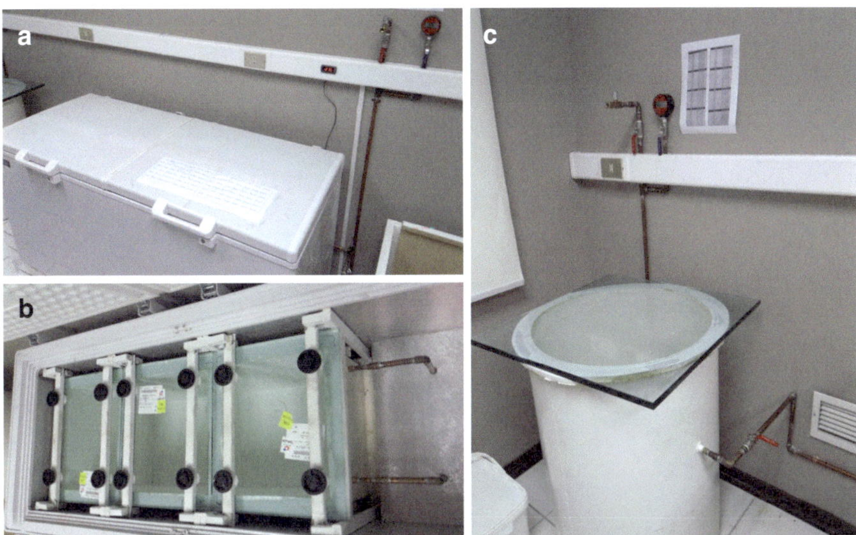

Fig. 5.7 Vacuum chambers for forced impregnation steps. (**a**, **b**) Vacuum chamber located inside a freezer (−25 to −15 °C) for cold-temperature plastination technique. There are three separate tempered glasses (20 mm thick) on silicone gasket, which ensure an adequate distribution of pressure inside the vacuum chamber, made of 7-mm iron, with connections through copper pipes to the digital manometer, air inlet control needle bypass valve, and the vacuum pump located in another room. (**c**) Vacuum chamber for room-temperature plastination, with tempered glass (20 mm thick), silicone gasket for support, and pressure control system (digital manometer, air inlet control needle bypass valve, and copper pipe connections to this system and the vacuum pump located in a separate room). Likewise, the acetone gas extraction system is visualized, located 20 cm above the ground, due to the greater weight of acetone gas compared to air

and through the surrounding silicone-catalyst mixture in the form of gaseous bubbles.

The impregnation speed depends on the sample size, the number of samples, and also the class of polymer used (silicone or S10). High-viscosity polymers require a longer impregnation time than lower-viscosity polymers. The greater the amount, the larger and more dense the samples, the slower the impregnation needs to be, and it is essentially preferable to use low-viscosity polymers. This is because a high-viscosity impregnation mixture (S10/S3) is more difficult to incorporate into the samples and therefore of the cell structure and the spaces in between, making impregnation of the samples remarkably difficult. To achieve impregnation, any small oil-driven blade vacuum pump is not sufficient; it must be suitable for continuous work and of considerable quality. This vacuum pump will be connected to the vacuum chamber, and in addition, there will be a connection from the vacuum chamber that will be controlled with a needle valve, bypass valve, to adjust and regulate the pressure reductions inside the vacuum chamber very finely and gradually. In addition, there will be a vacuum gauge connected to the vacuum pump at the opposite end (it is currently recommended that it be digital, since these equal the

Fig. 5.8 Samples submerged in the silicone-catalyst mixture (100:1) in a stainless steel container located inside the vacuum chamber, ready to start the forced impregnation process in cold-temperature plastination technique. It is visualized that samples must be totally submerged in the impregnating mixture and plastic grids are placed on which in turn weights will be placed to keep the samples submerged

accuracy of Bennett's mercury manometer), with which the pressures in the vacuum chamber can be viewed and controlled, and in this way they will be compared with what happens inside the vacuum chamber in relation to the presence or not of acetone bubbles. Regarding the impregnation speed, with the generation of the vacuum, and therefore, in search of the extraction of acetone, it must be controlled and adjusted via a needle valve that permits the incorporation of air to the interior of the vacuum chamber. The extraction of acetone can be viewed through the identification of bubbles inside the vacuum chamber, through the glass cover (Fig. 5.9). This extraction can be put into effect from 215 mmHg, the pressure at which the acetone becomes gas, and this way it can be effectively extracted from inside the samples. For this, it is very important to consider the speed at which this pressure is reached, because it is possible to see bubbles between 760 and 215 mmHg. However, these bubbles might be air or could be some other component incorporated into the samples as the catalyst. But it would not be acetone, which is extracted in the form of gas from 215 mmHg. Thus, reducing the pressure too quickly to obtain the acetone extraction will cause the samples to shrink, with a remarkable deterioration. For this reason, it is necessary to reduce the pressure in a controlled and gradual way, and it has been defined in general that the forced impregnation process in S10 plastination techniques lasts between 3 and 6 weeks. In this sense, it is recommended in the four first days to perform a controlled reduction to reach 215 mmHg on the fifth day.

Fig. 5.9 Visualization of bubbles during the forced impregnation stage in the cold-temperature plastination technique, corresponding to the extraction of acetone from the samples with the corresponding entry of the impregnation mixture (silicone-catalyst) into the interior of the samples

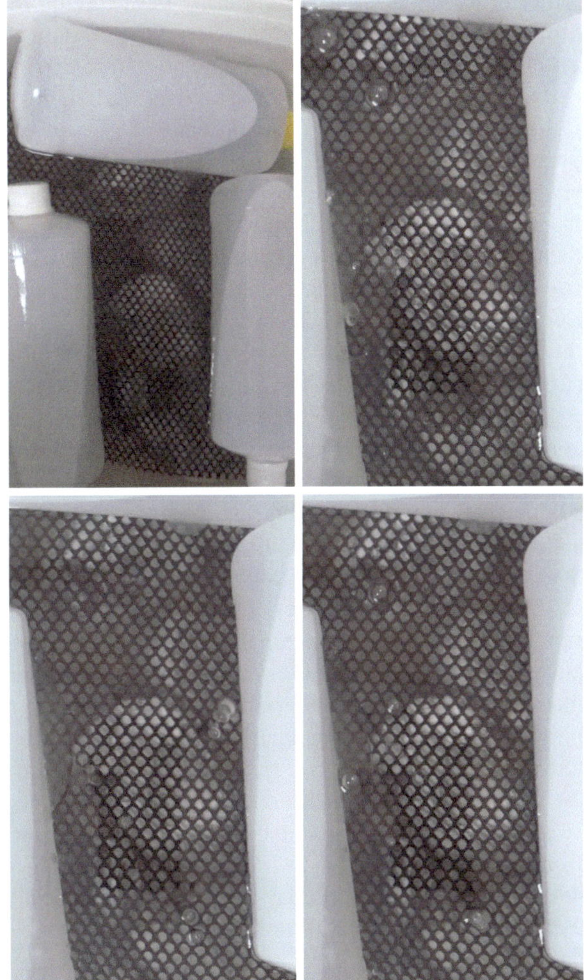

How Is the Pressure Reduced, Vacuum Generation, Inside the Vacuum Chamber?

There are several possibilities. In our experience, once the samples are placed inside the vacuum chamber, submerged in the silicone-catalyst mixture, they must be left for 24 h so the samples can stabilize inside the vacuum chamber with the silicone-catalyst mixture, before initiating the reduction in pressure (vacuum generation). On the second day, the vacuum pump can be turned on and begin with the vacuum generation inside the chamber in order to begin the forced impregnation process. Numerous air bubbles (<2–5 mm) will be noted on this first day of forced impregnation, when the vacuum is generated, and will correspond not to acetone bubbles, but rather to air bubbles produced during the agitation that occurs when pouring the impregnation mixture in the container of the vacuum chamber and then loading the samples. To eliminate this bubbling, and to reach the effective pressure for acetone extraction, in which the extraction of acetone bubbles will be seen (from 1 to 2 cm in size), which will be the steam pressure of the acetone at $-15\,°C$ (215 mmHg), the pressure must be lowered controlling for the appearance of bubbles. Once bubbles are observed, the pressure reduction must be stopped, and this is done by opening the needle valve that permits air to enter the chamber (increasing the pressure), and then seeking to make the pressure level at which bubbles are seen. The bubbles will rise slowly to the surface, without splashing. When the bubbles disappear at the reached pressure, the needle valve must be closed, and this way the entry of air to the interior of the system is avoided, which ensures the reduction in pressure. This is how to reduce the pressure, and this depends on the sight of the bubbles, which will determine or not whether the vacuum generation continues, and the pressure is consequently reduced. Otherwise, an overly accelerated reduction in pressure will cause the acetone to vaporize too fast, in turn leaving the sample too and causing the silicone-catalyst mixture, in its viscous state, to not adequately penetrate the sample and for the acetone to leave just as quickly, in gaseous form, causing a collapse at the cellular and interstitial level of tissues, thus impeding the silicone from access-ing the inside of the samples. This will also cause considerable shrinkage of the samples, resulting in an incomplete impregnation, so the plastination process will have failed. The decrease in pressure will be controlled by observing both the for-mation of bubbles and the device/manometer pressure record [59, 68]. As indicated previously, the pressure will have to be reduced slowly over a period of 3–6 weeks until reaching 5 to 7 mmHg. Therefore, to reach the effective extraction pressure of the intermediary solvent (acetone), 215 mmHg, it is recommended that this pressure be reached in 5 days, and in the following way. The second day, after 24 h of stabi-lization, the pressure must be taken to 600 mmHg. The third day, the idea is to reach 450 mmHg and the fourth day 300 mmHg. And the fifth day, it can be brought to 215 mmHg. From here, this process of pressure reduction must continue, observing the bubbles that from 215 mmHg will be from acetone, larger than the first ones (between 1 and 2 cm), as greater amounts of acetone are released from the samples and are pumped outside the silicone impregnation mixture. It is necessary to achieve a final vacuum pressure of 5 to 7 mmHg, and no bubbles are seen, or their appear-ance is much less than at the outset, i.e., they are reduced in number. Once this hap-pens, essentially from the significant reduction or total disappearance of bubbles at 5–7 mmHg, it means that complete extraction of the acetone inside the samples has been achieved, and thus the silicone-catalyst mixture has been fully incorporated into the impregnated samples. It is important to point out that throughout this pro-cess of pressure reduction, with the resulting vacuum generation, inside the vacuum chamber, from 760 to 7–5 mmHg final, the bubbling that is demonstrated in any of the stages (the initial, corresponding to air bubbles from the initial mixture of the silicone with the catalyst or those of impregnation, from 215 mmHg, corresponding to acetone bubbles), it does not have to be like boiling water, but rather it must be a slow and steady bubbling, with the observation of some bubbles on the surface of the impregnation mixture having around five to ten bubbles at a time. Once the impregnation is finished, the vacuum pump must be turned off and air allowed to enter the system in order to return the pressure to 760 mmHg (or the atmospheric pressure corresponding to the location of the laboratory), to then be able to extract the samples from the impregnation mixture. However, it is necessary to leave the samples submerged in the impregnation mixture for 24 h with no vacuum pressure (at 760 mmHg), so there is also a balance between the samples and the mixture, and in this way the samples can be removed from the silicone-catalyst mixture. Once this is done, the next step in the plastination technique must be carried out.

Drainage of Samples and Positioning

The next step is to open the vacuum chamber, removing the glass covers and lifting the samples above the level of the impregnation mixture so that the samples drain their excess polymer in the same mixture and inside the freezer at $-15\,^{\circ}$C (Fig. 5.10). Once it is observed that no more polymer is draining from the samples, they can be extracted from the freezer, and drainage of the excess polymer can continue with paper towels, in a room at room temperature, so the samples are as dry as possible. The drainage is going to increase as the polymers adapt to this new room temperature, since a greater amount of silicone inside the samples tends to drain with higher temperatures [60]. It is important to rotate the samples, both in the freezer and then at room temperature, so they can drain the excess polymer from every part.

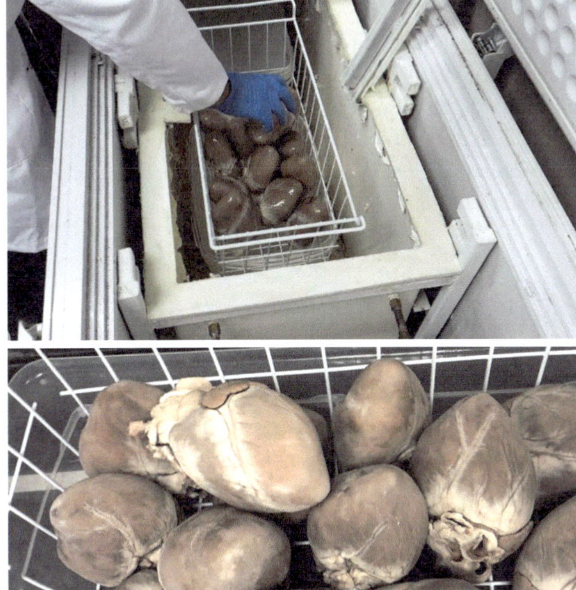

Fig. 5.10 Completion of the forced impregnation process. (**a**) Once the forced impregnation stage is over, the samples must be left inside the freezer ($-15\,^{\circ}$C) to favor the drainage of excess silicone into the container of the vacuum chamber. (**b**) The samples can then be extracted to continue with the positioning process and finally cured

Fig. 5.11 Positioning of
the anatomical structures
in the correct anatomical
position prior to the curing
(polymerization) process,
which will determine the
drying and (relative)
hardening of the samples.
Positioning can be done
with all kinds of
accessories (needles,
threads, separators, etc.)
that allow the anatomical
structures to be correctly
located

Therefore, the surface of the samples will continue being cleaned and these excess polymers eliminated. Once the extraction of the excess silicone is complete, and before moving to the curing (polymerization) stage, the anatomical positioning of the sample must be done (Fig. 5.11). This positioning will be fundamental and is done prior to the curing (polymerization) stage because the objective of this last stage of plastination is to dry the samples practically definitively, and in turn the samples will acquire a certain rigidity characteristic of the classic S10 plastination technique, which will not allow the anatomical structures to be moved or positioned again. This is why it is important at this point to correctly locate all the anatomical elements, in addition to being able to make windows in muscles, or cavities, for the viewing of deep levels; it is also possible to join the structures with threads and needles, to provide them with a correct anatomical location, depending on the anatomical region subjected to the S10 plastination technique. Thus, once the samples have been positioned, the final step in plastination may proceed: the curing (polymerization) stage.

Curing (Polymerization)

The polymerization or hardening process consists of the final curing and drying of the preparation [1–4, 6, 7, 9, 11, 59, 60, 69–72]. The silicone-catalyst mixture remains within the tissue. During this final phase, the main objective will be to obtain a completely dry sample and finish the hardening process initiated previously. The curing compound (S6, tetraethyl orthosilicate: hardener, cross-linker, or curing gas) is a liquid that will be gasified and that causes the silicone molecules to bind "side by side," interacting with the long molecules of S10/S3 of the samples, elongating and contributing to the binding and cross-linking of the long silicone molecules. The S3 contained in the impregnation mixture initiates the hardening of the silicone molecules by end-to-end polymerization. Due to the cross-linking during the final curing with gas, the silicone within the sample will be solidified and dried. The hardener will cause a three-dimensional cross-reaction, whereas the catalyst is responsible for the flexibility of the sample (extension of the molecule). The S6 will dry and harden the sample, but sometimes it can become fragile (cross-effect). Therefore, the more catalyst (S3) and the less hardener (S6), the more flexible the sample will become. While it acts as a cross-linker, the hardener (S6) will arrive at the center of the sampling bottles in a gaseous state.

Returning to the curing process, after eliminating the excess silicone, the sample will be placed in a hermetically sealed curing chamber and exposed to the curing agent in the form of a gaseous steam (S6, tetraethyl orthosilicate) (Fig. 5.12), which will generate an atmosphere continuously saturated with this gaseous hardener [1–4, 6, 7, 9, 11, 59, 60, 69–72]. It is important to add silica crystals to this curing chamber, which absorb the moisture from the atmosphere and keep the inside of the curing chamber dry. A small amount of S6 is placed in a container connected to a small membrane pump (like those used in aquariums) (Fig. 5.13), through which

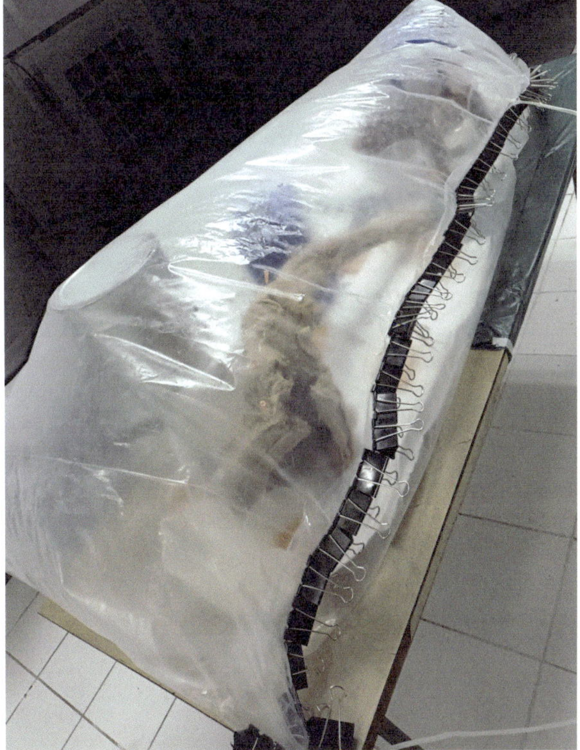

Fig. 5.12 Assembly of hermetic curing chamber to carry out the polymerization process of the silicone-catalyst mixture, subjecting the samples to the curing agent (tetraethyl orthosilicate) by means of vaporization

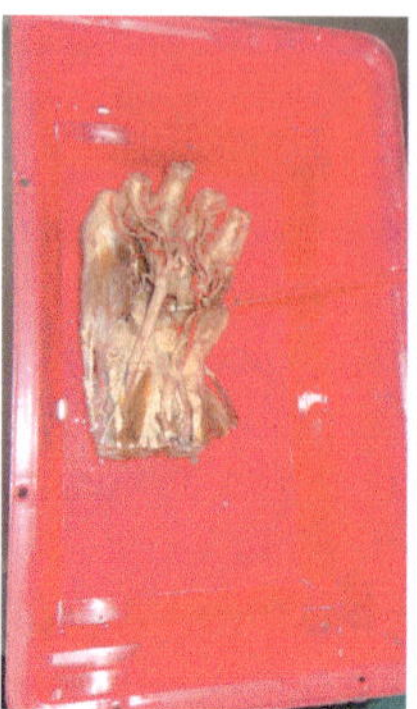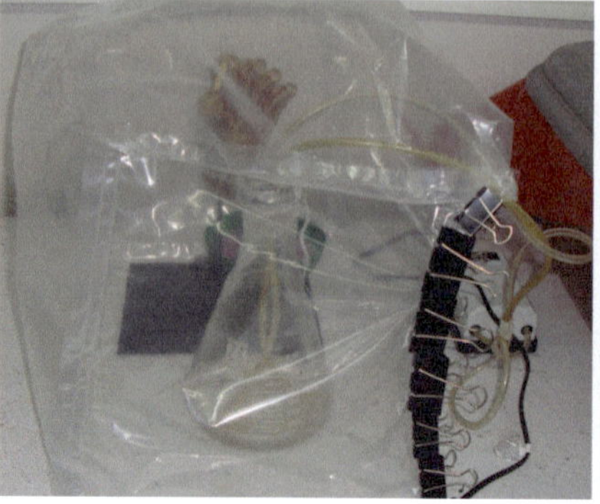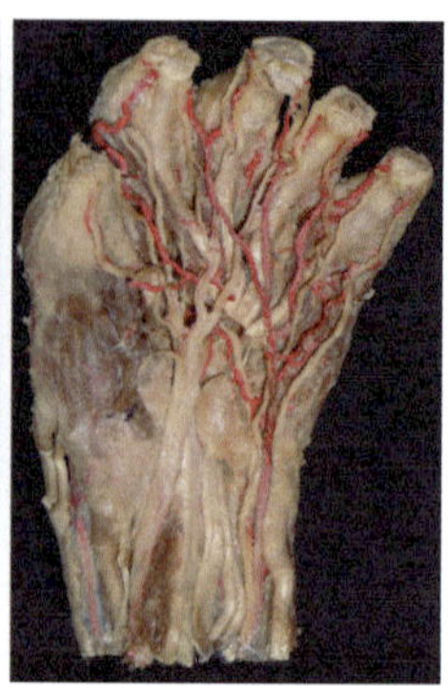

Fig. 5.13 Hermetic curing chamber with the aquarium motor that allows the curing agent to be vaporized and the silica crystals (violet) that allow the environment inside the curing chamber to be kept dry, which seeks to favor the elimination of humidity and the drying of the samples. (**a**) Sample completely impregnated with silicone, prior to draining the excess silicone and beginning the curing stage. (**b**) Sample inside the hermetic curing chamber. (**c**) Final result of the curing stage, and therefore, of the plastination process

bubbling of the S6 liquid is produced, and this will accelerate the vaporization of the agent, which will quickly be distributed throughout the curing chamber, coming into contact with the sample and therefore with the S10/S3 mixture on which it will act. This vaporization process must be maintained for 15–20 min, and then the motor can be turned off, repeating this process three times a day. This is why it is not necessary to keep the vaporization turned on throughout the day. The next day, before beginning the vaporization of the S6 again, the curing chamber must be opened, always with face protection to avoid the aspiration of gases or under an exhaust hood, and the samples must be dried with paper towel to avoid lumps or spots of S10/S3 being generated on their surface (Fig. 5.14). Once this is done, the vaporization of S6 can begin again on the sample to continue with the curing process. This polymerization dynamic must be repeated the number of times necessary for the samples to be totally dry on the surface. This can last between 5 and 10 days, or more, depending on the sample size as well as the amount. This way, the surface drying having been achieved; the rapid surface curing process is complete, continuing with the deep slow curing internally in the sample for a longer time, which, according to some authors, lasts 3 or 4 months. For this, the sample must be kept in a plastic bag to achieve the final polymerization inside the preparation, since it will allow a uniform distribution of the curing agent (S6). In addition, as room temperature accelerates the thickening of the silicone base by the S3 hardener, less and less S6 gaseous curing agent will be required to cross-link the silicone in the entire sample [1–3]. Once this curing process is finished, the sample can be stored at room temperature with minimal care.

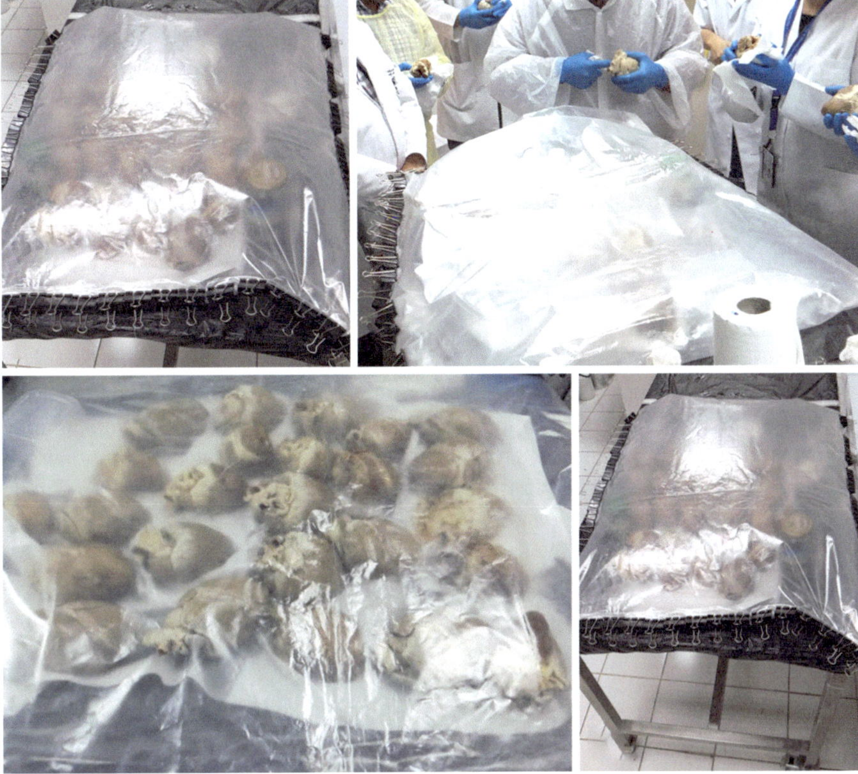

Fig. 5.14 Curing chambers and drying process. (**a, c, d**) Samples inside the curing chamber subjected to the vaporization of the curing agent. (**b**) When opening the curing chambers, if the samples are not yet dry, they must be dried with the help of a paper towel. Once dry, they are placed back inside the curing chamber to continue steaming until they are cured and superficially dried

Room-Temperature Plastination Technique

As indicated in Chap. 2, the first communication in scientific journals on plastination at room temperature was made in the *Journal of Plastination* (former *Journal of the International Society for Plastination*) in 1996 by Zheng et al. [47], to later, in 1998 [48], appear in the same journal, an original article by the same group of Zheng and a work abstract by Glover, Henry, and Wade, corresponding to an abstract presented at the ninth International Conference on Plastination held in Quebec City, Canada [49]. The primary objective of this process was to achieve a plastination technique that could achieve the same results but faster and, consequently, due to changes in the methodology, more economically for the researchers. In this sense, Zheng et al. [48] published the basic foundations for the development of the room-temperature plastination technique in the *Journal of the International Society for Plastination*, i.e., using vacuum chambers for forced impregnation without a freezer, at 20 °C.

The technique basically consists of combining the components differently: in forced impregnation, the impregnation mixture is between silicone and the curing agent (S10 and S6, respectively), and the forced impregnation takes place at room temperature, i.e., the vacuum chamber is not in a refrigerator, and the impregnation temperature is 20–22 °C [6, 9, 23, 47, 48, 61, 73–77]. When the cross-linker (S6) and the polymer (S10) are mixed, they stay stable and no reaction is produced. Therefore, there are temperature restrictions when performing the "room temperature process," just as there are for low-temperature impregnation [61, 73].

This way, the silicone-catalyst impregnation mixture is less viscose at room temperature and can be incorporated into the samples more quickly. This is why the forced impregnation process can last between 1 and 2 weeks. However, over time this room-temperature impregnation mixture will tend to harden, which is why it must be stored immediately in a refrigerator (at −25 °C), so it does not harden for a longer time and can be reused. Another difference with the classic low-temperature silicone plastination technique (S10 plastination) is in the curing step (polymerization). In this step, S3 (dimethyl dibutyltin) is used as a curing agent, which, due to its characteristics, cannot be vaporized, but can be sprinkled or painted onto the samples. Once this is done, the samples must be covered with plastic wrap to promote the action of S3 on the other components (S10 and S6). The curing process must be repeated until a dry sample is obtained.

Alternative Room-Temperature Plastination Technique

In 2015, we published new advances on an alternative plastination technique at room temperature [6] in which we communicated for the first time in a scientific journal the possibility of using the same combination of reagents suggested by von Hagens, but at room temperature, that is, carrying out the forced impregnation process in vacuum chambers at room temperature, without the need to resort to a freezer [47–49, 73].

Dehydration

This second phase endeavored to replace the formaldehyde solution by an intermediary solvent, miscible in water, for example, acetone. Dehydration took 4 weeks. The sample was placed in a container of the same size and covered in acetone 2 cm over the level of the specimen. The sample was submerged completely in the dehydrating liquid. Next, the samples received four baths with increasing acetone concentrations. The first bath began with an acetone concentration over 90% (it was possible to reuse the previous acetone baths). Successive baths were done with progressively higher concentrations (again, reusing previous acetone baths or using an acetone bath at 100%). The final bath was 100% acetone, the objective of which was to defat the preparation. Before changing the sample to a new acetone bath, the purity of the acetone was measured with an acetonometer/alcoholmeter. We performed all the dehydration processes at room temperature, except for the brain [6].

Forced Impregnation

This was the most important phase of the plastination process [6]. The intermediary solvent (acetone) was replaced with silicone. Forced impregnation extracts acetone from the structure of the cell and the interstitial space of the preparation and replaces it with silicone. Forced impregnation must be done slowly: as the polymer enters the sample, the acetone changes from liquid to steam and leaves the preparation. The speed of impregnation was carefully adjusted through the controlled addition of air to the vacuum pump through a bypass valve. The anatomical material was placed in a small plastic container, made to measure for the preparation, to avoid wasting the silicone. The mixture in which the preparation was submerged included silicone and catalyst in a ratio of 100:1. The preparation was completely submerged in the mixture, using elements to keep it submerged.

The essential elements of this stage were the vacuum chamber and the vacuum pump. We did not use a freezer (Figs. 5.7c and 5.15a). Therefore, once the dehydration phase was complete, the sample was placed in the vacuum chamber so the

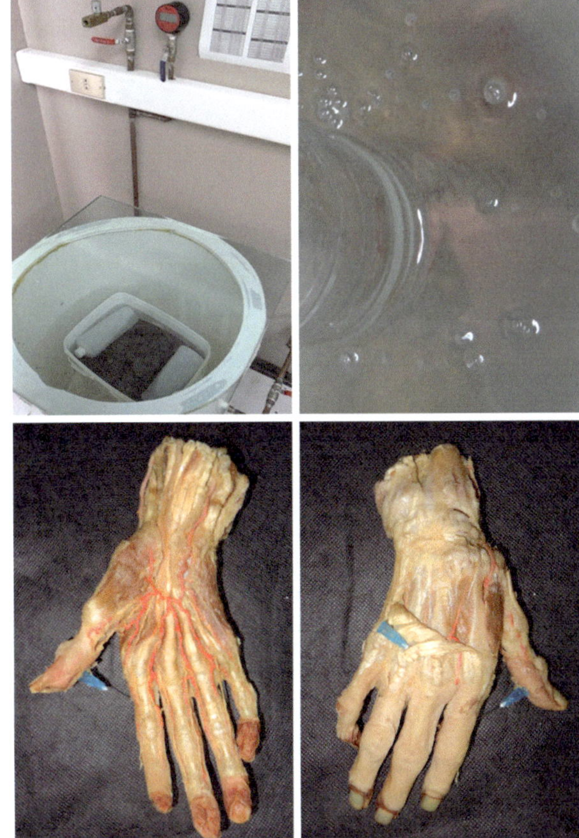

Fig. 5.15 Room-temperature plastination, alternative technique proposed by Ottone et al. [6]. (**a**) Vacuum chamber at room temperature, with samples (human hand) under "active" forced impregnation. (**b**) Bubbling corresponding to forced impregnation at room temperature. (**c**, **d**) Fully impregnated sample is positioned to continue the curing/polymerization process

forced impregnation phase could begin. The entire vacuum process was done at room temperature.

The forced impregnation became a fast process: from 2 to 4 days, depending mainly on the size of the preparation. The pressure was reduced by closing the valve (increasing the vacuum) until bubbling was observed (Fig. 5.15b). When the bubbling was produced, the pressure was maintained so the process could take place at that pressure; i.e., the bubbling was contained by partially opening the valve. The bubbling was acetone steam that left the sample, and the silicone-catalyst mixture occupied the remaining spaces (Fig. 5.16). Later, it was necessary to verify the presence of bubbles. When the bubbling decreased, the vacuum was increased (reducing the pressure) by closing the valve until the bubbles reappeared.

To be more precise, the reduction in pressure to achieve a greater vacuum to reach the final goal of 5 mmHg was not done continuously. We developed the forced impregnation in two phases: "active" forced impregnation and "passive" forced impregnation [6]. The preparation was under "active" forced impregnation for 8 h every day through the action of a vacuum pump. Once these 8 h were over, the valves of the vacuum chamber were closed, and the vacuum pump was switched off. The preparation was left in the chamber, and the "passive" forced impregnation phase began. This phase extended to the next day, when it began again from the pressure at which the process had paused. In the case of small and medium specimens, the first day the bubbling began at approximately 460 mmHg, and from there the pressure was reduced from the first day to 260 mmHg. The pressure was verified

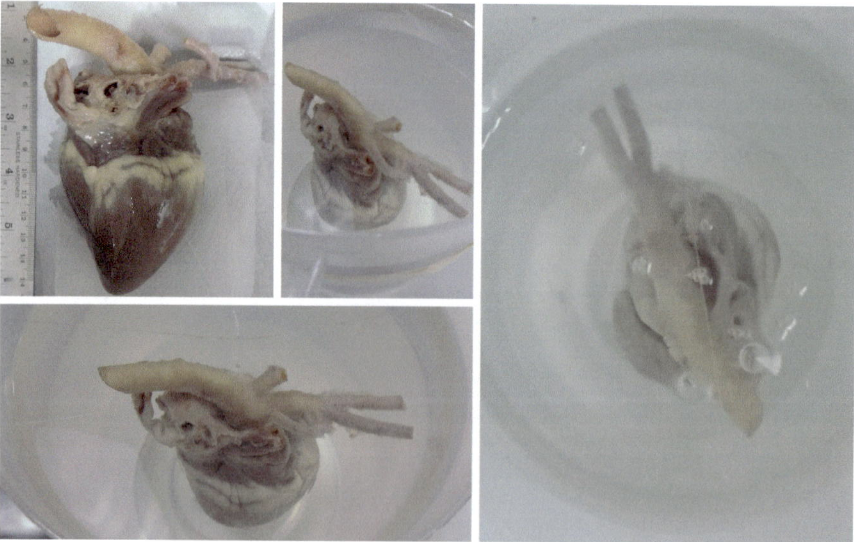

Fig. 5.16 Room-temperature plastination, alternative technique proposed by Ottone et al. [6]. (**a–c**) Sample (pig heart) subjected to the forced (passive) impregnation process, and positioned inside the resin so as not to affect the morphology of the sample. (**d**) "Active" forced impregnation process, with visualization of the bubbling corresponding to the extraction of acetone

every half an hour, and the vacuum was increased whenever the bubbling decreased. The second day, the pressure continued descending, from the value of 260 mmHg reached the day before to 60 mmHg. Finally, on the third day, the final pressure was 5 mmHg. Once this pressure was reached and the bubbling stopped, the forced impregnation phase was finished [6].

The active and passive forced impregnation process avoided the continuous operation of the vacuum pump and allowed the silicone to enter the sample slowly and uniformly, avoiding shrinkage, especially in preparations that typically undergo a high degree of shrinkage, like the central nervous system [6].

Drainage and Positioning

Immediately after the forced impregnation phase, the vacuum was interrupted, and the silicone-catalyst mixture preparation was removed and left to drip in the same container to eliminate the excess silicone. It is also possible to use paper towels to help eliminate the excess silicone, and, in the case of hollow organs, the paper can be placed inside the cavities of the sample to absorb the excess silicone and help it dry (Fig. 5.17). This can take 2 days, and depends on the size of the specimen. Once

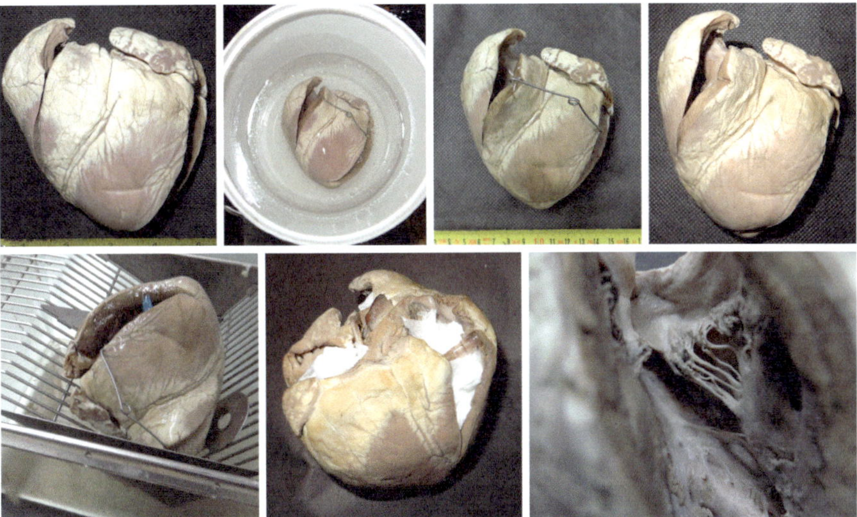

Fig. 5.17 Pig heart subjected to the room-temperature plastination proposed by Ottone et al. [6]. (**a**) Sample after the dehydration process. (**b**) Sample immersed in the silicone-catalyst mixture (S10:S3) subjected to forced impregnation at room temperature. (**c**, **d**) Drainage of excess silicone. (**e**) Placement of paper inside the cavities to favor drying. (**f**) Visualization of the exterior of the sample finished the plastination process. (**g**) Visualization of the interior of the sample finished the plastination process

the preparation was completely drained, the next step was to position the anatomical structures of the specimen according to what section of the dissected region was going to be viewed [6] (Fig. 5.15c, d).

Curing (Polymerization)

The curing process (polymerization) is similar to that applied in the classic S10 plastination technique, and which was explained previously (Fig. 5.18). In Table 5.2 compares Gunther von Hagens' plastination technique to protocols suggested by other researchers.

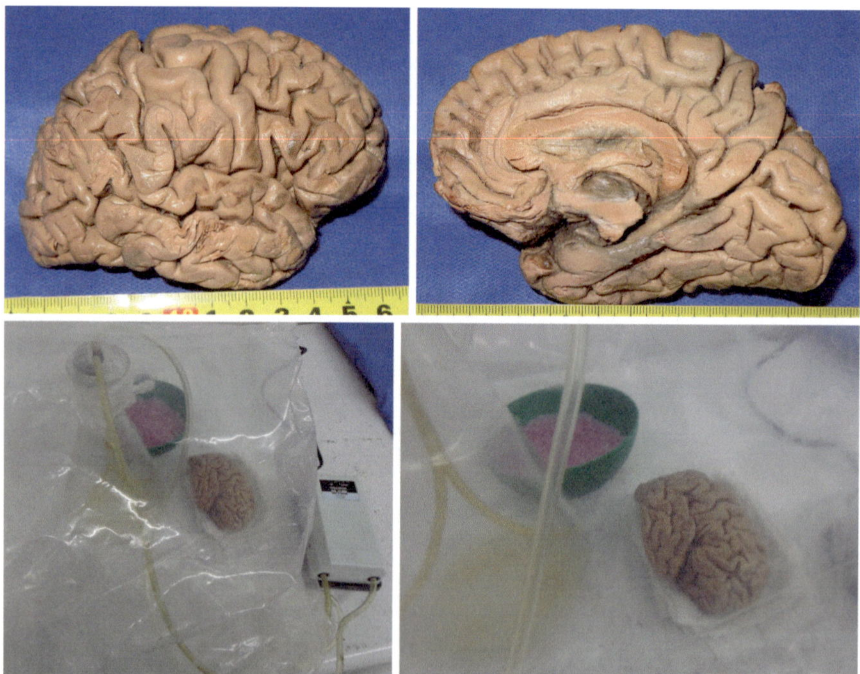

Fig. 5.18 Curing process in a human cerebral hemisphere. (**a, b**) Lateral and medial aspects of the cerebral hemisphere. (**c, d**) Sample in a curing chamber, with visualization of the curing agent, vaporized from the function of the fish tank motor, and silica crystals that keep the interior environment of the chamber dry

Table 5.2 Comparison between plastination techniques with silicone

Steps	von Hagens [1]	Zheng et al. [47, 48] Glover et al. [49]	Ottone et al. [6]
Dehydration	Acetone 100% (−25/−20 °C)	Acetone 100% (−25/−20 °C)	Acetone 100% (−25/−20 °C)
Forced impregnation	Cold temperature: −25/−20/−15 °C Mixture: S10 + S3 Continue forced impregnation	Room temperature: +20/+22 °C Mixture: S10 + S6 Continue forced impregnation	Room temperature: +20/+22 °C Mixture: S10 + S3 Active/passive forced impregnation
Curing	Vaporization of S6	Vaporization of S6	Spraying and/or brushing of S3

Discussion

With the aid of previous studies [1–3, 47–50, 62, 74–76, 78–83], and adapting processes developed in other laboratories to our situation, we achieved our objective of developing a fast room-temperature plastination technique with materials that can be reproduced in any laboratory. Introducing the concept of "active" and "passive" forced impregnation, and using economical materials and equipment, we developed a plastination technique that can be easily available for any institution and has results similar to those obtained with classic plastination [6].

Our plastination technique was developed with local materials and methods, thus radically reducing its cost [6].

First, we chose to perform the technique at room temperature [6]. This has two advantages: it is economical, because a freezer adapted for forced impregnation (which must be modified by placing the motor in a contiguous room, and using multiple connections to allow the hoses from the vacuum chamber to connect to the outside) is unnecessary. In addition, not using a freezer adds an element of biosafety, reducing the possibility of explosion due to the interaction between the acetone and the ignition of the freezer motor.

Second, we built an iron vacuum chamber (instead of stainless steel) [6]. It has a thickness of 7 mm and is treated against rust, which makes it very durable. Safe connections were placed. The opening of the chamber is sealed with silicone and 20-mm-thick glass is used as a lid.

As far as silicone, the catalyst, and the hardener, we found local generic products with chemical characteristics similar to the original silicones used (Biodur S-10 and NCS-10 from North Carolina, which we also used in this study to compare with generic silicone) [6]. Thus, we developed this technique using commonly available silicone, polydimethylsiloxane, a catalyst, dibutyltin dilaurate, and a hardener (gas curing agent, cross-linker), tetraethyl orthosilicate (TEOS). In addition, laboratory acetone was used for the dehydration phase, which also greatly helped the development of the technique.

We were able to obtain cadaver preparations of high quality from all points of view of the anatomical dissection to preserve them dried indefinitely, maintaining their texture and color, and, very importantly, without needing liquid preservatives and avoiding their toxic and irritating effects [6].

We ordered the construction of vacuum chambers for complete human bodies as well as for small- and medium-sized specimens [6]. The vacuum chambers were made of iron 7 mm thick, rather than stainless steel, which is more expensive and more difficult to obtain. The vacuum chamber necessary depends on the sample size. We were able to plastinate a large number of preparations, from hearts, isolated hands, kidneys, heads, and even complete limbs. These vacuum chambers worked correctly and served as the basis for the development of this plastination technique, which was done at low cost. O'Sullivan and Mitchell [79] described a low-cost plastination technique in 1995 using a rudimentary element, a kitchen oven, as a vacuum chamber. Later, Zheng et al. in 1998 [48] also described their own room-temperature plastination technique and reported on whole-body vacuum chambers in their work, which inspired us to build the whole-body vacuum chamber in our laboratory.

With respect to the coloration of the hands, the arteries were injected with latex before being dissected. Nevertheless, other authors have proposed the addition of different pigments after dissection and before the dehydration step [84]. In addition, the incorporation of dyes, such as imidazole, to the impregnation mixture in the classic S10 low-temperature plastination has been considered [82]. In such cases, addition of these chemical dyes both to the acetone baths and to the impregnation mixture colors the entire sample and not only its vascular structures, losing accuracy in the development of this step and staining structures that should not be colored [85].

Dehydration consists of eliminating any liquid (water, formalin) and fat from tissues. To achieve this, an organic solvent must be used to replace these tissue fluids. To act as a dehydrating agent, the solvent must be miscible in water (alcohol or ketones). It is important to record the acetone purity weekly, since if this process is not adequately recorded, the dehydration cannot be complete and will cause all the embalming or tissue fluid from the samples to be eliminated, knowing that the steam pressure from the remaining water in the samples will be near 0 in cold [60, 62, 74]. Therefore, it will not be vaporized or evaporated from the tissue, and this will prevent the S10/S3 impregnation mixture from penetrating the cells during the forced impregnation, and, therefore entry of the impregnation mixture will be blocked at cell level. Either ethanol or acetone can be used, but cold acetone is the preferred agent because it triggers dehydration and defatting [1–3, 69, 86, 87]. However, the main disadvantage of dehydration is tissue shrinkage, particularly in tissues with fat high levels, as in the central nervous system. According to several authors [1–3, 51, 55, 56, 68, 77], shrinkage can be avoided using cold acetone, in which the ice in the sample is replaced by acetone (through a process called "freeze substitution"), or using a lower percentage of ethanol. It is always important to use a suitable volume of dehydrating liquid. For this process, ten volumes of acetone per volume of tissue are recommended [47–49, 75, 76]. Nevertheless, we have found that placing the sample in a container adjusted to the preparation is enough to use a volume of acetone of at least 2 cm over the level of the sample, which must be completely submerged, to initiate dehydration. In addition, using the exact amount of acetone saves money. For musculoskeletal samples or other organs (heart, liver), acetone can be used at room temperature without causing tissue shrinkage. With respect to the defatting, its implementation is essential after dehydration. Lipids do not impregnate well and will maintain a slippery sensation, in addition to producing a rancid odor in the samples [60, 62, 74]. MCl is a maximum defatter, with good steam pressure for extraction, but it does not dehydrate samples; therefore, when MCl is used, it must only be used in this defatting step, having to dehydrate previously with acetone [60, 62, 74]. However, due to its toxicity, it is recommended that MCl be used in other plastination processes, for example, in the plastination of epoxy resin slices, a technique in which it is fundamental to achieve great transparency of tissues, due to the principles that the technique implies (and that will be seen in section E12).

Thanks to the work by Chaynes and Mingotaud [50], we have the chemical structure of the basic components used in the classic S10 plastination technique. From this starting point, we looked for generic components corresponding to the same chemical structure. Once obtained and used in the same suggested proportion, we obtained a suitable mixture of silicone and catalyst.

If we compare the silicone-catalyst mixture at room temperature and low temperature, at room temperature the mixture is less viscous, which enables a faster penetration of the sample and acetone gas bubbles to escape more easily [47–49]. On the other hand, at low temperatures, like the classic S10 plastination technique (−15 °C), the acetone is vaporized at 28 mmHg, being a sufficiently cold temperature to slow down the S3 catalytic reaction with S10 and, therefore, slow down the elongation of the silicone chains and thus slow down the increase in viscosity of the impregnation mixture [60]. An even colder temperature (−20 °C) makes the vaporization of the acetone require an even greater vacuum generation, i.e., a much lower pressure, close to 21 mmHg. According to Henry et al. [60], a more powerful vacuum pump could be needed for this, for colder impregnation temperatures, making a S10/S3 impregnation mixture even more viscous, which would complicate its penetration in the samples, also due to the acetone steam pressure being lower, and this would be added to the complications in the acetone-impregnation mixture exchange.

In this sense, Sora in 2017 [88] investigated the viscosity of Biodur silicone, which is normally used at cold temperature (S10), but using it to impregnate at room temperature (as we suggested in Ottone et al. [6]) and came to the conclusion that impregnation at room temperature is possible with S10 silicone, recommending a period of 2 weeks. Likewise, Adds also in 2017 [89] carried out an analysis of the viscosity of the Biodur silicone-catalyst mixture (S10/S3) at room temperature, indicating that although with plastination at room temperature, the same technical results can be obtained in comparison. In addition to the classic cold plastination technique, it is also more economical and safer; there is a notable reduction in the half-life of the S10/S3 components when used at room temperature, which is also affected by fluctuations in the temperature of the environment, and finally determining cost implications of having to acquire new inputs due to the speed with which they harden at room temperature, compared to the longer useful life of these components if they are used and kept at cold temperatures.

On the other hand, in relation to the silicone (S10) and catalyst (S3) impregnation mixture, the authors who have achieved plastination at room temperature mixed polydimethylsiloxane with tetraethyl orthosilicate (cross-linker) at a ratio of 100:10 [75] or 100:8 [74]. Nevertheless, we used the mixture that is normally used in low-temperature plastination [1, 2, 59, 62, 81] but at room temperature, i.e., silicone/catalyst (polydimethylsiloxane/dibutyltin dilaurate, respectively) and in the same 100:1 ratio, with very good results, as described in this document. And the use of silicone? The mixture of catalysts, using the latter at only 1%, saves a large amount of liquid. In the Chinese Su-Yi method [47, 48], only silicone was used, not mixed with anything. However, the run-off process of the pieces lasts several months, unlike the other techniques, including ours, where this step only lasts a few days (in our case, only 2 or 3 days). This is due to the absence of interaction between the silicone and its corresponding catalyst, which contributes to the increase in its viscosity, and which is ultimately accelerated during polymerization, giving rise to the final hardening of the sample with its different variants according to the technique.

Forced impregnation is the most important phase of the entire plastination process. During this phase, the acetone is replaced by the silicone-catalyst mixture (100:1). The high steam pressure of the acetone is very important. When does it interact with the silicone? Due to the low steam pressure of the catalyst mixture, the pressure within the chamber is reduced, and the intermediary solvent boils and leaves the tissue as steam. Then, does the silicone-catalyst mixture fill the space that remains? This process is classically described as something that occurs continuously, without stopping the operation of the vacuum pump [1, 2, 59, 62, 75].

Nevertheless, the studies conducted by Zheng et al. [48] and the tests performed in our laboratories have demonstrated that stopping forced impregnation is valid. In addition, we discovered that stopping the process and doing it slowly and in a controlled way reduced the level of tissue shrinkage and made it possible for the silicone to enter the tissue in a regulated way without needing a vacuum pump. We call this "passive forced impregnation," whereas the silicone that replaces the acetone using a vacuum pump is called "active forced impregnation." Nevertheless, we recognize that there is a disadvantage in the "passive" process: it requires a longer impregnation phase, which delays the achievement of the final results.

With respect to the duration of the forced impregnation process, the first great difference is between the low- and room-temperature plastination processes, the first being longer, 7 or more days [1, 2, 59, 62, 75] with the exception of Sakamoto et al. [82], whose impregnation process lasted from 1 to 4 days depending on the size of the specimen. With respect to the room-temperature plastination techniques, Raoof et al. [75] and Henry [74] coincided in a duration of 5 days or more, whereas in our experience forced impregnation only took 3 days. In this case, the exception to this remarkable reduction in impregnation times in the room-temperature plastination technique compared to the low-temperature technique is in the work of Zheng et al. [48], who reported that forced impregnation took from 18 to 25 days.

In this sense, and the pressures issue, it is worth noting the first work published on plastination in the city of La Paz, Bolivia, from Rodriguez-Torrez and Ottone [90], where the geographical characteristics of the city determined notable modifications for the beginning of the impregnation, especially in relation to work pressures, in which the atmospheric pressure in the city of La Paz is 462 mmHg, and this allowed reaching the final 20 mmHg in only 3 days. This factor determined a reduction in forced impregnation time, speeding up this protocol.

During the curing process, most authors [48, 74, 75, 82] perform room-temperature plastination painting or sprinkling the specimens with catalyst to then wrap the preparation in plastic and wait until it is completely dry. This process is done several times, while the sample is moist until it finally hardens and is dried as desired. Nevertheless, in the modified plastination process (UFRO), we imitated the polymerization process of low-temperature plastination [1, 59, 74, 75] but at room temperature. The preparation is exposed to the hardener (TEOS) in a curing chamber, quickly obtaining the ideal sample completely dry and hard, enabling safe manipulation without needing gloves and making the samples long-lasting. This way of using TEOS in the form of gas, and the requirement of only a few milliliters (depending on the size of the container where the TEOS is placed to gasify, the bottom of which is generally filled until the liquid barely reaches a level of 1 cm), also saves on the use of the catalyst employed in the traditional room-temperature plastination method. The relation between the curing times is similar to the relation of the forced impregnation times between the low- and room-temperature techniques. On the other hand, it must be noted that the slow curing process is particularly beneficial to providing a better flexibility of tissues. The thin-walled specimens (stomach, intestine, and thin muscles) can be given a relative flexibility with this slow curing process. For this, these samples, ready to begin the curing process, must be placed in plastic bags for several weeks, after which they can be subjected to the vaporized S6 to reach the final cross-linking that will eliminate the stickiness of the specimen and allow it to achieve its definitive drying, preserving a certain flexibility, as indicated.

An important detail in the room-temperature plastination technique by Zheng [47, 48] and Glover [49] is associated with the vaporization of the acetone during the forced impregnation process, which, at room temperature, occurs around 256 mmHg (1/3 of a pressure atmosphere), without needing to achieve an almost total vacuum, as happens in low-temperature plastination ($-15\ °C$). Another advantage of room temperature in plastination is that the impregnation mixture (S10 + S6) does not react, which the low-temperature impregnation mixture (S10 + S3) does, where long silicone molecules form and the viscosity increases. This way, at room temperature, the impregnation mixture stays very fluid, with very low viscosity, which ensures an excellent penetration and in a very short time (it takes days compared to the weeks of low-temperature force impregnation). Starchik and Henry [61] indicate that the impregnation time is reduced by 65%. This is corroborated by our studies [5–9].

Problems That Can Arise During the Plastination Process

Johnson et al. [91] described different procedures aimed at repairing damage in plastinated preparations, especially at the level of muscular structures, a product of manipulation, and the passage of time. In this sense, they used silicone adhesive (GE Silicone II* Household Glue) together with portions of fascia and muscle impregnated with precured S10/S3, to strengthen, replace, and restore the anatomical appearance of the samples. Regarding the latter, the impregnated repair tissues (fascia and muscle) were provided by Plastinarium in Guben, Germany. In this way, they were able to repair the damaged structures, corresponding to head and neck samples, prolonging the useful life of these plastinated preparations.

Regarding the forced impregnation stage, Domínguez-Calderón et al. [92] make a communication in this regard. In plastination, forced impregnation is required, with a pipe connection to a vacuum pump, through which the pressure reduction will be carried out with the consequent production of vacuum inside the chamber. It is in this process where drawbacks arise, being special care to completely avoid the return of the oil from the vacuum pump into the chamber with polymer [93]. This situation occurs in the laboratory when there is no own electric power plant, or when the electricity fails a lot for different reasons. Once the forced impregnation process has begun, a vacuum is achieved inside the chamber of at least 10–20 mmHg; therefore, once the electricity fails and if the circuit is not closed manually, the chamber can act by exerting a retrograde suction, stripping the pump of lubricating oil, which is stored in the pipeline or, worse, can reach the vacuum chamber and contaminate the silicone and the entire process. Once the oil is in the circuit, the process must be suspended, and all the connections must be disassembled immediately, as well as washed with detergent and later with acetone. It is very important to mention that if the oil has already reached the silicone, it can be carefully removed, in case the oil is floating on the polymer, but if it is seen that the oil has already mixed with the polymer, the entire of the mix must be discarded and replaced with a fresh mix. In these cases, to avoid these situations, the authors recommend the placement of an oil trap device in the vacuum pipe to avoid these retrograde flows in cases of electrical failures or special situations.

In the case of maintenance situations associated with curing, Aja-Guardiola et al. in 2012 [93] made a contribution in relation to this important topic. Biodur S6 (tetraethyl orthosilicate) is a gas in liquid carrier, used for polymerization and hardening of Biodur silicone S10 (polydimethylsiloxane) with a molecular weight of 27,200 and silane functions. Store at room temperature in closed plastic containers to prevent evaporation. After finishing the forced impregnation process inside the vacuum chamber, the specimens must be faced with a curing process, which can occur in the environment in very long periods of time or in shorter and programmed periods with the help of the curing gas S6. The specimens already impregnated with S10 polymer are placed inside a curing chamber and exposed to S6 gas, which acts by vaporization inside the chamber, through the use of a small membrane pump (small aquarium aerator), which through small hoses (one connected to the aerator and the lid of the S6 container, penetrating to the bottom of this, and another from another hole in the lid and toward the curing chamber), both hoses thus connected, the first will "bubble" inside the S6 and the other it will carry this vaporized gas to around the impregnated specimens, thus favoring the curing of the polymer. On certain special occasions, such as exhibitions, public demonstrations, storage of specimens, surprise trips with the material, etc., it may happen that some of the specimens have not been properly cured and that the polymer appears on the surface as drops, which is called "a specimen that sweats or perspires polymer." This can be solved very easily by taking the affected specimen, and immediately pour about 50 cm³ of S6 into a small resistant plastic container, cover it perfectly with a wide layer of plastic (plastic bag, like making a hat), and leave it that the S6 "vaporizes" naturally inside the plastic bag or hermetic container, since this gas, being outside its container, will dissolve in the air around the sample, curing and hardening the polymer of the specimen, but without leaving the sample. The plastic cover will be left there until curing is considered complete, and the specimen must be checked for drying or dried with absorbent paper each time it is checked and it is still wet. This simple procedure allows to correct the defect of the curing not well finished, or altered by climatic conditions, accidents, or special situations [92].

Regarding the appearance of fungi in plastinated samples, according to Adds [94] and Prinz et al. in 1999 [95], in Brazil, product of a tropical storm the first massive infestation of fungi in a collection of plastinated samples occurred. On the other hand, Alshery et al. in 2015 [96] described the appearance of a crystallization on plastinated samples, corresponding to fungi of the Aspergillus type, a fungus transported in the air, in brain sections plastinated with silicone, considering the changing parameters of temperature and humidity the possible causes of the biological changes in sample storage sites. In this sense, the damaged preparations corresponded to already plastinated samples acquired by the corresponding institution describing failures in the air conditioning system during the summer vacations, which determined that the samples were exposed to temperatures of 30–35 °C, in addition to having. There was a flood inside the anatomy laboratories, due to the breakage of a fire sprinkler, which, combined with the high temperatures, caused a considerable increase in humidity inside the storage areas of the plastinated samples, causing the appearance of fungi on plastinated samples. Faced with these situations, both Prinz et al. [95] and Alshery et al. [96] proposed methods to disinfect the samples. Prinz et al. [95] developed a chlorine alcohol-/formalin-based treatment, while Alshery et al. [96] developed the powder-based Virkon method. Regarding the method suggested by Prinz et al. [95], the samples must be brushed manually, while they are kept under running tap water, to remove the surface crystals, thus avoiding the propagation of the crystals in the air and their possible inhalation, which would be very dangerous for human health. Subsequently, they submerge the samples in pure formalin for 5 min, and then submerge the samples in an alcohol chlorine solution (100 mg of chlorine per 100 mL of absolute alcohol) for 20 min. Finally, the samples must be washed in tap water and then left to dry in a well-ventilated area. Regarding the Virkon method suggested by Alshery et al. [96], in this case, initially the samples should also be brushed under running tap water, removing the crystals, and then submerged in a fresh, recently prepared solution of 3% Virkon (3 g in every 100 mL of water) several times. Finally, the samples should be allowed to dry in an area with good ventilation. Personal protection is essential when carrying out these interventions, both with gloves, apron, and eye and/or facial protection, to avoid any skin or eye irritation and inhalation. For his part, Adds in 2021 [94] described the appearance of a large growth of fungus in a sagittal section of the head plastinated with S10 silicone. Adds indicates that this preparation was mistakenly stored cold, in the body storage chamber, at a temperature of 4 °C. Upon microbiological analysis, they observed abundant growth on the agar plates after incubation at 37 °C, allowing to find colonies of *Penicillium* sp. (specifically finding two: *Penicillium corylophilum* and *Penicillium solitum*, in different samples). As Adds [94] indicates, the environment in which the plastinated sample was left is cold and humid, with a strong smell of formaldehyde, and must be ventilated every time personnel enter; therefore, it would be thought that the appearance of fungi on the samples, however, *Penicillium corylophilum* colonizes cold and humid environments, while *Penicillium solitum* survives at low temperatures, being found even in Antarctic marine sediments [94]. Finally, Barros, Navarini, and Masuko [97] described the finding of a thin layer of white matter on the surface of a plastinated kidney cut in half. After the corresponding analysis, they discovered the presence of *Penicillium* sp. In this case, Barros et al. [97] identified that the problem could have been caused by incomplete fixation of the sample, which caused the appearance of fungi, unlike the cases of Prinz et al. [95], Alshery et al. [96], and Adds [94], in which fundamentally temperature and humidity problems seem to have been responsible for the appearance of these fungal colonies. Finally, Komarnitki et al. [98] performed microbiological analyses on samples taken from plastinated specimens, as well as bones and plastic models, and also took samples from surfaces (dissection tables, chairs). In this sense, the results were interesting, since the plastinated samples did not present bacterial or fungal growth, unlike the rest of the tests carried out, in which they found growths of *Staphylococcus epidermidis* (door handle, table top) and *Micrococcus* species (skull, plastic, and silicone models).

Distribution of Research on Silicone Plastination According to the Regions or Anatomical Structures Preserved and Analyzed

Before concluding this chapter, a brief detail of some silicone plastination applications is shared, according to the region or anatomical structure plastinated.

Heart Among these works, we must highlight the first on the subject, developed in 1982, a few years after the invention of plastination, and published by Tiedemann and von Hagens [4]. Later, in 1987, and on the occasion of the launch of the first issue of the *Journal of the International Plastination Society*, the works of Oostrom [99] and Henry [100], also, highlight the application of imaging studies in plastinated cardiac samples: stereolithographic reproduction [101], computerized 3D anatomical modeling [102], echocardiographic in dog plastinated hearts [103], cone beam computed tomography [104], and three-dimensional X-ray computed tomography [105]. On the other hand, the aortic arch and its clinical implications in aortic regurgitation from plastination have also been studied [106]. In addition, the silicone plastination technique is also applicable to large specimens, as we will see later, and in this case also highlighting the plastination of "giant" hearts: Miller et al. in 2017 [107] plastinated the heart of a blue whale (*Balaenoptera musculus*), while Miller et al. in 2021 [108] and Latorre et al. in 2022 [109] reported silicone plastination of the heart of a killer whale (*Orcinus orca*). Finally, in 2023, highlighting the first work on plastination from Ecuador, published by the group of Dr. Revelo-Cueva, from the Central University of Ecuador and published by Toaquiza et al. [110] in *International Journal of Morphology*, in 2023, corresponding to the plastination of dog hearts, donated for science to the university.

Vascular Variations Regarding the description of vascular variations in the upper limbs, in 1996, Grondin and Olry [111] identified arterial variations of the palmar arch in samples of plastinated human hands, while Ottone et al. [26] identified two vascular variations of the upper limbs, specifically a superficial ulnar artery, originating from the axillary artery, and a supernumerary branch of the radial artery, which was provided at the level of the lower third of the forearm, with a trajectory superficial through the anatomical snuffbox, both variations with important surgical implications and their descriptions are necessary, both those of Grondin and Olry [111] and those of Ottone et al. [26], to explain the relevance of anatomical knowledge, both standardized and that of variations, when performing surgical approaches in the different anatomical regions.

Parasites All kinds of specimens can be plastinated, including parasites: *Ascaris* [112], *Oestrus ovis* larvae [113], various species of parasites [114], and *Toxocara vitulorum* [115].

Larger and Massive Specimens In relation to the plastination of large and massive specimens, as Nader, Henry, and Sui rightly indicate in their 2019 work [116], they represent a notable infrastructure challenge, but commenting that despite these large structures that must be mounted at the time of carrying out the plastination of these specimens, in addition to the need for a large amount of economic resources, not only for the purchase of inputs but also for the transfer of the samples, the basis of the plastination technique is the same as the described since its invention by Gunther von Hagens [1–3], and considering all the modifications and complements incorporated to it in all these years. Thus, in 2015, Yu et al. [117] describe plastination of a whole horse, while Jian et al. [118], in 2022, describe the plastination of a sperm whale.

Some Clinical and Surgical Applications of Silicone Plastination The plastination technique has been applied in multiple studies with clinical correlation: radiographic and magnetic resonance studies of the tibiofibular syndesmosis [119, 120], Dupuytren's disease [121], development of plastinated human placentas [77, 122], analysis of the calcaneal tendon and its rupture [123], surgical approaches to the infratemporal fossa [124], vascular studies of the ethmoid region [125, 126], plastinated models of fingertip flaps [127], development of lung phantoms from specimens plastinated [128], plastinated models of the maxillary sinus [129], application of plastination for the development of surgical practice in temporal bone [130], and base of the cranium microsurgical approaches [131].

Cultural Heritage In this unique work of Buendía, Latorre, and Lopez-Albors [132], from the University of Murcia, Spain, the validation of plastination is highlighted as a technique for the conservation of flooded archaeological materials, highlighting in this work the singularity of having preserved elephant tusks, obtained from an archaeological excavation corresponding to a Phoenician shipwreck in Bajo de la Campana, San Javier, Murcia, Spain. The characteristic of this work, and as the authors highlight, was the possibility of preserving ivory by means of a silicone plastination technique, especially indicating the inefficiency of traditional conservation procedures given the extreme density of ivory.

Autopsy Samples The works presented by Porzionato et al. in 2004 [133] and 2018 [134] correspond to human samples from autopsies. In the first case [133], corresponding to an aneurysm of the great cerebral vein (Galen's vein), the authors performed plastination of brain slices with silicone, and were able to identify the communication of the aneurysm with the right thalamoperforating artery and, bilaterally, with the posterior choroidal and mesencephalic arteries. And in the 2018 paper [134], the authors demonstrated the application of the plastination technique in the field of forensic sciences (see also DNA identification in plastinated samples by our Laboratory, in Chap. 8), by plastinating the brain of a victim of homicide by firearm and its presentation in court as documentary evidence. In this work, the possibility of the brain of the victim in three dimensions allowed the judge to analyze the trajectory of the projectile and to compare this information with the statements, postmortem information, and police investigations, thus allowing to demonstrate the usefulness of plastination in this type of court.

Neuroanatomy and Neurosciences. In this section there are also various contributions that suggest new plastination protocols for preserving human brains, modifications to existing techniques, and analysis of possible failures in their implementation [135]. Additionally, a publication compares different room temperature methods (15-18°C) for preserving human brains [136] and another research compared different forced cold impregnation processes [137]. Other notable works include the application and comparison of different stains on brain sections, which are then plastinated [138]. There are also publications that guide the use of plastination for the development of neurosciences [139] and teaching in neuroanatomy [140]. Moreover, brain plastination techniques are also applied in human fetuses, combining them with different staining techniques [141]. Comparisons are made with different types of polymer resin (S10 and P35) [142], and finally a publication that present the possibility of making ventricular models in situ with S10 [143].

Conclusions

Finally, plastination is a widely accepted technique due to the durability and high research value that such specimens offer. However, none of the plastination techniques described here will improve the characteristics of a cadaver preparation. For that reason, it is fundamentally important to develop a suitable embalming technique and a technique of precise, meticulous dissection, care, and excellent identification of the anatomical structures of interest. The correctly executed plastination technique will make it possible to preserve biosafe cadaver material for an indefinite period without the toxicity of formaldehyde, preserving its size and shape. This is beneficial at times when there are few cadavers for research and they are difficult to acquire. These preparations can be used several purposes, for example, in morphology and macroscopic anatomy, and in different places: anatomy in medical practice, essential for the education of students, and for the higher education of professionals in surgical, clinical, and diagnostic imaging practices, as well as for research.

References

1. von Hagens G. Impregnation of soft biological specimens with thermosetting resins and elastomers. Anat Rec. 1979;194(2):247–55. https://doi.org/10.1002/ar.1091940206.
2. von Hagens G, editor. Heidelberg plastination folder. Collection of technical leaflets of plastination. Heidelberg: Biodur Products GmbH; 1986.
3. von Hagens G, Tiedemann K, Kriz W. The current potential of plastination. Anat Embryol (Berl). 1987;175(4):411–21. https://doi.org/10.1007/BF00309677.
4. Tiedemann K, von Hagens G. The technique of heart plastination. Anat Rec. 1982;204(3):295–9. https://doi.org/10.1002/ar.1092040315.
5. Ottone NE, Baptista CAC, Del Sol M, Muñoz Ortega M. Extraction of DNA from plastinated tissues. Forensic Sci Int. 2020;309:110199. https://doi.org/10.1016/j.forsciint.2020.110199.
6. Ottone NE, Cirigliano V, Bianchi HF, Medan CD, Algieri RD, Borges Brum G, Fuentes R. New contributions to the development of a plastination technique at room temperature with silicone. Anat Sci Int. 2015;90(2):126–35. https://doi.org/10.1007/s12565-014-0258-6.
7. Ottone NE. Plastination: techniques fundamentals and implementation at Universidad de La Frontera. J Health Med Sci. 2018;4(4):293–302.
8. Ottone NE, Cirigliano V, Lewicki M, Bianchi H, Aja-Guardiola S, Algieri RD, Cantin M, Fuentes R. Plastination technique in laboratory rats: an alternative resource for teaching, surgical training and research development. Int J Morphol. 2014;32(4):1430–5. https://doi.org/10.4067/S0717-95022014000400048.
9. Ottone NE. Unified plastination protocol with silicone at cold and room temperature. Int J Morphol. 2021;39(2):630–4. https://doi.org/10.4067/S0717-95022021000200630.
10. Bravo H. Plastination an additional tool to teach anatomy. Int J Morphol. 2006;24(3):475–80. https://doi.org/10.4067/S0717-95022006000400029.

11. Bickley HC, Walker AN, Jackson RL, Donner RS. Preservation of pathology specimens by silicone plastination. An innovative adjunct to pathology education. Am J Clin Pathol. 1987;88(2):220–3. https://doi.org/10.1093/ajcp/88.2.220.
12. Dawson TP, James RS, Williams GT. Silicone plastinated pathology specimens and their teaching potential. J Pathol. 1990;162(3):265–72. https://doi.org/10.1002/path.1711620314.
13. Grondin G, Grondin GG, Talbot BG. A study of criteria permitting the use of plastinated specimens for light and electron microscopy. Biotech Histochem. 1994;69(4):219–34. https://doi.org/10.3109/10520299409106291.
14. Hawley DA, Marlin DC, Cook DC, Becsey D, Clark MA, Pless JE, Standish SM. Specimens for teaching forensic pathology, odontology, and anthropology. I. Soft tissue. Am J Forensic Med Pathol. 1991;12(2):164–9. https://doi.org/10.1097/00000433-199106000-00015.
15. Hawley DA, Marlin DC, Cook DC, Becsey D, Clark MA, Pless JE, Standish SM. Specimens for teaching forensic pathology, odontology, and anthropology. II. Teeth and bone. Am J Forensic Med Pathol. 1991;12(2):170–4. https://doi.org/10.1097/00000433-199106000-00016.
16. Janick L, DeNovo RC, Henry RW. Plastinated canine gastrointestinal tracts used to facilitate teaching of endoscopic technique and anatomy. Acta Anat (Basel). 1997;158(1):48–53. https://doi.org/10.1159/000147910.
17. Mohamed AMA, Ahmed AA, Adam Asia I, Ali AM, Taha AAM. Establishing for the first time the use of the standard S10 technique for plastination in the Sudan. J Plastination. 2018;30(2):JP-18-03. https://doi.org/10.56507/BAXU9458.
18. Monteiro YF, Juvenato LS, Bittencourt APSV, Siqueira BMM, Monteiro FC, Baptista CAC, Bittencourt AS. Influence of the temperature on the viscosity of different types of silicone. J Plastination. 2018;30(1):JP-18-02. https://doi.org/10.56507/HETT9088.
19. Neha SL, Dhingra R. Plastinated knee specimens: a novel educational tool. J Clin Diagn Res. 2013;7(1):1–5. https://doi.org/10.7860/JCDR/2012/5142.2657.
20. Pendovski L, Petkov V, Popovska-Percinic F, Ilieski V. Silicone plastination procedure for producing thin, semi-transparent tissue slices: a study using the pig kidney. J Int Soc Plastination. 2008;23:10–6.
21. Pond KR, Holladay SD, Luginbuhl JM. Technical note: preservation of tissues and gastrointestinal tract portions by plastic coating or plastination. J Anim Sci. 1992;70(4):1011–4. https://doi.org/10.2527/1992.7041011x.
22. Ramos ML, de Paula TAR, Zerlotti MF, Silva VHD, Carazo LB, de Paula MF, Silva FFR, Santana ML, Silva LC, Ferreira LBC. A comparison of different de-plastination methodologies for preparing histological sections of material plastinated with Biodur® S10/S3. J Plastination. 2018;30(1):JP-17-09. https://doi.org/10.56507/OHLF5315.
23. Ruthig VA, Labrash S, Lozanoff S, Ward MA. Macroscopic demonstration of the male urogenital system with evidence of a direct inguinal hernia utilizing room temperature plastination. Anatomy. 2016;10(3):211–20. https://doi.org/10.2399/ana.16.036.
24. Shanthi P, Singh RR, Gibikote S, Rabi S. Comparison of CT numbers of organs before and after plastination using standard S-10 technique. Clin Anat. 2015;28(4):431–5. https://doi.org/10.1002/ca.22514.
25. Steinke H, Rabi S, Saito T, Sawutti A, Miyaki T, Itoh M, Spanel-Borowski K. Light-weight plastination. Ann Anat. 2008;190(5):428–31. https://doi.org/10.1016/j.aanat.2008.02.005.
26. Ottone NE, Guzmán D, Bianchi HF, del Sol M. Anatomical variations of radial and ulnar arteries in plastinated upper limbs. Int J Morphol. 2023;41(2):548–54. https://doi.org/10.4067/S0717-95022023000200548.
27. Bickley HC, Conner RS, Walker AN, Jackson RL. Preservation of tissue by silicone rubber impregnation. J Int Soc Plastination. 1987;1(2):30–9.
28. Tiedemann K. A silicone-impregnated knee joint as a natural model for arthroscopy. J Int Soc Plastination. 1988;2(1):13–7.
29. Oostrom K. Plastination of the human placenta. J Int Soc Plastination. 1988;2(1):18–23.
30. Oostrom K. Plastination of the human kidney. J Int Soc Plastination. 1988;2(2):21–4.
31. Riepertinger A. Plastination of the brain with attached spinal cord. J Int Soc Plastination. 1989;3:22–8.

32. Holladay SD. Plastination of inflated hollow gastrointestinal organs from large animals. J Int Soc Plastination. 1989;3:34–7.
33. Baptista CAC, Conran PB. Plastination of the heart: preparation for the study of the cardiac valves. J Int Soc Plastination. 1989;3:3–7.
34. Baptista CAC, Skie M, Yeasting RA, Ebraheim N, Jackson WT. Plastination of the wrist: potential uses in education and clinical medicine. J Int Soc Plastination. 1989;3:18–21.
35. Bore P, Boyes R. Plastination down under. J Int Soc Plastination. 1994;8:19–21.
36. Cook P, Dawson B. Plastination methods used in Auckland, New Zealand. J Int Soc Plastination. 1996;10:32–3.
37. Grondin G, Olry R. Dissection and plastination of the human cerebral dura mater with the base of skull. J Int Soc Plastination. 1998;13(1):23–5.
38. Asadi MH. Plastination of sturgeons with the S10 technique in Iran: the first trials. J Int Soc Plastination. 1998;13(1):15–6.
39. Baker JA. COR-TECH PR-10 silicone: initial trials in plastinating human tissue. J Int Soc Plastination. 1999;14(2):13–9.
40. Pendovski L, Ilieski V, Nikolovski G. Silicone plastination of a malpositioned long-term formalin-fixed green Iguana. J Int Soc Plastination. 2004;19:40–2.
41. Alpár A, Glasz T, Kálmán M. Plastination of pathological specimens—a continuing challenge. J Int Soc Plastination. 2005;20:8–12.
42. Oostrom K. Fixation of tissue for plastination: general principles. J Int Soc Plastination. 1987;1:3–11.
43. Mohsen SM, Esfandiari E, Rabiei AA, Hanaei MS, Rashidi B. Comparing two methods of plastination and glycerin preservation to study skeletal system after Alizarin red-Alcian blue double staining. Adv Biomed Res. 2013;2:19. https://doi.org/10.4103/2277-9175.108003.
44. Monteiro YF, Juvenato LS, Bittencourt APSV, Baptista CAC, Bittencourt AS. Plastination of glycerin-fixed specimens. J Plastination. 2021;33(2):JP-21-06. https://doi.org/10.56507/RWCV2838.
45. Cannas M, Fuda P. Plastination of old formalin-fixed specimens. J Int Soc Plastination. 1991;5:11–5.
46. Ripani M, Bassi A, Perracchio L, Panebianco V, Perez M, Boccia ML, Marinozzi G. Monitoring and enhancement of fixation, dehydration, forced impregnation and cure in the standard S-10 technique. J Int Soc Plastination. 1994;8:3–5.
47. Zheng TZ, Weatherhead BL, Gosling J. Plastination at room temperature. J Int Soc Plastination. 1996;11(2):33.
48. Zheng T, Liu J, Zhu K. Plastination at room temperature. J Int Soc Plastination. 1998;13(2):21–5.
49. Glover R, Henry R, Wade R. Polymer preservation technology: Poly-Cur, a next generation process for biological specimen preservation. In: Abstracts of the 9th International Conference on Plastination held in Quebec City, Canada. J Int Soc Plastination. 1998;13(2).
50. Chaynes P, Mingotaud AF. Analysis of commercial plastination agents. Surg Radiol Anat. 2004;26(3):235–8. https://doi.org/10.1007/s00276-003-0216-9.
51. Henry RW, Janick L, Salmos FP. Specimen preparation for silicone plastination. J Int Soc Plastination. 1997;12(1)
52. Henry RW. Principles of plastination—specimen preparation. J Int Soc Plastination. 1995;9:13.
53. Schwab KH, von Hagens G. Freeze substitution of macroscopic specimens for plastination. Abstract, Sixth European Anatomical Congress. Acta Anat. 1981;111(12):399.
54. Henry RW. Principles of plastination—dehydration of specimens. J Int Soc Plastination. 1995;9:27.
55. Sagoo MG, Adds PJ. Low-temperature dehydration and room-temperature impregnation of brain slices using Biodur TM S10/S3. J Plastination. 2013;25(1):3–8. https://doi.org/10.56507/LYMW8377.
56. Parsai S, Frank PW, Baptista CAC. Silicone plastination of brain slices: using sucrose to reduce shrinkage. J Plastination. 2020;32(1):JP-20-05. https://doi.org/10.56507/IDZK2594.

57. Arnts H, Kleinnijenhuis M, Kooloos JG, Schepens-Franke AN, van Cappellen van Walsum AM. Combining fiber dissection, plastination, and tractography for neuroanatomical education: revealing the cerebellar nuclei and their white matter connections. Anat Sci Educ. 2014;7(1):47–55. https://doi.org/10.1002/ase.1385.
58. Baeres FM, Møller M. Plastination of dissected brain specimens and Mulligan-stained sections of the human brain. Eur J Morphol. 2001;39(5):307–11. https://doi.org/10.1076/ejom.39.5.307.7377.
59. DeJong K, Henry RW. Silicone plastination of biological tissue: cold-temperature technique BiodurTM S10/S15 technique and products. J Int Soc Plastination. 2007;22:2–14.
60. Henry RW, von Hagens G, Seamans G. Cold temperature/Biodur®/S10/von Hagens'-silicone plastination technique. Anat Histol Embryol. 2019;48(6):532–8. https://doi.org/10.1111/ahe.12472.
61. Starchik D, Henry RW. Comparison of cold and room temperature silicone plastination techniques using tissue core samples and a variety of plastinates. J Plastination. 2015;27(2):JP-16-006. https://doi.org/10.56507/NTQJ7764.
62. Henry RW. Silicone plastination of biological tissue: cold-temperature technique North Carolina technique and products. J Int Soc Plastination. 2007;22:26–30.
63. Nicoll S. Plastination of specimens using the silicone (S10) technique following embalming with Dodge© products results in improved color retention and tissue differentiation. J Plastination. 2021;33(1):21–8. https://doi.org/10.56507/TNNI3327.
64. Satte MS, Ali TO, Mohamed AHY. The use of vacuum forced impregnation of gum Arabic solution in biological tissues for long-term preservation. J Plastination. 2017;29(1):JP-17-03. https://doi.org/10.56507/NSSX1797.
65. Henry RW, Nel PPC. Forced impregnation for the standard S10 method. J Int Soc Plastination. 1993;7:27–31.
66. Henry RW. Principles of plastination—forced impregnation. J Int Soc Plastination. 1995;9:26.
67. Smodlaka H, Latorre R, Reed RB, Gil F, Ramírez G, Váquez-Autón JM, López-Albors O, Ayala MD, Orenes M, Cuellar R, Henry RW. Surface detail comparison of specimens impregnated using six current plastination regimens. J Int Soc Plastination. 2005;20:20–30.
68. Pereira-Sampaio MA, Marques-Sampaio BP, Sampaio FJ, Henry RW. Shrinkage of renal tissue after impregnation via the cold Biodur plastination technique. Anat Rec (Hoboken). 2011;294(8):1418–22. https://doi.org/10.1002/ar.21432.
69. Bickley HC, von Hagens G, Townsend FM. An improved method for the preservation of teaching specimens. Arch Pathol Lab Med. 1981;105(12):674–6.
70. Parel SM, Bickley H, Holt GR, Shuler S. Prosthetic use of plastinated facial structures: a feasibility study. J Prosthet Dent. 1983;49(4):529–31. https://doi.org/10.1016/0022-3913(83)90317-7.
71. Weiglein AH, Henry RW. Curing (hardening, polymerization) of the polymer—Biodur S10. J Int Soc Plastination. 1993;7:32–5.
72. Henry RW. Principles of plastination—gas curing (hardening). J Int Soc Plastination. 1995;9:13.
73. Starchik D, Henry RW. Room temperature/Corcoran/Dow Corning™-Silicone plastination process. Anat Histol Embryol. 2019;48(6):539–46. https://doi.org/10.1111/ahe.12505.
74. Henry RW. Silicone plastination of biological tissue: room-temperature technique North Carolina technique and products. J Int Soc Plastination. 2007;22:15–9.
75. Raoof A, Henry RW, Reed B. Silicone plastination of biological tissue: room-temperature technique Dow/Corcoran technique and products. J Int Soc Plastination. 2007;22:21–5.
76. Raoof A. Using a room-temperature plastination technique in assessing prenatal changes in the human spinal cord. J Int Soc Plastination. 2001;16:5–8.
77. Prieto R, Vargas CA, Veuthey C, Aja-Guardiola C, Ottone NE. Fundamental concepts of the modified room temperature plastination protocol with silicone, with subsequent pigmentation, and its application for the conservation of human placenta. Int J Morphol. 2019;37(1):375–6. https://doi.org/10.4067/S0717-95022019000100369.
78. Gubbins RBG. Design of a plastination laboratory. J Int Soc Plastination. 1990;4:24–7.

79. O'Sullivan E, Mitchell BS. Plastination for gross anatomy teaching using low cost equipment. Surg Radiol Anat. 1995;17(3):277–81. https://doi.org/10.1007/BF01795063.
80. Briggs CA, Robbins SG, Kaegi WH. Development of an anatomical technologies laboratory. J Int Soc Plastination. 1997;12(2):8–11.
81. Smodlaka H, Henry RW, Daniels GB, Reed RB. Correlation of computed tomographic images with anatomic features of the abdomen of ringed seals (Phoca hispida). Am J Vet Res. 2004;65(9):1240–4. https://doi.org/10.2460/ajvr.2004.65.1240. Erratum in: Am J Vet Res. 2004;65(12):1738–9.
82. Sakamoto Y, Miyake Y, Kanahara K, Kajita H, Ueki H. Chemically reactivated plastination with shin-Etsu Silicone KE-108. J Int Soc Plastination. 2006;21:11–6.
83. Latorre RM, García-Sanz MP, Moreno M, Hernández F, Gil F, López O, Ayala MD, Ramírez G, Vázquez JM, Arencibia A, Henry RW. How useful is plastination in learning anatomy? J Vet Med Educ. 2007;34(2):172–6. https://doi.org/10.3138/jvme.34.2.172.
84. Steinke H, Spanel-Borowski K. Coloured plastinates. Ann Anat. 2006;188(2):177–82. https://doi.org/10.1016/j.aanat.2005.10.001.
85. McCreary J, Iliff S, Hermey D, McCreary K, Henry RW. Silicone-based coloration technique developed to highlight plastinated specimens. J Plastination. 2013;25(2):13–20.
86. Baptista CAC, Cerqueira EP, Conran PB. Impregnation of biological specimens with resins and elastomers: plastination with Biodur S10 resin. Rev Bras Cienc Morfol. 1988;5(1):60–2.
87. Baptista CAC, Bellm P, Plagge MS, Valigosky M. The use of explosion proof freezers in plastination: are they really necessary? J Int Soc Plastination. 1992;6:34–7.
88. Sora MC. Room temperature impregnation with cold temperature Biodur® silicone: a study of viscosity. J Plastination. 2017;29(1):JP-17-02. https://doi.org/10.56507/QIDA3140.
89. Adds PJ. Biodur® S10/S3 and S15/S3 at "room temperature": a viscosity study. J Plastination. 2017;29(1):JP-17-04. https://doi.org/10.56507/RNBU9088.
90. Rodriguez-Torrez VH, Ottone NE. First plastination experience at 4,150 meters above sea level, in the height of La Paz, Bolivia. Int J Morphol. 2023; In Press
91. Johnson JH, Baker EW. Rehabilitation of plastinated anatomical prosections using silicone adhesive and pre-cured S10/S3—impregnated fascia and muscle. J Plastination. 2017;29(2):JP-17-10. https://doi.org/10.56507/YVPS5638.
92. Domínguez-Calderón RG, Aja-Guardiola S, Mayoral-Robles S, García-Jarquín J, Romero-Ramírez M, Ottone NE. Elimination of oil in the vacuum chamber caused by backflow accidents. In: Presentation. XXIV National Congress of Anatomy, Mexican Society of Anatomy, Zacatecas, Mexico. October 2 to 5, 2012.
93. Aja-Guardiola S, Figueroa Gutiérrez J, Grajeda López S, Barreto Oble D, Domínguez Calderón RG, Ottone N. Emergent curing by vaporization of Biodur-S6 in special situations of the plastination technique. In: Presentation. XXIV National Congress of Anatomy, Mexican Society of Anatomy, Zacatecas, Mexico. October 2 to 5, 2012.
94. Adds PJ. Fungal growth on a plastinated anatomical specimen. J Plastination. 2021;33(2):JP-21-13. https://doi.org/10.56507/CAXM7316.
95. Prinz R, Correia J, Moraes A, da Silva A, Queiroz S, Pezzi L. Fungal contamination of plastinated specimens. J Plastination. 1999;14(2):20–4.
96. AlShehry MA, AlObaysi MM, Al-Hamdan N. Cleaning excessive cross-linker crystallization on S10 plastinated brain slices. J Plastination. 2015;27(1):13–7.
97. Barros M, Navarini A, Masuko TS. Fungal contamination of plastinated specimen. In: Experimental biology 2015 meeting abstracts. FASEB J. 2015;29(S1). https://doi.org/10.1096/fasebj.29.1supplement.547.12.
98. Komarnitki I, Skadorwa T, Dziedzic D, Grzegorczyk M, Ciszek B. Are anatomical specimens plastinated using cold-temperature S10 silicone technique microbiologically safe? J Plastination. 2019;31(1):JP-19-05. https://doi.org/10.56507/GYMH8431.
99. Oostrom K. Plastination of the heart. J Int Soc Plastination. 1987;1(2):12–9.
100. Henry RW. Plastination of an integral heart-lung specimen. J Int Soc Plastination. 1987;1(2):20–4.

101. Greil GF, Wolf I, Kuettner A, Fenchel M, Miller S, Martirosian P, Schick F, Oppitz M, Meinzer HP, Sieverding L. Stereolithographic reproduction of complex cardiac morphology based on high spatial resolution imaging. Clin Res Cardiol. 2007;96(3):176–85. https://doi.org/10.1007/s00392-007-0482-3.
102. Tunali S, Kawamoto K, Farrell ML, Labrash S, Tamura K, Lozanoff S. Computerised 3-D anatomical modelling using plastinates: an example utilising the human heart. Folia Morphol (Warsz). 2011;70(3):191–6.
103. Gómez A, Del Palacio JF, Latorre R, Henry RW, Sarriá R, Albors OL. Plastinated heart slices aid echocardiographic interpretation in the dog. Vet Radiol Ultrasound. 2012;53(2):197–203. https://doi.org/10.1111/j.1740-8261.2011.01880.x.
104. Chang CW, Atkinson G, Gandhi N, Farrell ML, Labrash S, Smith AB, Norton NS, Matsui T, Lozanoff S. Cone beam computed tomography of plastinated hearts for instruction of radiological anatomy. Surg Radiol Anat. 2016;38(7):843–53. https://doi.org/10.1007/s00276-016-1645-6.
105. Hirasaki Y, Tomita T, Yanagisawa M, Ueda K, Sato K, Okabe M. Heart anatomy of Rhincodon typus: three-dimensional X-ray computed tomography of plastinated specimens. Anat Rec (Hoboken). 2018;301(11):1801–8. https://doi.org/10.1002/ar.23902.
106. Khubulava GG, Marchenko SP, Starchik DA, Suvorov VV, Krivoshchekov EV, Shikhverdiev NN, Naumov AB. Geometric and morphological features of the aortic root in norm and aortic regurgitation. Khirurgiia (Mosk). 2018;5:4–12. https://doi.org/10.17116/hirurgia201854-12.
107. Miller JR, Henry RW, Nader P, Engstrom S, Iliff S, Chereminskiy V, von Hagens G. The challenges of plastinating a blue whale (Balaenoptera musculus) heart. J Plastination. 2017;29(2):22–9.
108. Miller JR, Lopez Albors O, Raverty S, Cottrell P, Engstrom MD, Latorre R. Plastination of an endangered southern resident killer whale heart (Orcinus orca). J Plastination. 2021;33(1):29–41. https://doi.org/10.56507/JWUL5276.
109. Latorre R, Graïc JM, Raverty SA, Soria F, Cozzi B, López-Albors O. The heart of the killer whale: description of a plastinated specimen and review of the available literature. Animals (Basel). 2022;12(3):347. https://doi.org/10.3390/ani12030347.
110. Toaquiza AB, Gómez C, Ottone NE, Revelo-Cueva M. Conservation of organs (heart, brain and kidney) of canine by cold-temperature silicone plastination done at an animal anatomy laboratory in Ecuador. Int J Morphol. 2023. In Press;41:1004.
111. Grondin G, Olry R. Vascular patterns of plastinated human hands with special reference to abnormalities of the arterial palmar arches. J Int Soc Plastination. 1996;10:19–21.
112. Asadi MH, Mahmodzadeh A. Ascaris plastination through S10 techniques. J Int Soc Plastination. 2004;19:20–1.
113. Gonzalves M, Ortiz J, Latorre R. S10 plastination technique for preservation of parasites: the case of Oestrus ovis larvae. J Plastination. 2017;29(2):JP-17-08. https://doi.org/10.56507/MUBJ3876.
114. Gonzálvez M, Ortiz J, Navarro M, Latorre R. Preservation of macroparasite species via classic plastination: an evaluation. Folia Parasitologica (Praha). 2018;65:019. https://doi.org/10.14411/fp.2018.019.
115. Kumar N, Solanki JB, Shil P, Patel DC, Meneka R, Chaurasia S. Dry preservation of Toxocara vitulorum by plastination technique. Vet World. 2019;12(9):1428–33. https://doi.org/10.14202/vetworld.2019.1428-1433.
116. Nader PB, Henry RW, Sui HJ. Plastination of larger and massive specimens-with silicone. Anat Histol Embryol. 2019;48(6):547–51. https://doi.org/10.1111/ahe.12481.
117. Yu SB, Zhang ZJ, Chi YY, Gao HB, Liu J, Sui HJ. Plastination of a whole horse for veterinary education. J Plastination. 2015;27(1):29–32.
118. Jiang WB, Han J, Ma XW, Liu H, Yu SB, Sui HJ. Plastination of a sperm whale. J Anat. 2022;240(4):669–77. https://doi.org/10.1111/joa.13581.
119. Beumer A, van Hemert WL, Niesing R, Entius CA, Ginai AZ, Mulder PG, Swierstra BA. Radiographic measurement of the distal tibiofibular syndesmosis has limited use. Clin Orthop Relat Res. 2004;423:227–34. https://doi.org/10.1097/01.blo.0000129152.81015.ad.

120. Hermans JJ, Wentink N, Kleinrensink GJ, Beumer A. MR-plastination-arthrography: a new technique used to study the distal tibiofibular syndesmosis. Skelet Radiol. 2009;38(7):697–701. https://doi.org/10.1007/s00256-008-0631-4.
121. Meinel A. Morbus Dupuytren. New aspects in pathogenesis and principle of surgery. Handchir Mikrochir Plast Chir. 1999;31:339–45. https://doi.org/10.1055/s-1999-13549.
122. McRae KE, Davies GA, Easteal RA, Smith GN. Creation of plastinated placentas as a novel teaching resource for medical education in obstetrics and gynaecology. Placenta. 2015;36(9):1045–51. https://doi.org/10.1016/j.placenta.2015.06.018.
123. Neusel E, Graf J, Jochem C, Rompe G. Long-term results following subcutaneous Achilles tendon rupture. Sportverletz Sportschaden. 1990;4(1):36–40. https://doi.org/10.1055/s-2007-993595.
124. Prades JM, Timoshenko A, Merzougui N, Martin C. A cadaveric study of a combined trans-mandibular and trans-zygomatic approach to the infratemporal fossa. Surg Radiol Anat. 2003;25(3–4):180–7. https://doi.org/10.1007/s00276-003-0126-x.
125. Musumeci D, Lang FJW, Duvoisin B, Riederer BM. Plastinated ethmoidal region: II. The preparation and use of radio-opaque artery casts in clinical teaching. J Int Soc Plastination. 2003;18:23–8.
126. Musumeci D, Lang FJW, Duvoisin B, Riederer BM. Plastinated ethmoidal region: II. The preparation and use of radio-opaque artery casts in clinical teaching. J Int Soc Plastination. 2003;18:29–33.
127. Alpar A, Gal A, Patonay L, Kalman M. Local flaps for fingertip injuries: a plastinated model. J Int Soc Plastination. 2001;16:42–5.
128. Yoon S, Henry RW, Bouley DM, Bennett NR, Fahrig R. Characterization of a novel anthropomorphic plastinated lung phantom. Med Phys. 2008;35(12):5934–43. https://doi.org/10.1118/1.3016524.
129. Durand M, Rusch P, Granjon D, Chantrel G, Prades JM, Dubois F, Esteve D, Pouget JF, Martin C. Preliminary study of the deposition of aerosol in the maxillary sinuses using a plastinated model. J Aerosol Med. 2001;14(1):83–93. https://doi.org/10.1089/08942680152007936.
130. Maeta M, Uno K, Saito R. The potential of a plastination specimen for temporal bone surgery. Auris Nasus Larynx. 2003;30(4):413–6. https://doi.org/10.1016/s0385-8146(03)00089-0.
131. Resch KDM. Plastinated specimens for demonstration of microsurgical approaches to the base of the cranium. J Int Soc Plastination. 1989;3:28–33.
132. Buendía M, Latorre R, Lopez-Albors O. Plastination applied to the conservation of cultural heritage. J Plastination. 2017;29:11–21.
133. Porzionato A, Macchi V, Parenti A, De Caro R. Vein of Galen aneurysm: anatomical study of an adult autopsy case. Clin Anat. 2004;17(6):458–62. https://doi.org/10.1002/ca.20008.
134. Porzionato A, Russo M, Macchi V, Aprile A, De Caro R. The utility of plastinates in court: a case of firearm homicide. Forensic Sci Med Pathol. 2018;14(2):216–20. https://doi.org/10.1007/s12024-018-9958-x.
135. Miklosová M, Miklos V. Plastination with silicone method S 10—monitoring and analysis causes of failure. Biomed Pap Med Fac Univ Palacky Olomouc Czech Repub. 2004;148(2):237–8. https://doi.org/10.5507/bp.2004.048.
136. Mooncey MS, Sagoo MG. Comparative staining methods with room temperature plastination (15–18°C) of brain specimens, using BiodurTM S10/S3. J Plastination. 2014;26(2):21–9. https://doi.org/10.56507/TFQH5165.
137. Owolabi JO, Adeteye OV, Fabiyi OS, Olatunji SY, Olanrewaju JA, Obaoye A. Comparative study of the outcome of forced impregnation of whole brains at cold temperature, and an alternative diffusion/impregnation process. J Plastination. 2019;31(1):JP-18-06. https://doi.org/10.56507/RTIG2240.
138. Suriyaprapadilok L, Withyachumnarnkul B. Plastination of stained sections of the human brain: comparison between different staining methods. J Int Soc Plastination. 1997;12(1):27–32.
139. Weiglein AH. Plastination in the neurosciences. Keynote lecture. Acta Anat (Basel). 1997;158(1):6–9. https://doi.org/10.1159/000147902.

140. Holladay SD, Hudson LC. Use of plastinated brains in teaching neuroanatomy at the North Carolina State University, College of Veterinary Medicine. J Int Soc Plastination. 1989;3:15–7.
141. Ulfig N. Staining of human fetal and adult brain slices combined with subsequent plastination. J Int Soc Plastination. 1990;4:33–8.
142. Weiglein AH. Preparing and using S-10 and P35 brain slices. J Int Soc Plastination. 1996;10:22–5.
143. Grondin G, Sianothai A, Olry R. In situ ventricular casts of S10 plastinated human brains. J Int Soc Plastination. 2000;15(1):18–21.

Chapter 6
Epoxy Sheet Plastination Technique

General Description

Epoxy sheet plastination, called the E12 plastination technique by von Hagens [1, 2], makes it possible to produce thin and ultra-thin slices of tissues that permit the study of the anatomy by sections, without deformation, collapse, or alteration, in their anatomical position [3–12]. It is the technique of choice to produce dry, odorless, and transparent body sections [1–31]. Sheet plastination for the production of transparent sheets with E12 epoxy resins was patented in 1982 [13] by Gunther von Hagens, the inventor of plastination. Figure 6.1 shows the basic steps of the classical technique of sheet plastination with epoxy resin.

The plastination process causes a certain shrinkage of tissues, and for that reason it is fundamental to achieve minimum shrinkage of the samples, and this can occur mainly by performing the plastination technique incorrectly, either by excessive dehydration or a very rapid forced impregnation. In addition, the previous fixation phase could notably affect the size and shape of tissues, if this was done incompletely. Nevertheless, one of the advantages of epoxy sheet plastination is the almost total absence of shrinkage, and this is why it is recommended for studies in the area of anatomy and morphological sciences. The method halts the decomposition of the sample and embeds it in epoxy, obtaining great transparency thanks to the refractive index of epoxy resin that facilitates microscopic study, creating a high-quality optical interface with the tissue [32]. In addition, the sheets obtained can be easily processed morphometrically [33].

Sheet plastination with epoxy resin (E12) can achieve plastinated sections from any region of the body with a thickness between 2 and 5 mm [1, 2]. It was first introduced by Gunther von Hagens in 1980. There is also an ultra-thin sheet plastination technique, between 0.3 and 1 mm [34]. In general, the 3–5 mm sections are used for the descriptive and morphometric analysis, and the ultra-thin sections (<1 mm) are used for three-dimensional reconstruction or histological examinations [33, 35–39]. We describe an ultra-thin sheet plastination or microplastination

© The Author(s), under exclusive license to Springer Nature Switzerland AG 2023

N. E. Ottone, *Advances in Plastination Techniques*,
https://doi.org/10.1007/978-3-031-45701-2_6

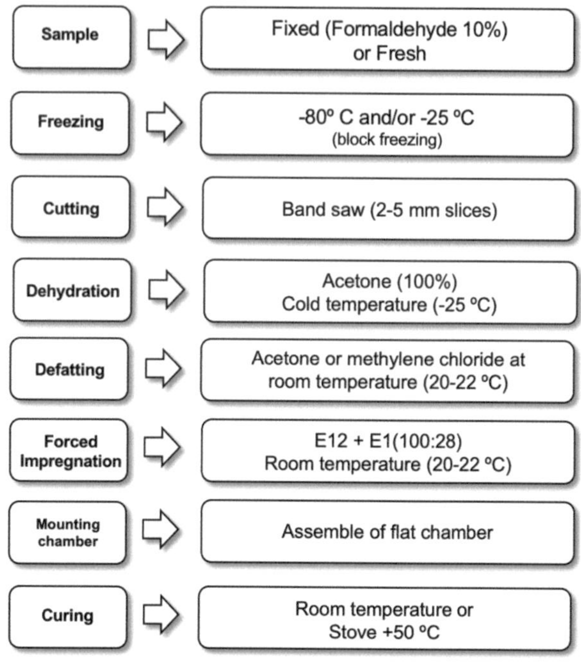

Fig. 6.1 Basic steps of the classical technique of sheet plastination with epoxy resin

technique [10], which achieves plastinated sections between 130 and 230 μm, being reduced still further with the use of polishers.

Table 6.1 provides three sheet plastination (E12) processes: the classic one, developed by Gunther von Hagens et al. [2], and then two techniques developed by one of the researchers, who has the largest number of publications on this technique, Prof. Constantin Sora, who described two protocols, for 3–5 mm sections [3], and a technique for obtaining ultra-thin sections (<1 mm) [15].

Slices of organs or transparent organs are for education and research purposes, as they permit the study of the topography of all the body structures in a state that is neither collapsed nor dislocated [4, 37, 40–45], preserving the anatomical position "in situ" of all the tissues for examination at macro- and microscopic levels [37, 46, 47].

In addition, the specimens are useful in preparatory training programs (computed tomography and magnetic resonance imaging), providing, from the student to the specialist, a much clearer understanding of the anatomical structures due to the deeper knowledge of the anatomy by sections in terms of morphological recognition and clinical diagnosis [3–5, 8–12, 24, 31, 37, 42, 48–71].

In the traditional version of this technique by von Hagens [1, 2], the fundamental steps of plastination are respected, but it requires the previous sectioning of the samples, since those will be the sections to be impregnated. On the other hand, in the ultra-thin sheet technique [15] or microplastination [10], the sample is impregnated, creating an epoxy resin block with the sample inside, and then this block will be cut with a diamond blade. Thus, returning to the traditional version, the sample must be placed for cutting inside a container, filling it with polyurethane foam or gelatin as well, which will enable the most suitable manipulation of the sample

Table 6.1 E12 sheet plastination technique general protocols (von Hagens et al. [2]; Sora and Cook [3]; Sora [15]) (modified and from Ottone et al. [9])

Sheet plastination of transparent slices, filling method with epoxy resin (E 12) (von Hagens et al. [2])		General protocol for epoxy slices (2–3 mm) (Sora and Cook [3])		Ultra-thin epoxy slice (<1 mm) protocol (Sora [15])	
Day 0	Freeze unfixed or fixed specimens	Day 0	Freeze specimen −75 °C	Day 1	Immerse specimens in n°1 −25 °C acetone bath (>90 °C)
Day 2	Saw into 2–5 mm slices	Day 1	Slice and clean sawdust from slices	Day 14	Check and record purity of acetone bath n°1; immerse in n°2, −25 °C acetone (100%)
Day 2	Stack between plastic nettings and grids; avoid thawing and drying	Day 2	Immerse specimens in n°1 −25 °C acetone bath (>90 °C)	Day 28	Check and record purity of acetone bath n°2; immerse in n°3, −25 °C acetone (100%)
Day 2	Dehydrate by freeze substitution with acetone followed by degreasing in methylene chloride	Day 5	Immerse in n°2 −25 °C acetone bath (100 °C). Check and record purity of bath n°1	Day 42	Degrease slices in MeCl at room temperature (RT) (minimum 2 weeks)
Day 10	Immerse in E 12/ E1, reaction mixture	Day 8	Immerse in n°3 −25 °C acetone bath (100 °C). Check and record purity of bath n°2	Day 56	Immerse in E12 resin-mix (+30 °C)
Day 11	Impregnate in vacuum to under 2 mmHg	Day 11	Degrease slices in acetone or MeCl at room temperature (RT)	Day 57	Impregnate in E12 resin-mix (+30 °C)
Day 12	Place slices onto glass plate; assemble flat chamber and fill	Day 18	Impregnate in E12 resin-mix (+5 °C or RT)	Day 61	Impregnate in E12 resin-mix, and increase temperature to +60 °C
Day 13	Cure at room temperature or at + 50 °C	Day 20 or 27	Cast or sandwich slices. Lay cast 15° from horizontal at RT	Day 62	Cure specimen +65 °C oven
Day 14	Open flat chamber	Day 21 or 28	Cure upright in +45 °C oven	Day 66	Remove block and slice when convenient
		Day 25 or 33	Open flat chamber or the sandwich, cover slice with foil, saw, and sand		

when cutting with the circular saw. The block must be cooled previously to ensure adequate cutting. Once cold, the block is cut, and the sections obtained must be immediately placed in 100% acetone (at −25 °C) for dehydration. Once the dehydration is complete, they must be defatted, which will eliminate the fat, ensuring greater transparency. This can be done with acetone or dichloromethane (more toxic component, and if used, it must be done under a gas extraction hood, with face protection). Once defatted, the sections must be placed in the epoxy resin and catalyst mixture (E12/E1), there being a myriad of protocols in relation to the mixture percentages and components used [8, 9]. This stage is done in a vacuum chamber at room temperature (20 °C), reducing the pressure from 760 to 10 mmHg, over 24 h. Finally, the curing stage is reached, which consists of the assembly of a curing chamber ("sandwich method"), formed by two glass plates and acetate sheets between which the sections are placed with a new E12/E1 mixture, and this "sandwich" is placed in an oven at 50 °C, ensuring the hardening and drying of the sections in 48–72 h. Table 6.2 lists the equipment, instruments, accessories, and chemical products used in the plastination techniques of sheets with epoxy resin.

Table 6.2 Equipment, instruments, accessories, and chemicals used in sheet plastination techniques with epoxy resin

Steps[a]	Equipments and accessories	Chemicals
Fixation[b]	Cannulas Basic dissection instruments Containers Injection pump	Formalin
Sectioning	Band saw Low-speed diamond band saw Deep freezer (−80 °C) Freezer (−25 °C)	Liquid nitrogen Gelatin Polyurethane foam
Dehydration	Freezer (−25 °C) Containers (HDPE plastic or stainless steel) Acetonometer/s Grids (plastic or metal)	Acetone
Defatting	Containers (HDPE plastic or stainless steel) Grids (plastic or metal)	Acetone Dichloromethane Extraction hood
Forced impregnation	Vacuum chamber (RT)[c] Vacuum pump Vacuometer Valves (vacuum control) Grids (plastic or metal)	Impregnation mixture (see Table 6.3)
Curing	Tempered glass (4–5 mm thickness) Acetate sheets Silicone gasket Clamps or paper clips Stove	Impregnation mixture (see Table 6.3)

[a]The order of the steps can vary according to the sheet plastination technique with epoxy resin implemented (see in this chapter)
[b]The sample can be fresh, without fixation
[c]There are protocols that use low temperatures (see in this chapter and Table 6.7). *RT* room temperature

Standard Method of Epoxy Sheet Plastination

Sectioning Block Preparation

To facilitate the cutting of the sample, it will be placed in a container (wood or cardboard), and then the sample will be covered with polyurethane foam, which hardens in 24 h (Fig. 6.2). Some also use gelatin to generate this section block [15, 22, 72–79]. After this, the entire section block, comprised of the container that holds the sample completely covered by the polyurethane foam or gelatin, will be placed in a freezer at a temperature between −25 and −20 °C for at least 2 weeks, followed by a 1-day deep-freezing process at −80 °C. This process of cooling the section block is fundamental, since it will prevent the block and above all the slice to be obtained from defrosting, which will make the cutting of the required measurement precise (up to 2 mm thick with a conventional saw) [8, 9].

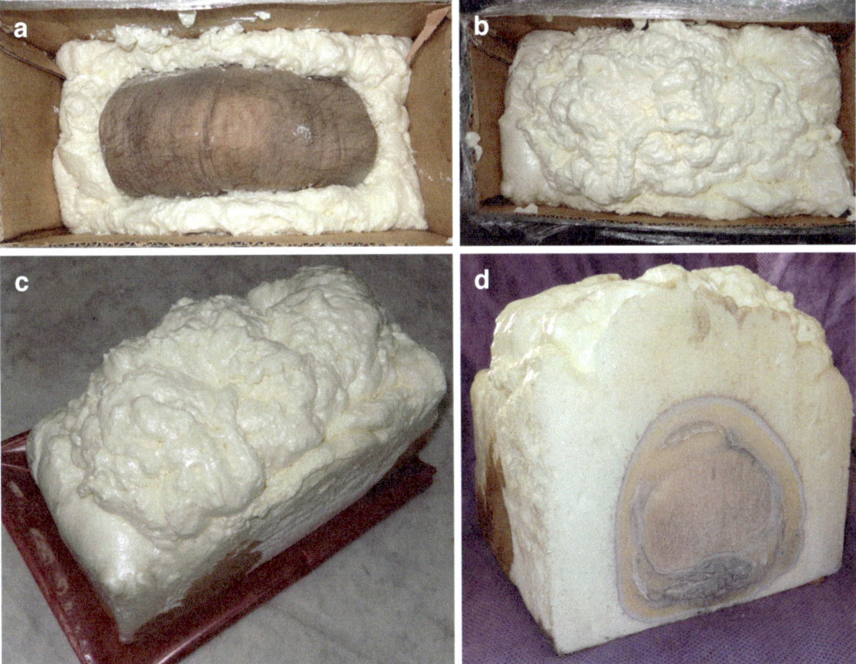

Fig. 6.2 Preparation of the polyurethane foam block for cutting a fixed sample (knee). (**a**) The elaboration of the block is shown in a cardboard box of appropriate measurements for the selected sample, which rests on a polyurethane foam mattress. (**b**) Complete coverage of the sample with polyurethane foam, pending expansion. (**c**) After 24 h at room temperature, the polyurethane foam hardens and must be placed in freezers at low temperatures (−25and −80 °C) to proceed with cutting (see details in the text). (**d**) Visualization of the sample in section

Sectioning the Sample

The polyurethane block will be extracted, with the sample inside, from the cardboard box to make the sections [8, 9]. The block must be placed on the cutting table, which will have a cutting guide adapted to determine the thickness. This guide must also be kept cold using liquid nitrogen, which will be brushed on the cutting guide as it advances with the block to be cut (Fig. 6.3). This, added to the frozen block, will ensure it does not defrost, the temperature on the cutting guide does not increase, and the cut can be made as precisely as possible. Moreover, the cutting machine must have a thin saw blade, with fine teeth, that can make thin cuts with the smallest amount of material lost when making the section. The goal is to make sections between 2 and 5 mm thick. Once the sections have been made, they will be cleaned on one side, where dirt from the cutting remains (with brush and the same cold acetone that soon will be used for dehydration), and then they must be placed immediately in pure (100%) and cold ($-25\ °C$) acetone to begin dehydration [8, 9].

Dehydration

Dehydration will consist of passing the sections through consecutive acetone baths of 100% purity, beginning with the first bath with an acetone more than 90% pure and continuing with purities of increased concentration, with the last bath being at 100%. It is recommended that the dehydration be done in cold, with a temperature between -25 and $-20\ °C$ being suitable [8, 9]. The appropriate volume is when the amount of acetone is ten times greater than the volume of the samples. However, experience shows that this proportion is not necessary and can be adapted to a suitable amount according to the sample size, using a container large enough to contain the samples and submerging them in acetone, covering them up to 4 cm from the sample closest to the surface. In addition, it is recommended that separation grills be used in order to keep the anatomical sections separate and organized, and to keep everything together, threads can be used to form a kind of "package" so all the sections can move at the same time, without being at risk of coming apart and ensuring their anatomical integrity (Fig. 6.4a–c). The acetone density must be checked daily throughout the dehydration process. A particular duration for this process cannot be guaranteed, but it is fundamental to know when the acetone is due to be changed and renewed. For that reason, daily control of the purity of the acetone with an acetonometer or alcoholometer is important. For this, to take the measurements, when it is stable (with the same purity) for at least 2 consecutive days, this is indicative of the need to change the acetone for one of greater purity than the previous one or 100 %. In general, four acetone changes are required to reach the purity of 99.5 % or greater. Some authors also indicate that this one must be 99 %. However, reaching these final concentrations (for us greater than or equal to 99.5%) for at least 2 consecutive days can already occur once the dehydration process is finished. The

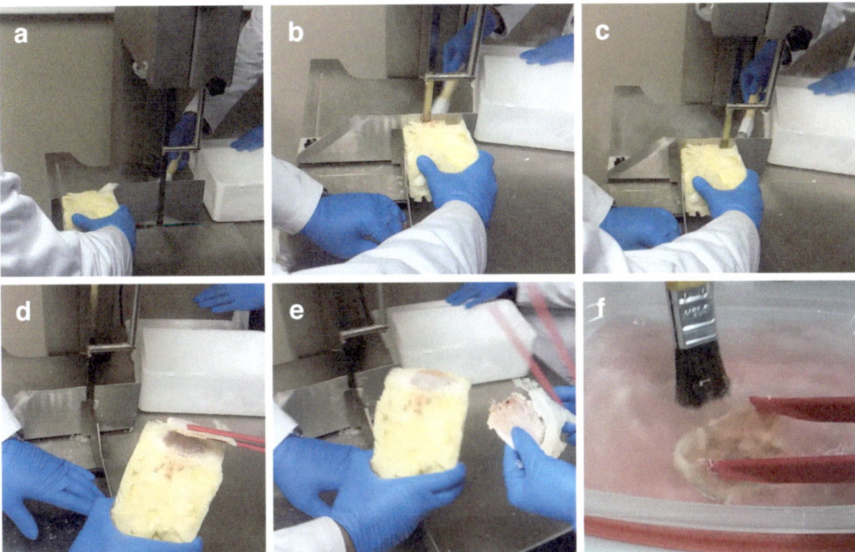

Fig. 6.3 Polyurethane foam block cutting process. (**a–c**) Positioning of the block on the cutting guide, which is brushed with liquid nitrogen to keep the cutting guide cold, thus avoiding heating the guide, which would prevent making thin slices, in this case, 2 mm thick. (**d, e**) Once the cut is finished, it is taken, and the surrounding polyurethane foam is removed. The frozen section is visualized, since the block is immediately extracted from the freezer (−80 °C) and the liquid nitrogen contributes to keep it cold. (**f**) The slices are immediately placed in cold acetone (−25 °C), and previously the cut must also be cleaned in cold acetone, especially on the side of the cut where dirt from the saw remains

dehydration process, in sections between 2 and 5 mm, will last 7–14 days, depending on the size and number of sections [8, 9].

Once dehydration is complete, it is essential to continue with the defatting process, which requires that the sections be placed at room temperature. A hundred percent acetone can be used (Fig. 6.4d); methylene chloride or dichloromethane (MeCl) can also be used [15]. This component has the advantage that at room temperature it is a powerful defatting agent, particularly because being immiscible in water, it favors the elimination of fat and above all it favors the transparency of the connective tissue, something essential in this epoxy sheet plastination process. In addition, given that its boiling point is 40 °C (vapor pressure), it is a highly appropriate compound to be extracted from the sample in the plastination process (forced impregnation), ensuring progressive and total penetration of the resin inside the sample. However, it is a highly toxic chemical component (as will be described in ultra-thin sheet plastination), and therefore its use and handling must be done safely, both for the environment and for the personnel (extraction hoods and protection masks with a built-in breathing filter). In this sense, it is essential to consider that sections of brain or samples that include the central nervous system must not be subjected to the defatting process because they will suffer a lot of shrinkage. In

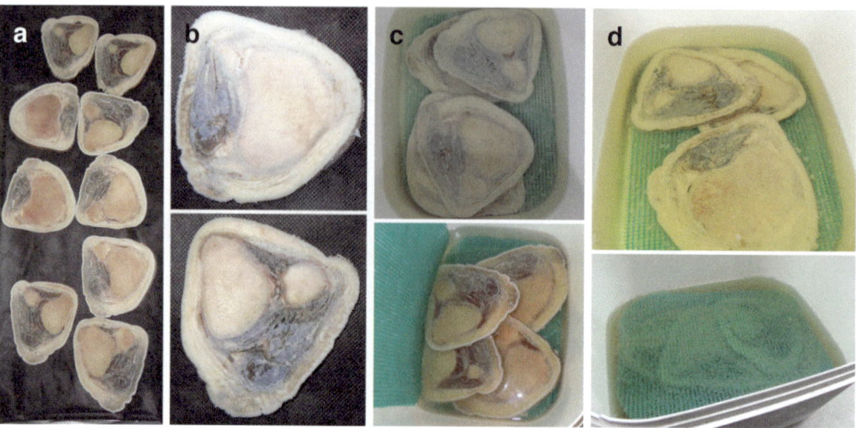

Fig. 6.4 Slices of human samples (knee) and their dehydration and defatting process. (**a, b**) Slices obtained from the polyurethane foam block, visualized in (**b**) the frozen samples thanks to the block cooling process and the cutting guide with liquid nitrogen. (**c**) Slices subjected to the dehydration process. (**d**) Slices subjected to the defatting process at room temperature, keeping the cuts separated with perforated grids

these cases, the dehydration process should only be done at low temperatures (−25 °C) [8, 9].

Forced Impregnation

Forced impregnation will be carried out for 24 h maximum, placing the samples in an epoxy resin mixture with its corresponding catalyst. When the mixture is ready, the polymerization process of the mixture begins [8, 9]. The classic proportion for this impregnation mixture is 100:28 (epoxy resin: catalyst) [1], but there are many proportions in which the samples can be submerged, as well as the use of other components (e.g., accelerants, although this component is usually used for ultra-thin sheet plastination). In addition, the forced impregnation process is done in a vacuum chamber at room temperature (20–22 °C). Therefore, once the samples have been placed in the impregnation mixture, the process must be left to rest, without subjecting it to vacuum pressure, for 24 h. After this time, the vacuum pump will be turned on to begin the forced impregnation stage. At this stage, an extraction of the acetone will be done, verified by viewing, through tempered glass, bubbles on the surface (Fig. 6.5), and this way, the impregnation speed and incorporation of resin into the samples can be controlled. At the beginning, the bubbling will be intense, and the bubbles will be small (Fig. 6.6). And as the process advances, i.e., as the pressures are reduced, larger and larger bubbles will be noted. The balance of

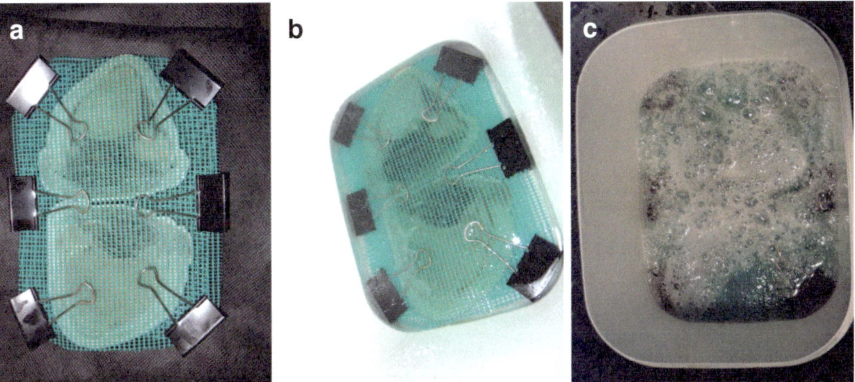

Fig. 6.5 Forced impregnation process in sheet plastination with epoxy resin. (**a**) Assembly and preparation of the package with samples between perforated grids and paper clips, which allows adequate handling of the samples, maintaining their anatomical integrity. (**b**) Placement of the package in the impregnation mixture, prior to the start of the forced impregnation process. (**c**) Intense and characteristic bubbling of the sheet plastination with epoxy resin

Fig. 6.6 Intense bubbling with large bubbles, characteristic of the final phases of the forced impregnation stage in plastination of sheets with epoxy resin

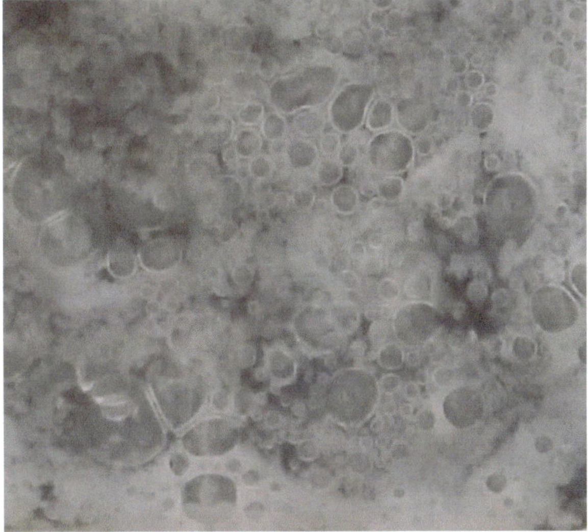

pressures, which will occur between the acetone extraction and the incorporation of the impregnation mixture, will reduce the pressure until reaching the final 5 mmHg. Once this pressure has been reached, and with no visible bubbles, this will indicate the end of the forced impregnation process and will mean that the section is fully impregnated. Other authors have developed a more prolonged forced impregnation process with the use of low temperatures: 5 °C [3, 80] and 0 °C [72].

Curing

Immediately after the forced impregnation process, the resin samples will be extracted, and the "sandwich"/"flat chamber" curing chambers will be assembled, within which the samples will be placed to carry out their polymerization (curing). The curing or polymerization of the samples takes advantage of tissue amines contained in the specimens from the impregnation with epoxy resin (E12) [23]. The action of these amines justifies their use in the technique, because they act as reaction accelerants, and by combining them with the hardener, made up of anhydrides, they reinforce the polymerization reaction (complete curing) of the mixture when they are subjected to heat (+ 50 °C, classic technique) [8, 9].

Assembling the curing chambers requires the following materials: rectangular glass plates 2–3 mm thick (the size of the plates will vary according to the samples to be cured), acetate sheets (the same size as the glass plates), and paper clips (Fig. 6.7).

An acetate sheet must be placed on one glass plate, on which a small amount of the epoxy-catalyst mixture (100:40) is poured. The sections are placed on the mixture in an orderly fashion, separated from each other by at least 15 mm. Once all the sections are placed, the epoxy resin-catalyst (100:40) mixture is poured on them again. Then, a new acetate sheet is placed over the sections, and a new glass plate is placed over all of it. The glass plates must be pressed to obtain a uniform distribution of the resin between the sections, until all are fully covered. Then, the glass plates must be adjusted with paper clips, which will be located on the sides of the glass plates [8, 9].

Once the curing chamber is assembled, it will be placed in the oven at 45–50 °C for at least 5 days, continuously (Fig. 6.8). After these 5 days, the curing chamber will be removed from the oven, and the chamber will be disassembled, undoing the clips and separating the glass plates from the sections contained between the acetate sheets. Then, the acetate sheets must be separated to access the sections totally embedded in resin, already hardened and dry (Fig. 6.9). If necessary, the excess

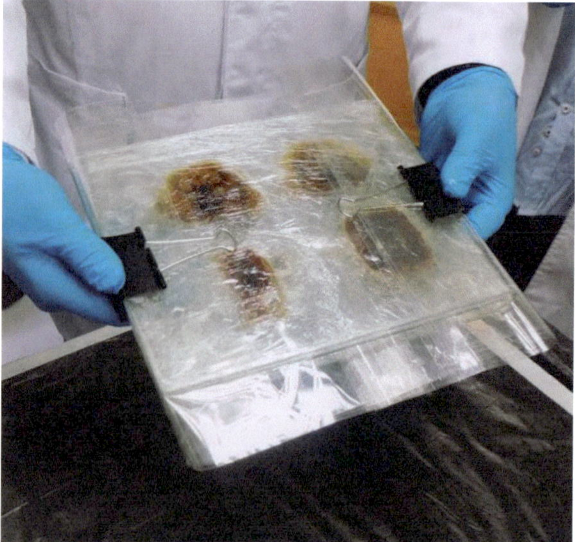

Fig. 6.7 Assembly of the curing chamber (sandwich type), with glass plates, acetate sheets, paper clips (see detail in text)

Fig. 6.8 Placement of the curing chambers in oven to develop the curing/polymerization stage

Fig. 6.9 Curing step of sheet plastination with epoxy resin. (**a**, **b**) Curing chambers inside the oven. (**c**) Polymerization completed and chamber ready for disassembly. (**d**) Slice (knee) completely plastinated in epoxy resin

resin between the sections can be eliminated by using scissors to cut the sections to their shape [8, 9]. Examples of final results of the technique are shown in Figs. 6.10 and 6.11.

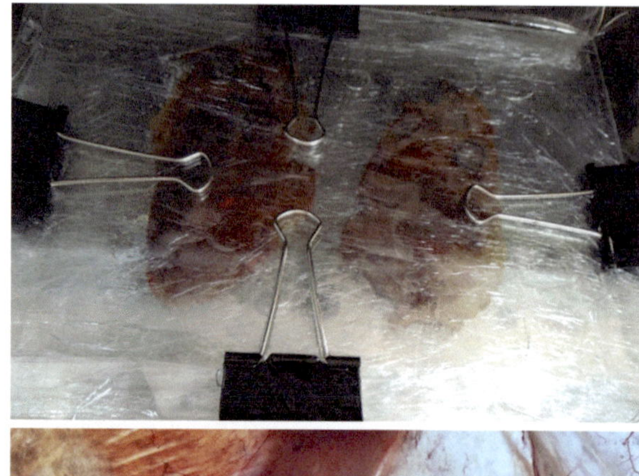

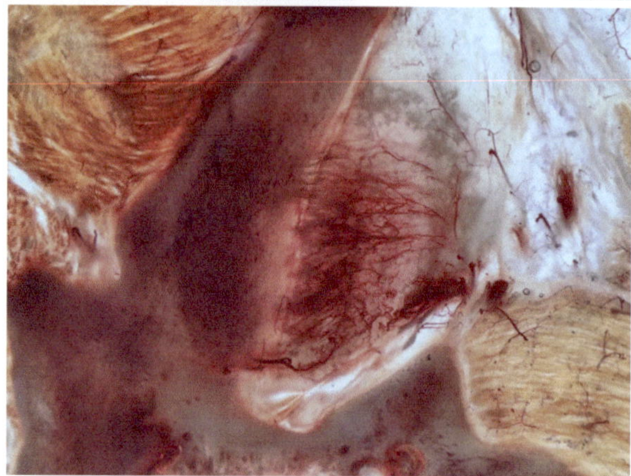

Fig. 6.10 Slices in the curing chamber prior to disassembly and enlarged view of the final result, corresponding to an area injected with epoxy resin from the fresh sample (Biodur E20), prior to starting the sheet plastination technique

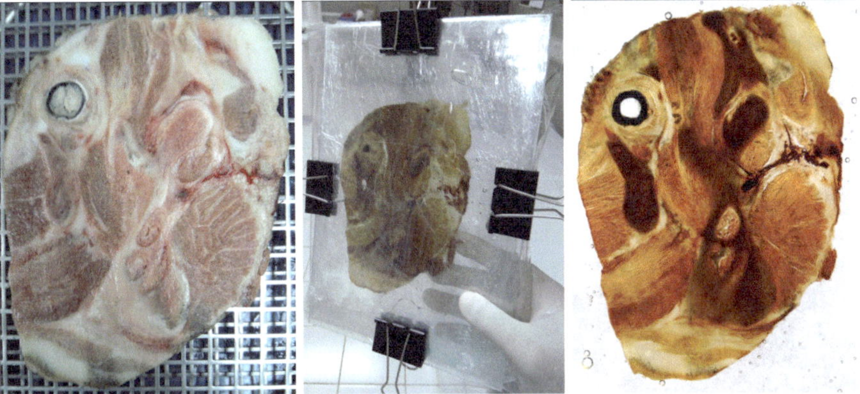

Fig. 6.11 Visualization of a fresh slice (pig's head), prior to sheet plastination with epoxy resin, the same slice inside the curing chamber, and the final result, after disassembling the curing chamber

Analysis of Various Epoxy Sheet Plastination Protocols

Generation of the Block and Its Sections

When opting for epoxy sheet plastination, which technique to use must first be defined, since it must be decided if the cadaveric sample will be sectioned, and then these sections will be plastinated or if, on the contrary, the cadaveric piece is to be plastinated in its entirety, from the block already impregnated in epoxy resin, to make the sections that will ultimately be subjected to curing. This is directly associated with the desired thickness of the sections. When ultra-thin (less than 2 mm) sections are desired, the whole piece will be subjected to the plastination technique to then make the ultra-thin sections with a diamond saw [6, 10, 11, 15, 16, 23, 44, 50, 81–125]. On the other hand, for 2, 2.5, to 5 mm sections, the cadaveric piece is cut after subjecting it to a very low temperature (-25 °C and then -80 °C), to then make the sections which will be subjected to the plastination technique [8–12, 17, 27, 37, 41, 42, 46–48, 72–75, 80, 82, 89, 90, 113, 121, 126–156]. With respect to data on the temperature, there are differences among researchers: Steinke [157], -35 °C; Porzionato et al. [158], -20 °C; Sora and Cook [3], -75 °C; Macchi et al. [159], -20 °C; Scali et al. [160, 161] and Steinke et al. [162], -85 °C; and Bernal-Mañas et al. [135], -20 and -80 °C.

In addition, before making the section from the anatomical block, as indicated previously, its inclusion in various components is recommended. For example, Soal et al. [43] described, like us, the use of polyurethane foam, whereas Liang et al. [74, 75] and Zhang and An [22] prepared the sectioning block with gelatin, like other authors [15, 72–79].

Already with the anatomical block preparation, it is necessary to proceed to sawing the sections, and for this it is important to consider the type of saw blade. In

1997, Alston et al. [163] analyzed the type of blade necessary to cut; they compared several circular saw blades, including a "standard" blade (10 teeth per inch) and also a saw blade identified as a "Shark Band," with fewer teeth per inch (three teeth per inch) and greater depth of separation between the teeth. One of the outstanding characteristics of the "Shark Band" blade is that the sawdust generated when cutting is eliminated more easily from the surface of the slices obtained, presenting a smoother and cleaner appearance than the slices obtained with the standard blade. On the other hand, the disadvantage of the "Shark Band" is that the greater quality of the section obtained occurs at the expense of the working life of the blade, which is shorter than with standard blades.

To obtain the thin slices (2–5 mm), Fasel et al. [45] added liquid nitrogen by injection, while the sample was being processed in the saw, thereby cooling the saw blade and the cutting guide on which the sample was supported to make the section. It is also important to have acetone at −25 °C to clean the sections of the sawdust produced by the saw and immediately place them for dehydration. The use of liquid nitrogen is crucial to keep the sample from thawing while it is being cut [8, 9].

Dehydration

In relation to dehydration, most authors do this at a temperature of −25 °C [3, 6, 8–12, 27, 33, 39, 41–44, 72, 73, 80, 81, 84, 88–90, 98, 99, 107, 119, 120, 126–129, 131, 132, 137, 139, 143, 146, 158–161, 164–168] with the exception of Zhang and Lee [82], Leaper et al. [46], Chen et al. [133], Liu et al. [134], Liang et al. [74, 75], Thorpe Lowis et al. [47], and Zhang and An [22], who do this at −30 °C and Eckel et al. [111, 169] who do dehydration at −23 °C. On the other hand, Pavkov et al. [145] and Aigner et al. [170] do dehydration at −20 °C. In every case, the acetone volumes at which dehydration occurs are greater than 90% at the beginning until reaching the 100% needed for the last acetone bath (see Table 6.7). There are only a few authors who did not indicate the dehydration temperature [37, 82, 83, 86, 87, 130, 135, 171, 172], which supposes that, according to the references mentioned in their respective publications, mainly with reference to the publication by Prof. Gunther von Hagens, it is done at −25 °C (see Table 6.7).

In general, in terms of the dehydration times, the thinner the sections, the less time the dehydration step requires. For example, Sittel et al. [171, 172] made larynx sections between 4 and 80 μm thick, with the dehydration process being 48 h and 2 h of defatting at room temperature with MeCl. Soal et al. [43] made 1 mm sections (in specimens longer than 150 mm), and these were dehydrated in 10 days. In addition, we can see in the work of Sha et al. [81], who made 1.2 mm ankle sections, that these were dehydrated in a total of 35 days. On the other hand, in the works of Qiu et al. [37] and Qiut et al. [173], in 4 weeks they dehydrated sections of 1 mm and 700 μm. However, Porzionato et al. [158] made sections between 2 and 3 mm, and the dehydration was done in 2 weeks. The remaining researchers took between 4 weeks and 4 months to complete the dehydration process (see Table 6.7).

By contrast, Steinke [157] added 8 % peracetic acid acetone to produce a mild whitening during freeze substitution (according to a personal recommendation by von Hagens).

Defatting

The defatting stage makes it possible to eliminate the fat in the specimens at room temperature. At this stage, there are authors who, taking advantage of the defatting power of acetone, defatted their specimens with 100% acetone [43, 44, 46, 47, 59, 72–75, 82, 133, 134, 157, 167]. Other authors chose MeCl as the defatting agent, known for its powerful characteristics, but with greater toxicity than acetone [6, 33, 39, 42, 80–90, 120, 126, 128–131, 139, 144, 165, 166, 171, 174, 175]. On the other hand, in one study the choice was made to combine the use of acetone and MeCl, always beginning with acetone, as a continuation of the previous dehydration process [37]. There were studies in which acetone or MeCl was used interchangeably and with no further indications [3, 158]. In every case, the defatting was done at room temperature, with these temperatures varying between 15 and 24 °C. Cook and Al-Ali [37] indicate that the best optical quality obtained in the sections occurs when the defatting is done at room temperature, with acetone first and then with MeCl, to increase the extraction of lipids.

Forced Impregnation

This is the crucial stage in the development of the different plastination techniques and is the one in which resin is incorporated into the samples with the extraction of acetone in the form of gas using the action of a vacuum pump. The mixture in which the samples are submerged is an epoxy resin, catalysts, and accelerants, mainly produced by Biodur, in their different versions (see Table 6.3). In some studies, the use of epoxy resin and its corresponding catalysts and accelerants were mentioned, without naming the commercial brand [37, 85–88, 160, 161]. With respect to the forced impregnation process times, the authors who performed this stage in 24 h predominate [43, 46, 72–75, 82, 90, 171, 172]; these are followed by authors who, as classically described from the works by von Hagens [1, 2], perform the process in 48 h [37, 44, 47, 131, 159, 161]. There are other authors who perform the process over 2 weeks [88, 89], agreeing with the studies in relation to dealing with fetal pelvis samples and making ultra-thin sections of between 300 and 700 μm with a diamond saw. The need to perform a slower than normal forced impregnation in this case to avoid shrinkage of the fetal tissues can be understood. Eckel et al. [6] make similar indications, performing the forced impregnation process over 10 days, corresponding to larynx sections, in adult specimens, of 0.8 mm. It is the same in this case, performing ultra-thin sections and the need to avoid tissue shrinkage. Wegman

Table 6.3 Forced impregnation mixtures (modified and updated from Ottone et al. [8])

Products of mixture	Mixture proportions	Authors
Biodur® E12/E6/E600	100:70 pbw, 0.15%	Eckel et al. [6]
	100:50 pbw, 0.2%	Sora [15]; Sora et al. [33]; Sora and Matusz [34]; Kürtül et al., [44]; Adds and Al-Rekabi [18]; Vargas et al. [11]; Texeira et al. [190]
	100:50 pbw, 0.1% no information	Koebke et al. [143]; Elsner et al. [120]; Zhang and Lee [82] Zhang et al. [83]; Kaulhausen et al. [154]; Chen et al. [133]; Liu et al. [55]; Arredondo et al. [125]; Thorpe Lowis et al. [47]; Adds et al. [190]; Thorpe Lowis et al. [136]; Liang et al. [190]; Singh et al. [76]; Wu et al. [191]; Shi et al. [192]; Ma et al. [193]
Biodur® E12/AE30/ AE10/E1	95:5:20:26 pbw	Sora and Cook [3]; Cook and Al-Ali [37]
Biodur® E50/E7/AE10/ E700	20:16:5 pbw; 0.1%	Sittel et al. [171]; Qiu et al. [172]
Biodur® E12/E6/E500	100:50 pbw, 0.1%	Sha et al. [81]; Qiu et al. [84]; Sora et al. [173]
Biodur® E12/E6	72–18%	Windisch and Weiglein [186]
Biodur® E12/E1/AE10	95:26:10 pbw	Sora, Brugger and Strobl [174]; Elnady and Sora [131]
	100:28:10 pbw	Hammer et al. [194]
	No information	Sora et al. [80]; Sora and Genser-Strobl [128]; Eckel et al. [174]
Biodur® E12/ E1/AE10/AE30	100:28:20:5, pbw	Zhang and An [22]; Phillips et al. [79]; Zhang and Lee [82]; Leaper et al. [46]; Nash et al. [73, 146]; Diao et al. [122]
	95:26:20:5, pbw	Basa et al. [77]
Ph	69.6:19.4:13.9:3.4 pbw	Soal et al. [43]; Liang et al. [74, 75]
	No information	Xu et al. [26]; Wegmann et al. [167]; Liugan et al. [195]; Xu et al. [196]; Zwirner et al. [197]
Biodur® E12/E1/AE20		Bond et al. [198]; Shi et al. [192]
Biodur® E12/E1	No information	Fritsch [126]; Fritsch et al. [89, 127]; Macchi et al. [159]; Sora et al. [129]; Al-Ali et al. [130]; Villamonte-Chevalier et al. [199]; Konschake and Fritsch [90]; Steinke et al. [200]; Stegman et al. [78]; Pieroh et al. [201]; Noriego et al. [202]
Epoxy resin + catalyst	No information	Cook [179]; Sebe et al. [85–87]; Lunacek et al. [88]; Koslowsky et al. [165, 166]; Scali et al. [160, 161]
Epoxy resin (YD-128) + catalyst (C −403) (20:1)		Starchik et al. [26]
Commercial epoxy resin pre-accelerated (without hardener)		Ottone et al. [19]

References: [*E12, E50* epoxy resin; *E1, E6, E7* catalyst; *AE3* glass separator (for the separation of the cured specimen from the glass walls of the flat chamber); *AE10* plasticizer (for epoxy resins); *E500, E600, E700* epoxy accelerators] et al. [167] applied forced impregnation for only 10 h for lower limb sections of between 3 and 4 mm thick. The temperature at which this process is performed, from the classic descriptions by von Hagens [1, 2], is at room temperature. Nevertheless, researcher Mircea-Constantin Sora, in his works [3, 4, 16, 33, 34, 41, 42, 80, 128, 129, 174, 176], performs the forced impregnation process at 5 °C for sections from 1.6 to 3.8 mm. There are other researchers who apply lower temperatures, from 0 °C [46, 47, 72–75, 82, 133, 134]. Scali et al. [160, 161] used temperatures between −8 and 0 °C. On the other hand, Sora et al. [33] made ultra-thin sections (1 mm) with a diamond saw from the block plastination of ankles, the forced impregnation of which was done in a vacuum drying oven at 30 °C; the sections were made after curing.

In relation to forced impregnation, Reed [177] conducted experiments in which E12, E1, and AE30 were combined in different ways, determining the levels of bubbling and final hardening of the mixtures. Thus, it was identified that the first bubbles produced appear with a slight decrease in pressure and come mainly from air trapped in the mixtures. The second appearance of bubbles occurs when the pressure gauge reaches 103 mmHg, which are bubbles from the epoxy resin (E12). And the third generation of bubbles occurs when 10 mmHg is reached. These last bubbles come from the hardener (E1). In Table 6.3 are shown various possibilities of combination of the impregnation resins.

Latorre et al. [177] conducted an interesting study on this technique, and essentially they propose that forced impregnation be performed only with epoxy resin and without hardener. The aim of this work was to define whether this modification to the impregnation mixture could result in a decrease in the acquisition of a yellowish tone in the plastination sections days after having obtained them. They also wanted to verify if the resin mixture alone (without hardener) could remain longer without catalyzing, i.e., in a liquid state so as to be fit for reuse. And in addition, they sought to demonstrate if the casting time of the sections could be extended indefinitely after being impregnated with epoxy resin and thus obtain better quality sections. They managed to demonstrate a reduction in the acquisition of a yellowish tone by the plastination sections, extending the usage time of the epoxy resin for not being mixed with the hardener, as well as the completion time of the sections, giving the researchers more time to finish assembling the curing chambers to finish the sheet plastination process.

Curing

The curing stage consists of the polymerization of the epoxy resin mixture with its corresponding catalyst, seeking to obtain its final hardening to obtaining sample slices. Curing is classically done in an oven at 45 °C [1, 2] and consists of introducing the slices into curing chambers built with glass and acetate sheets and in which the epoxy-catalyst resin mixtures (E12-E1, respectively) according to von Hagens [1] are placed to achieve hardening and drying with the heat. The curing chambers can be assembled as a "flat chamber," i.e., with a silicone gasket between two glass plates held together with a paper clip, creating a pressurized compartment, inside which the epoxy resin and catalyst mixture is placed and then the tissue slice is already fully impregnated, and which this curing stage will seek to polymerize [8–10, 12, 41, 42, 80, 89, 90, 126, 128–131, 171, 172, 174, 178–180]. The other method for curing the slices is the "sandwich" method, which consists of positioning the sample sections one on top of the other, separated by acetate sheets, and the entire contents between two glass plates. This way, in the sandwich method, several tissue sections can be cured at the same time [3, 8–10, 43, 46, 72–75, 82, 157, 160, 161].

Borzooeian and Enteshari [179] proposed a silicone gasket, designed with a steel center covered by 1.5 mm silicone. This produced sections that had smooth, square edges and a rectangular shape, which ensured a simpler casting, favoring the positioning of the specimens, due to the absence of glass, and also facilitating the removal of the bubbles.

With respect to curing times, Konschake and Fritsch [90] subjected the samples to only 2–3 days as a minimum, whereas other researchers subjected them to the heat of the oven for a maximum of 2 weeks [46, 81–84, 173]. The temperatures were usually around 30–65 °C (see Table 6.7).

On the other hand, there is also the possibility of plastinating the entire sample with epoxy resin, generating an epoxy resin block and then cutting this block [6, 34, 44, 81, 84, 165, 166, 173]. Sora et al. [33] made this modification to the technique, but also developing the forced impregnation process within a vacuum oven, at 30 °C, ultimately managing to obtain ultra-thin (<1 mm) sections as is described next.

Ultra-thin Sheet Plastination [23]: Microplastination [10]

Epoxy sheet plastination can be divided into the thin technique, where the sections have a thickness of up to 2 mm, and an ultra-thin technique, where the sections have a thickness less than 2 mm, using diamond blade saws in this case (Table 6.4). Then, the sections can be scanned to perform a detailed analysis of the microstructure, allowing morphometric quantification as well as the possibility of implementing a three-dimensional reconstruction. We define the concept of microplastination, an ultra-thin sheet plastination technique, to obtain ultra-thin sections, less than 250

Table 6.4 Steps of the microplastination technique

Steps	Temperature (°C)	Times (days)
Dehydration	−25	10
Defatting	22 (room temperature)	4
Forced impregnation	30–65	5
Cutting (200–250 µm)	22 (room temperature)	12
Assembly and polishing (50–100 µm)	22 (room temperature)	2

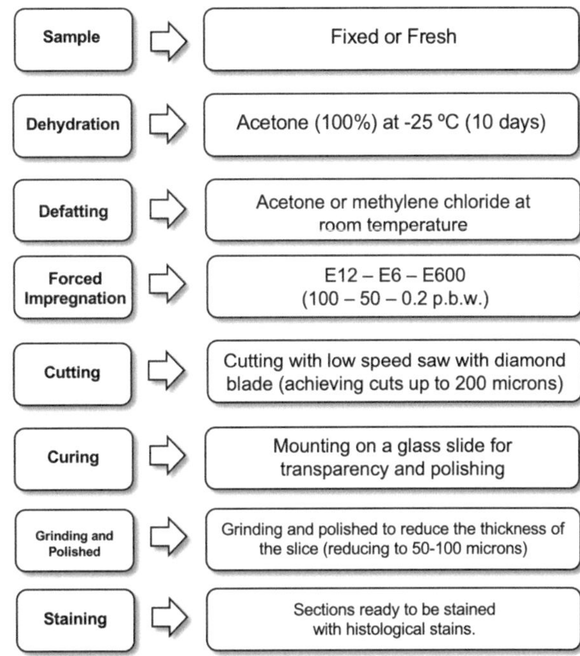

Fig. 6.12 Basic steps of the microplastination/ ultra-thin sheet plastination

μm thick, for the identification and visualization of microscopic characteristics of any anatomical region or organ applied in morphology and pathological experimental protocols [10]. Figure 6.12 shows the steps of microplastination/ultra-thin sheet plastination.

Stage of the Samples

Samples can be fresh or fixed; formalin is the traditionally used fixative, in a final concentration from 5% to 10%, reaching this point after increasing fixations of 2%, 3%, 4%, 5%, and 7%.

Dehydration

Dehydration of the samples must be done in 100% acetone, and to avoid tissue shrinkage, it is recommended that it be done at temperatures below −25 °C. The time will depend on the sample size, and the purity of the acetone should be verified daily with an acetonometer, which may indicate renewal of the acetone every 4 days. When the acetone concentration remains constant, with a value equal to or greater than 99% for 2 consecutive days, the dehydration stage can be considered complete.

Defatting

Once dehydration is complete, the samples must be defatted to promote greater transparency in the tissues. In this technique, the ideal procedure is immersion in methylene chloride or dichloromethane (MeCl) at room temperature; being a compound immiscible with water, it is not a dehydrating fluid, only a defatting agent, thereby favoring the elimination of fat and benefitting the transparency of the connective tissue. In addition, as the boiling point of MeCl is 40 °C (vapor pressure), it is a highly appropriate compound to be extracted from the sample in the plastination process (forced impregnation), ensuring progressive and total penetration of the resin inside the samples. In this sense, it is a powerful defatting agent. A defatting time of between 1 and 3 weeks is recommended, depending on the sample size. However, it is highly hazardous to health, and its handling requires strict safety measures, since its unsafe handling can cause headaches, dizziness, nausea, feeling of intoxication, and irritation in the eyes, nose and throat. Prolonged contact with the skin can cause irritation and even burns. Therefore, the use of safety masks and handling of the MeCl under an extraction hood are recommended.

Due to these disadvantages of MeCl, the use of acetone as a defatting agent is also recommended, rather than MeCl, also at room temperature. In this case, the defatting process can be extended to 2–5 weeks depending on the sample size.

Forced Impregnation

The forced impregnation process will be done in a vacuum chamber oven (AccuTemp-09s, Across International) at temperatures between 30 and 65 °C (Tables 6.5 and 6.6) (Fig. 6.13a). The reason for performing the impregnation at high temperatures is that it increases the fluidity of the impregnation mixture, and this facilitates its penetration within the specimen.

Table 6.5 Forced impregnation pressures and temperatures

Days	Pressure (mmHg)	Temperature (°C)
1	760	22
2	760–20	30–65
3[a]	760–20	65
4[b]	760–20	65
5	760	22
6	"Remove the blocks for cutting"	

The pressure reductions are gradual during the day, checking every half hour for the presence or absence of bubbling, which determines the maintenance or reduction of pressure, respectively
[a] On the third day, the mixture of epoxy resin, catalyst, and accelerator must be renewed
[b] On the fourth day, when 20 mmHg was reached, the bubbling ended (the acetone extraction ended), so the impregnation of the sample is considered complete

Table 6.6 Procedure to calculate the proportions of the components for the mixture of the "Forced Impregnation" step

Forced impregnation mixture
E12, epoxy resin (100 pbw); E6, catalyst (50 pbw); E600, accelerator (0.2 pbw)
1. $100 + 50 + 0.2 = 150.2$
2. $E12 = 100/150.2 \times 100 = 66.6\%$
3. $E6 = 50/150.2 \times 100 = 33.3\%$
4. $E600 = 0.2/150.2 \times 100 = 0.1\%$
5. For example, to prepare 1000 g of mixture E12:E6:E600 (100:50:0.2 ppp):
– 66.6% of 1000 g (666 g of E12)
– 33.3% of 1000 g (333 g of E6)
– 0.1% of 1000 g (1 g of E600)

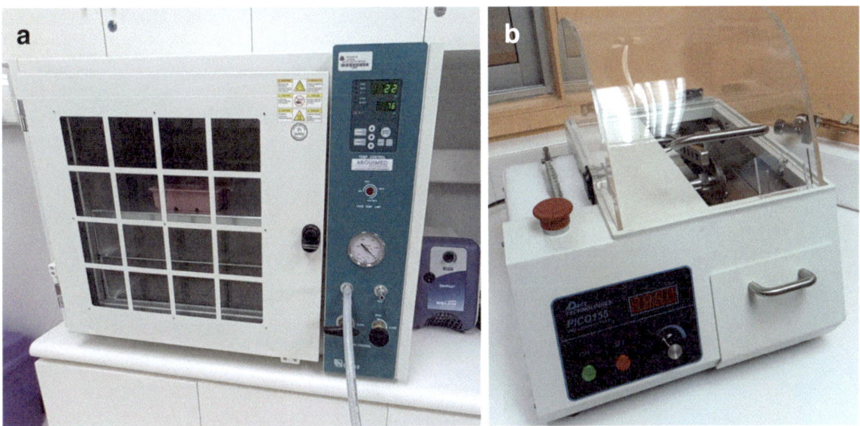

Fig. 6.13 Microplastination/ultra-thin sheet plastination equipments. (**a**) Vacuum chamber oven. (**b**) Low-speed band saw with diamond blade

The impregnation mixture will consist of three components: epoxy resin (100 pbw) (E12 Biodur); hardener (50 pbw) (E6, Biodur); and catalyst (0.2 pbw) (E600 Biodur).

Epoxy resin will be the main component in quantity. As can be seen in the previous distribution, its proportion in the mixture is 100 parts, in grams. Over time it tends to precipitate so that if crystals of precipitate are observed, their dissolution using the oven at 50 °C is recommended. On the other hand, the hardener is added in half the proportion of the resin (50%) corresponding to 50 parts, also measured in grams. Finally, the catalyst must be added to the impregnation mixture, together with the resin and the hardener, in an amount corresponding to 0.1% of the total volume, in cm^3, and a syringe or micropipette can be used for its incorporation. Some researchers incorporate it in a proportion that goes from 0.2% to 0.5%. From the proportions indicated, first the epoxy resin must be mixed (e.g., 100 g) with the hardener (according to the example, 50 g), and then the catalyst is added with a syringe or micropipette (0.1 mL).

The impregnation mixture will be placed in a container of a size suitable for the later positioning of the samples (Fig. 6.14). If dealing with small samples, a plastic,

Fig. 6.14 After the dehydration step, the samples are submerged in the impregnating mixture in a container of suitable sizes

Fig. 6.15 Visualization of the acetone extraction bubbles during the forced impregnation stage at high temperature (60–65 °C)

square container could be used, similar to an ice-cube tray. To avoid the samples coming into contact with the bottom of the container, grills will have to be placed at the bottom. Then the samples will have to be added, and they will also be covered by the same grills to keep them from floating. It is also necessary to place weights to keep the samples submerged in the impregnation mixture. Finally, it is recommended that the container be covered with cling film, perforated in three or four places.

Beginning of the forced impregnation process: Once in the impregnation mixture, the samples will be left for 24 h at 760 mmHg and at room temperature (day 1). After this time, the forced impregnation will begin at 30 °C. The entire forced impregnation, which will last 5 days, can be seen in Table 6.1 with some special details. The second day, as impregnation begins, at the fifth hour of impregnation, the temperature will be increased to 60 °C, reaching a final pressure of 18 mmHg. This process can be defined as a forced impregnation at high temperature. Should the bubbling continue (Fig. 6.15), impregnation will continue on the third day, at which point the resin will be changed for a new mixture and the pressure will be

Fig. 6.16 Epoxy resin
block with the sample
inside, obtained after the
forced impregnation
process at high temperature
(microplastination), which
will later be subjected to
cutting

reduced again to 18 mmHg. On the fourth day, if the bubbling ends, at a maximum
pressure between 2 and 5 mmHg, the forced impregnation process can be considered
complete. To finalize this stage, on the fifth day, the samples will be left for 24 h at
room temperature, and the pressure will again reach 760 mmHg. On the sixth day,
the samples can be removed from the vacuum chamber. This forced impregnation
process is finally combined with the curing/polymerization of the mixture, through
which the impregnation mixture hardens, forming a square epoxy resin block with
the sample inside, without requiring specific extra days for the curing (Fig. 6.16).

Cutting of the Blocks

The blocks must be cut with a low speed band saw (Fig. 6.13b), with a diamond
blade of 0.35 mm thick, which will create ultra-thin cuts between 130 and 230 μm
thick (Fig. 6.17). The sections obtained will not be transparent due to the dirt
adhered to the cuts from working the diamond blade (Fig. 6.18). To solve this, and
to obtain the transparency of the sections, a curing stage of the ultra-thin sections is
included. Thus, the sections must be placed in an E12:E1 (96:26 pbw) mixture
between two plates of glass and acetate sheets, which make up the curing chamber
(Fig. 6.19). It is suggested that the curing be done in the same oven with a vacuum
chamber so that while the sections harden and become transparent, the smallest
bubbles that are very difficult to see with the naked eye can be extracted. This process must be done at an increasing temperature until reaching 50 °C, which will
allow the resin to harden and the total transparency of the sections.

Fig. 6.17 Epoxy resin block subjected to the cutting process on a low-speed band saw with diamond blade

Grinding, Polishing and Slice Preparation for Histological Stains

Once the slices were obtained, they were opaque due to the passage of the diamond blade; in this way it is necessary to achieve their transparency and a thinning of the cuts to subsequently carry out histological staining. In this way, a grinding and polishing process is developed. To do this, the cut obtained must be mounted on a slide, and for this, a mixture of E12:E6:E600 (100:50:0.5, pbw) will be used as glue, which will be placed on the slide, and subsequently the cut will be applied on the

Fig. 6.18 Epoxy resin slices between 250 and 200 µm. The slices look opaque as a result of the cutting process. The slices must be mounted, not only to achieve transparency but also to reduce thickness by grinding and polishing

Fig. 6.19 Epoxy resin slice mounted on a slide, on a mixture of epoxy resins and catalyst (see detail in text), achieving the characteristic transparency

glue. The cut will be kept attached to the slide, to favor sticking, with a tweezer and placed in an oven at 60 °C for 4 h to speed up drying and gluing. Once this is done, you can proceed to polish the slice already adhered to the slide, from its free surface. To do this, six wet sandpapers (grades 240, 360, 600, 800, 1000, and 1200) will be used consecutively, under abundant irrigation with water, to achieve the necessary transparency and thinness. In this way, with this process it will be possible to reduce the thickness of the cuts, reaching between 100 and 50 µm. Each specimen is given a final touch using a P5000 silicon carbide disc (TrizactTM, 3M) and glycerin on an EcoMetTM 30 (Buehler) hand-operated polisher. After polishing each slice, the surface is sequentially treated with acetone, 30% hydrogen peroxide, and 10% EDTA. Following this, the samples can be stained using different histological stains, like Hematoxylin and Eosin, 0.5% Toluidine blue, or Goldner's Trichrome, among others. In sequence, the stained samples are immersed in descending concentrations of alcohol (from 50% to 100%), followed by the application of a neutral mounting solution (Entellan® new, Merck) and topped with glass cover slides [181, 182]. The ultra-thin sections visualized and scanned using the TissueFAXS i PLUS Cytometer TissueGnostics Axio Observer 7 System Carl Zeiss GmbH (TissueGnostics GmbH, Vienna, Austria) (ANID FONDEQUIP EQM200228) [183]. Fluorescence images can be obtained at 20x magnification with TissueFAXS 7.1.139 software. Acquisition objective: EC Plan-Neofluar 20x/0.50 M27 (20x, Air) [183]. Finally, in these epoxy resin microplastinated samples, endogenous autofluorescence is visible when excited with 488 nm light, revealing collagen, elastin, myofilaments, and neurofilaments (Fig. 6.20).

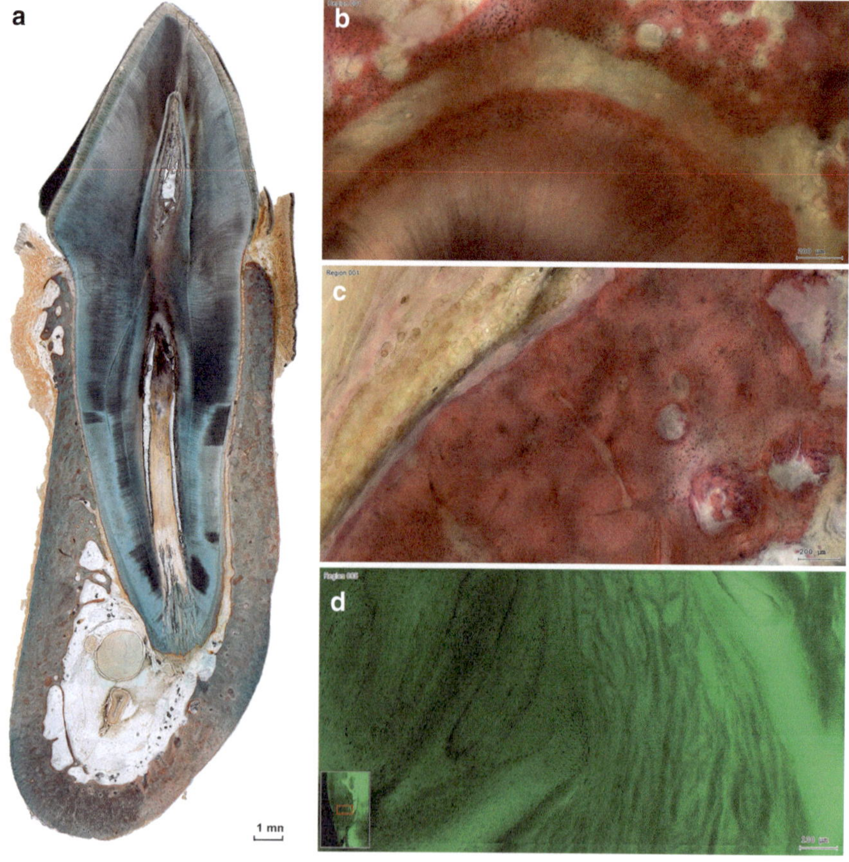

Fig. 6.20 The figure displays the outcomes of the micro-plastination technique in mandible blocks with teeth [181–183]. (**a**) Complete ultra-thin and microplastinated slices of the mandible and tooth region is shown. (**b**) and (**c**) Micro-anatomical visualization of the periodontal elements at the histological level after histological stain. (**d**) Appreciation of the endogenous autofluorescence of the same periodontal structures when excited at 488 nm

Details Related to the Ultra-thin Sectioning Technique

Fritsch [126] used a saw with diamond blade (Well@, Mannheim, Germany) to make sections of 300–500 μm of feet from newborns, which also underwent histological techniques, staining with methylene blue, blue IU, and contrasting with basic fuchsin [89, 126, 127].

Steinke [157] made adaptations to the diamond blade saw by cooling it with CO_2 liquid gas in addition to a continuous cleaning of the blade during cutting. A spout was fixed under the saw table and connected to a bottle of CO_2 and liquid nitrogen. This allowed the researcher to make sections of the human body at 800 μm.

Soal et al. [43] developed the ultra-thin sheet plastination technique by generating a block of tissue embedded in polyurethane foam, producing 1 mm slices with a less than 2% loss of tissue. In this technique, tissue shrinkage did not exceed 2% in each linear dimension, producing a maximum 6% potential loss of volume [43, 174]. However, the technique of ultra-thin slices by Soal et al. [43] requires decalcification of the bone due to the use of commercial saws, because otherwise the saw blade would be affected; moreover, the saw blade must be sharpened regularly.

In the literature, the work by Sittel et al. [171, 172] stands out for obtaining the thinnest sections (10 μm) in larynx samples compared to what has been reported by other authors: Fritsch [126], 300–500 μm; Sora [15], 400–500 μm; Eckel et al. [6]; Steinke [157], 800 μm; and Soal et al. [43], 1 mm. All the studies were based on normal anatomical samples. However, there are no major explanations in relation to the work by Sittel et al. [171, 172] regarding the sectioning, other than being larynx samples, which supposes a greater ease during the cutting process. The technique developed in this proposal makes it possible to obtain ultra-thin sections between 130 and 230 μm as a real option for the development of experimental morphological studies. Later, it will be necessary to polish these sections to reduce the thickness of the sections even further and this way be able to obtain the application of histological stains, immunofluorescence, and/or immunohistochemistry, something not yet very developed in this area.

The thin slices are more easily impregnated with resin than large tissue blocks, which may require higher temperatures (30–60 °C) during the treatment [33, 80]. But to achieve sections less than 1 mm thick, the complete block of the sample must be impregnated, before making the sections and doing so, as described previously, at a high temperature (during the 4 first days of impregnation, the temperature is 30 °C, and on the fifth day, it is raised to 60 °C), since at these temperatures the polymer becomes sufficiently fine as to deeply penetrate the block of the sample [33]. Ottone et al. [12] describe another protocol, as indicated previously, in which the complete impregnation can be done in 3 days, beginning on the first day with a temperature of 30 °C, but taking it to 60 °C when the third hour of forced impregnation is reached. On the following days, the 60 °C is maintained, and finally on the third or fourth day, the final pressure of 5 mmHg is reached without the presence of bubbles (absence of acetone extraction), completing the forced impregnation.

In general, the plastination techniques for ultra-thin sections require the initial impregnation of the complete sample, generating a resin block that will then be sectioned. Nevertheless, there are authors who make the sections at the beginning, as in the traditional plastination technique (3–5 mm) [43, 165, 166, 171, 172].

There are authors who consider that with the sections 1 mm thick they can, on their own, make suitable histological examinations, with or without staining [33, 38] or create good-quality three-dimensional reconstructions [33].

It is crucial to indicate that one of the great advantages of ultra-thin sheet plastination is that the sample does not require decalcification prior to cutting [6, 8, 10, 18, 33, 39, 41, 47, 84, 184].

Ultra-thin plastinated sections can be analyzed using low-power light microscopy or, even as in our case, to apply the use of tissue cytometers, fluorescence microscopes (TissueFaxs i Extra, of TissueGnostics GmbH) (ANID Fondequip EQM200228, 2020). We can also state, like Johnson et al. [180] and Marks et al. [32], that with this microplastination or ultra-thin sheet plastination the microscopic characteristics of the tissues are preserved, without jeopardizing the morphological arrangement of the samples [8, 10, 33, 41, 161, 184].

This way, transparent, dry, and resistant plastinated sections were obtained, as described by all the researchers who have undertaken this technique [3, 6, 8, 9, 19, 32, 37, 41, 43, 45, 84, 89, 128, 131, 135, 173, 174, 185]. Thanks to the transparency, the microscopic characteristics of the ultra-thin plastinated sections can be identified down to the histological level, and for example, in rat shoulder samples in which osteoarthritis was generated experimentally, and after applying another anatomical technique (arterial injection of epoxy resin) in the samples, we were able to view the neovascularization produced in the compact and cancellous bone tissue, the irrigation of which reaches the joint cartilage. According to Windisch and Weiglein [185] and Sora et al. [173], these findings are one of the characteristics of OA [11], and we were able to identify them for the first time using microplastination [10] or ultra-thin sheet plastination [15].

It is therefore worth noting that ultra-thin sheet plastination or microplastination is a method for the macro- and microscopic study of large specimens [37, 41, 43, 128, 173], and we agree with Leaper et al. [46], who stated that with histology only very small areas of a sample can be analyzed, unlike what occurs with the ultra-thin technique described here, through which the microscopic anatomy to the histology can be viewed in a macroscopic section.

We conclude that ultra-thin epoxy sheet plastination is a technique that offers the advantage of being applied to large samples, of any size, without needing decalcification, with the possible application of histological stains, and allowing their microscopic analysis through microscopy, contributing a new perspective in morphometric and histomorphological studies.

Discussion

The plastination technique created by Gunther von Hagens in Heidelberg, Germany, in 1978 is still a novel technique in its application and development. This technique holds the anatomical position of the structures and the adjacent elements "in situ," without alteration, similar to the living state [1, 2, 8, 12, 13, 19, 39, 41, 46, 47, 80, 83, 89, 129, 160, 161, 184, 186, 187]. This makes plastination a valuable tool in the evaluation of anatomical relationships [5, 8–12, 129].

From this, it may be said that the sections make it possible to study the topography of the body, human and animal, in a state free of collapse and dislocation [1–3, 8–12, 39, 83, 89, 174, 188]. The anatomical proportions are presented with minimal evidence of shrinkage [6, 41, 83] thanks to the low temperatures (-25 °C) [1]. According to Sora et al. [16], this property of the plastination technique makes it easy to take morphological measurements. On the other hand, Sora, Brugger, and Strobl [173] identified a shrinkage per section of 6.65% (±1.123 SD), with the average shrinkage after the first acetone bath being 1.33% (96% acetone) and during the second bath 0.8% (99% acetone), whereas from the defatting process with MeCl and the rest of the plastination process (forced impregnation and curing), the shrinkage was 4.52%. This shrinkage may be due to the shrinkage of the epoxy resin in itself and the shrinkage of the sections throughout the plastination process [174].

In addition, this sheet plastination technique offers the characteristic of not needing decalcification to make the sections [6, 19, 39, 41, 47, 84, 184].

Viewing them can be done under low-power magnification without jeopardizing the morphological arrangement [33, 41, 161, 184]. In this sense, Leaper et al. [46] consider that only very small areas of a sample can be seen with histology. Nevertheless, the sheet plastination technique is an excellent methodology to investigate the fine architecture of muscle and connective tissues in large specimens at macro- and microscopic level [37, 41, 43, 128, 173]. In addition, it is optimal for histological staining [35, 36, 38, 85–89, 126, 127, 189]. Adds and Al-Rekabi [18] indicated that the advantage of plastinated sections embedded in epoxy resin lies in their ability to be stained without needing to rehydrate the samples prior to applying the dye.

Plastinated sections are transparent, dry, resistant, odorless, of unlimited durability, and easy to maintain [3, 6, 32, 37, 41, 43, 45, 89, 131, 135, 174, 185]. And this property of transparency, particularly of the connective and fat tissue, makes the arterial and venous vessels visible and easy to measure [174, 186]. This way, these characteristics of plastinated sections are directly related to the ease they offer in their handling, storage at room temperature, and maintenance [3, 6, 41, 174].

The plastinated slices can be used for the three-dimensional reconstruction of any anatomical region [39, 41, 80, 81, 188], providing a resolution higher than photographs obtained from frozen samples [84].

According to several authors, it allows the conservation of the microscopic characteristics of tissues [8, 10, 11, 32, 184]. Marks et al. [32] establish that the infusion of cured epoxy polymers in tissues is also used in ultrastructural electron microscopy to preserve the details on a nm scale, which suggests that polymer embedding could also be useful for microscopy studies of tissue microstructure.

Anatomical dissection or the appearance of artifacts in conventional histological treatment can change the topographic anatomy; however, this does not occur with the epoxy sheet plastination, because all the tissues and structures, including the bone and cartilage, can be studied in sections [89].

Yet there are some limitations to the technique, such as the possible shrinkage of soft tissues like muscle and connective or fat tissue [129, 160, 161]. Scali et al. [160, 161], however, point out that this limitation could become an advantage, by allowing, in the case of the alar fascia and its adjacent spaces (primary objective of their research), a better examination by demonstrating a separation of the fascial planes and a better delineation of the aponeurotic structures due to this shrinkage effect.

With regard to histological processing, the development time of the plastination technique is greater than the embedding in paraffin of the histological techniques [6]. And another characteristic that can be an obstacle is that over time the plastinated sections can lose their transparency, acquiring a yellowish tone [37].

With respect to the application of the sheet plastination technique, its implementation in head and neck, pelvis, and lower limb mainly predominates. Table 6.7 displays information for comparing various sheet plastination protocols with epoxy resin.

Table 6.7 Detail of the techniques of plastination of sections with epoxy resin in publications that described the steps of the applied protocols (modified and updated from Ottone et al. [8])

Authors	Specimens (fixation/vascular injection)	Blocks (freeze/cutting)		Dehydration		Defatting	Forced impregnation		Curing		
				Bath	Time		Mixture	Time	Flat chamber/ sandwich	Oven (t °C)	Time
Eckel et al. [6]	Larynge/15% formalin	Block/cutting after impregnation and curing (no freeze)	0.8 mm/25–30 slices	Acetone/−25 °C/n = 4 or 5 ^	3 weeks	MeCl	Biodur® E12/E6/E600 (100:70 pbw; 0.15%)	10 days	Block	50 °C	1 week
Fritsch [126]	n = 7 (ft)/h	−80 °C	3–4 mm	Acetone/−25 °C	5 weeks	MeCl/RT/2 weeks	Biodur® E12[a]	–	Flat chamber	50 °C	–
Cook [178]	n = 1 (tronco)/h	−80 °C for 3 days	2.5 mm (0.5 mm loss)/60 slices	Acetone/n = 3: 93%, 97%, 99%	93%, 97%: 2 weeks each/99%: 5 days	Acetone (1 week) & MeCl (3 weeks)	Resin + catalyst	48 h	Flat chamber (5 mm glass; silicone 8 mm)	45 °C	5 days
Cook and Al-Ali [37]	n = 2 (whole body)/h/cut in regions	−80 °C/4 days	2.5–2.8 mm	Acetone/−25 °C/n = 3	5 weeks	Acetone & MeCl/ RT (18 °C)/8 weeks	Biodur® E12/AE30/ AE10/E1 (95:5:20:26 pbw)/0–5 °C	48 h	Flat chamber/ Biodur® E12/E1	45 °C	5 days
Sittel et al. [170, 171]	Larynge/15% formalin	Deep-frozen	4 µm (ordinary slicing machine)/80 µm (diamond wire-saw)	Acetone/n = 4 ^	48 hs total	MeCl (RT)/2 h	Biodur® E50/E7/AE10/ E700 (20:16:5 pbw; 0.1%)	24 h	Flat chamber		3 days
Skalkos, Williams, and Baptista [40]	Hearts/h/coronary arteries injected with 30% gelatin and 0.1 mL of gadopentetate dimeglumine (with eosin for color)/10% formalin fixation	Block (gelatin and polyethylene glycol)	3 mm	Acetone/−25 °C/n = 3	–	MeCl (RT)/2 h	Biodur® E12/AT30/ AT10/E1 (95:5:20:26 pbw)	24 h	Flat chamber/ Biodur® E12/ AT30/E1 (95:5:26 pbw)	(1) RT (15°) (2) 40 °C	(1) 2–3 days (2) 10 days

(continued)

Table 6.7 (continued)

Authors	Specimens (fixation/vascular injection)	Blocks (freeze/cutting)		Dehydration			Forced impregnation		Curing		
				Bath	Time	Defatting	Mixture	Time	Flat chamber/ sandwich	Oven (t °C)	Time
Sha et al. [81]	$n = 4$ (ankle)/h	Block/cutting after impregnation and curing	1.2 mm/diamond wire saw	Acetone/$n = 4$/−25 °C	1–3: 21 days/3–4: 14 days	–	Biodur® E12/E6/E500 (100:50 pbw, 0.1)	–	Block	50 °C	2 weeks
Steinke [157]	–	−35 °C	Diamond wire saw cooled with CO_2 and liquid nitrogen	(1) Acetone + 8% peracetic acid/−25 °C (2) Acetone/−25 °C	(1) 2 months (2) 2 weeks	Acetone/RT/2 weeks	Biodur® E12/AT30/E1 (95:8:26 pbw)	–	Sandwich[b]	–	–
Windisch and Weiglein [185]	$n = 5$ (feet)/h/ Thiel's method fixation/ Xantopren® blue for tendon sheaths	Pre-Block at 70 °C for 3 days/ cutting after impregnation and curing, final sheets	200 µm	Acetone/−25 °C	3 weeks	Acetone/RT/2 weeks	Biodur® E12/E6 (72–18%)/−4 °C	24 h	Block/flat chamber/Biodur® E12/E6 (72–18%)	45 °C	48 h
Sora et al. [41]	$n = 1$ (thorax)/h	−80 °C/1 week	3.8 mm (1 mm loss)/47 slices	Acetone/−25 °C	–	MeCl/1 week	Biodur® E12/E1/ AE10[b]/+5 °C	–	Flat chamber	(1) RT (15°) (2) 45 °C	(1) 1 day (2) 4 days
Sora et al. [42]	$n = 1$ (orbit)/h	−80 °C/1 week	3.8 mm (1 mm loss)/10 slices	Acetone/−25 °C	–	MeCl/1 week	Biodur® E12/E1/ AE10[b]/+5 °C	–	Flat chamber	(1) RT (15°) (2) 45 °C	(1) 1 day (2) 4 days
Sora, Brugger, and Strobl [173]	$n = 1$ (pelvis)/h	−80 °C/1 week	3.5 mm	Acetone/−25 °C/n = 2	(1) 4 days (2) 3 days	MeCl/1 week	Biodur® E12/E1/AE10 (95:26:10 pbw)[b]/+5 °C	2 days	Flat chamber/ Biodur® E12/E1/ AE10 (95:26:5 pbw)	(1) RT (15°) (2) 45 °C	(1) 1 day (2) 4 days
Zhang et al. [83][c]	$n = 3$ (head and neck)/h	Cutting after impregnation and curing (no freeze)	1 mm (0.3 mm loss)	–	–	–	Biodur® E12/E6/E600	–	–	–	–

Zhang and Lee [82]	$n = 5$ (cadaver)/10% NBF	−80 °C for 24 h	2.5 mm (2 mm loss)	Acetone/−30 °C/^A90–100%	6 weeks	Acetone/22–24 °C/2 weeks	Biodur® E12/E1/AE10/AE30 (100:28:20:5, pbw)/0 °C	24 h	Sandwich	32 °C/45 °C	1 week each
Qiu et al. [84]; Qiu et al. [172]	2003: $n = 3$ (head – temporal block)/2004: $n = 3$ (head – lateral skull base)/ formalin fixed/h	Cutting after impregnation and curing/epoxy block	2003: 1 mm/ 2004: 700 µm/ diamond wire saw	Acetone/−25 °C/100%	4 weeks	–	Biodur® E12/E6/E500 (100:50 pbw, 0.1)	–	Block	50 °C	2 weeks
Sora et al. [187]	$n = 20$ (lower limb)/h	−80 °C	1.6 mm (0.4 mm loss)/20 slices	Acetone/−25 °C	–	MeCl/1week	Biodur® E12/E1/ AE10/5 °C	–	Flat chamber/ flexible gasket 3 mm	(1) RT (15°) (2) 45 °C	(1) 1 day (2) 4 days
Sebe et al. [85, 86]	$n = 15$ (2005a), $n = 28$ (2005b) (fetus pelvis)/4% formalin/h	Cutting after impregnation and curing (no freeze)/epoxy block	200–600 µm	Acetone	–	–	Epoxy resin	–	–	–	–
Leaper et al. [46]	$n = 2$ (larynge)/10% NBF/h	−80 °C for 24 h	2.5 mm	Acetone/−30 °C/^A95–100%	6 weeks	Acetone/22–24 °C/2 weeks	Biodur® E12/E1/AE10/AE30 (100:28:20:5, pbw)/0 °C	24 h	Sandwich	+32 °C/45 °C	1 week each
Lunacek et al. [83]	$n = 29$ (fetal pelvis)/4% formalin (12 weeks)	Cutting after impregnation and curing/epoxy block	300–700 µm/ diamond wire-saw	Acetone/−25 °C	6 weeks	MeCl (RT)/1–2 weeks	Epoxy resin, hardener, accelerator	2 weeks			
Nash et al. [72]	$n = 13$ (head and neck)/10% NBF/h	−80 °C for 24 h/20% gelatin	2.5 mm (1.6 mm loss)	Acetone/−25 °C/^86.5–100%	8 weeks	Acetone (18–24 °C)/4 weeks	Biodur® E12/E1/AE10/AE30 (100:28:20:5 pbw)/0 °C	24 h	Sandwich (thick plastic sheets)	(a) RT (b) +30–40 °C oven	(a) 1 week (b) 3 days
Nash et al. [73]	$n = 7$ (cadavers)/h	−80 °C for 24 h/20% gelatin	2.5 mm (1.6 mm loss)	Acetone/−25 °C/^86.5–100%	8 weeks	Acetone (18–24 °C)/4 weeks	Biodur® E12/E1/AE10/AE30 (100:28:20:5 pbw)/0 °C	24 h	Sandwich (thick plastic sheets)	(a) RT (b) +30–40 °C oven	(a) 1 week (b) 3 days
Porzionato et al. [158]	–	−20 °C	+/2–3 mm	Acetone/−25 °C	2 weeks	Acetone/7 days	–	–	–	+50 °C	–

(continued)

Table 6.7 (continued)

Authors	Specimens (fixation/vascular injection)	Blocks (freeze/cutting)		Dehydration			Forced impregnation		Curing		
				Bath	Time	Defatting	Mixture	Time	Flat chamber/ sandwich	Oven (t °C)	Time
Sora and Genser-Strobl [128]	$n = 12$ (hands)/h	−80 °C	1.7 mm (0.4 mm loss)/20 slices	Acetone/−25 °C	–	MeCl/1 week	Biodur® E12/E1/ AE10[a]/5 °C	–	Flat chamber	(1) RT (15°) (2) +45 °C	(1) 1 day (2) 4 days
Fritsch et al. [89]	$n = 29$ (fetal pelvis)/4% formalin/3 months/h	Block/cutting after impregnation and curing	300–700 µm/ diamond wire saw	Acetone/−25 °C	5–6 weeks	MeCl/RT/1–2 weeks	Biodur® E12/E1/E600	2 weeks	–	–	–
Sora [15]	$n = 2$ (pelvis/ lower limb with endoprosthesis)/h	Cutting after impregnation and curing/epoxy block	1 mm or less (0.4 mm loss)/contact point diamond blade saw	Acetone/−25 °C	6 weeks	MeCl/RT/2 weeks	(1) Biodur® E12/E6/ E600 (100:50:0.2 pbw)/+30 °C (2) Biodur® E12/E6/ E600 (100:50:0.2 pbw)/+60 °C	(1) 4 days[d] (2) 1 day[d]	Block	+65 °C	4 days
Sora et al. [33]	$n = 1$ (ankle)/ unfixed/h	Cutting after impregnation and curing/block/−80 °C/1 week	1 mm (0.4 mm loss)/diamond blade saw	Acetone/−25 °C/ (1) 92% (2) 97% (3) 99%	(1) 4 weeks/ (2) 3 weeks/ (3) 2 weeks	MeCl/15 °C/4 weeks	Biodur® E12/E6/E600 (100:50 pbw, 0.2)/ vacuum drying oven 30 °C	5 days	Block	65 °C	4 days
Sora and Cook [3]	Limb sections	−75 °C	2–3 mm	Acetone/$n =$ 3/−25 °C/^A90–100%	3 days each	Acetone or MeCl (RT)/7 days	Biodur® E12/AE30/ AE10/E1 (95:5:20:26 pbw)/+5 °C or RT	2–9 days	Sandwich and flat chamber/ Biodur® E12/ AE30/E1 (95:5:20:26 pbw)	(1) RT (15° inclination) (2) 45 °C	(1) 1 day (2) 4 days
Macchi et al. [159]	$n = 4$ (pelvis)	−20 °C	2–3 mm	Acetone/−25 °C	2 weeks	Acetone (RT)/1 week	Biodur® E12 resin mix	–	–	+50 °C/ ultraviolet light	–
Sora et al. [129]	$n = 12$ (lower limbs)/h	−80 °C	1.5 mm (0.4 mm loss)	Acetone/−25 °C	–	MeCl/1 week	Biodur® E12 resin mix	–	Flat chamber	(1) RT (15° inclination) (2) 45 °C	(1) 1 day (2) 4 days

Al-Ali et al. [130]	$n = 2$ (pelvis)/h	−80 °C for 4 days	+/2.5 mm	Acetone	–	MeCl	Biodur® E12	–	Flat chamber	–	–
Elnady and Sora [131]	$n = 1$/10% formalin/a	−80 °C for 1 week	2.8 mm/44 slices (1 mm loss)	Acetone/$n = 2$/−25 °C/96 & 99%	4 & 3 days	MeCl (+15 °C)/1 day	Biodur® E12/E1/AE10 (95:26:10 pbw)/+5 °C	2 days	Flat chamber [E12/E1/AT30 (95:26:5)]	(a) 15 °C out oven (b) 45 °C oven	(a) overnight (b) 4 days
Soal et al. [43]	Specimens up to 150 mm in length width	Block of polyurethane foam (no freeze)	1 mm	Acetone/$n = 2$/−25 °C/100%	5 days each	Acetone (RT)/1 week	Biodur® E12/E1/AE10/AE30 (69.6:19.4:13.9:3.4 pbw)/0–1 °C	24 h	Sandwich[c]	40 °C	4 days
Koslowsky et al. [165, 166]	2011: $n = 15$ (forearm) 2015: $n = 11$ (elbow)/4% formalin fixation/ red epoxy resin vascular injection/h	Cutting after impregnation and curing/epoxy block	4 mm	Acetone/−25 °C/99%	4 months	MeCl/several weeks	Epoxy resin	3 weeks	Block	–	–
Kürtül et al. [44]	$n = 1$ (hip)/h/ ethanol	Cutting after impregnation and curing	diamond band saw	Acetone/−25 °C	2 weeks	Acetone/RT/1 week	Biodur® E12/E6/E600 (100:50 pbw, 0.2)[b]/21 °C	2 days	Block	50 °C	1 week
Sora et al. [39]	$n = 1$ (pelvis)/h	Cutting after impregnation and curing/−80 °C/1 week	1.6 mm/diamond blade saw	Acetone/−25 °C	–	MeCl	Biodur® E12/E6/E600 (100:50 pbw, 0.2)[f]	–	Block	65 °C	4 days
Chen et al. [133]; Liu et al. [134][g]	Chen et al.: $n = 4$ (cadaver) Liu et al.: $n = 16$ (cadaver)	–	–	Acetone/−30 °C/95–100%	–	Acetone/22–24 °C	Biodur® E12/E6/ E600/0 °C	–	–	32 °C & 45 °C	–

(continued)

Table 6.7 (continued)

Authors	Specimens (fixation/vascular injection)	Blocks (freeze/cutting)	Dehydration		Defatting	Forced impregnation		Curing		
			Bath	Time		Mixture	Time	Flat chamber/ sandwich	Oven (t °C)	Time
Wegmann et al. [167][h]	$n = 3$ (lower limb)/2% formalin/vascular injection of epoxy resin	–	3–4 mm/ diamond wire saw	Acetone/−25 °C/	2 months	Acetone/3 months	Epoxy resin	10 h	–	–	–
Adds and Al-Rekabi [18]	$n = 2$ (orbit)/4% formalin for 1 week/vascular injection of red silicone resin (Biodur® KEM 06)	Cutting after impregnation and curing	0.3 mm/diamond wire saw	Acetone/−20 °C/100%	4 days	Acetone/RT	Biodur® E12/E6/E600 (100:50 pbw, 0.2)/30 °C	–	–	–	–
Konschake and Fritsch [90][i]	$n = 6$ (midface)/h	Cutting after impregnation and curing/epoxy block	500 µm	Acetone/−25 °C	5 weeks	MeCl (RT)/2 weeks	Biodur® E12	–	–	–	–
Liang et al. [74, 75]	$n = 16$ (head)	20% gelatin/−80 °C for 24 h	2.5 mm	Acetone/−30 °C/^A95–100%	6 weeks	Acetone/22–24 °C/2 weeks	Biodur® E12/E1/AE10/ AE30 (100:28:20:5 pbw)/0 °C	24 h	Sandwich	32 °C and 45 °C	1 week each
Scali et al. [160, 161]	$n = 3$ (2015a) 13 (2015b) (cadaver)/h	−85 °C for 48 h	2 mm	Acetone/−25 °C/^	20 – 22 weeks	+	Epoxy resin/−8 °C to 0 °C	2 days	Sandwich	35 °C oven	24 hs
Thorpe Lowis et al. [47]	$n = 16$ (spine)/h	−80 °C for 5 days	2.5 mm	Acetone/−30 °C	3 weeks	Acetone (22–24 °C)/2 weeks	Biodur® E12/E6/ E600/0 °C	2 days	–	45 °C	5 days
Ottone et al. [19]	$n = 1$ (knee)/h/4% formalin fixation	−25 °C for 2 weeks	5 mm	Acetone/−25 °C/^90–100%	10 days	Acetone/RT/1 week	Commercial epoxy resin pre-accelerated (without hardener)	2 days	Sandwich/Epoxy resin—Hardener (50%:50%)	RT	24 h
Xu et al. [194]	$n = 6$ (pelvis, male cadavers)	−80 °C for 4 days	2.5 mm	Acetone/−30 °C	4 weeks	Acetone/RT (22–24 °C)/3 weeks	Biodur® E12/E1/AE10/ AE30/0 °C	2 days	–	45 °C	5 days

Xu et al. [26]	$n = 6$ (pelvis, male cadavers)	$-80\ °C$ for 7 days	2.5 mm	Acetone/$-30\ °C$	4 weeks	Acetone/RT (22–24 °C)/3 weeks	Biodur® E12/E1/AE10/AE30/0 °C	2 days	–	45 °C	5 days
Adds et al. [189]	$n = 9$ orbits (from 6 cadavers, 4 female, 2 males)	Cutting after impregnation and curing/epoxy block	0.3 mm	Acetone/$-20\ °C$	6 days	Acetone/RT/24 h	Biodur® E12/E6/E600/30–60 °C	2 days	Epoxy block	65 °C	6 days
Hammer et al. [203][k]	$n = 2$ triangular fibrocartilage complex (1 cadaver)	$-85\ °C$ (shock frozen in 85% acetone) Cutting after impregnation	250 µm	Acetone	2 weeks	–	Biodur® E12/E1/AE10 (100:28:10)	–	Epoxy block	–	–
Hedderwick et al. [204][l]	$n = 4$ knee joints	$-80\ °C$	2.5 mm	–	–	–	–	–	–	–	–
Liugan et al. [195][m]	$n = 9$ human heads (6 females, 3 males)	$-80\ °C$ for 7 days	2.5 mm 0.3 mm	Acetone/$-35\ °C$	4 weeks	Acetone/RT (22–24 °C)/3 weeks	Biodur® E12/E1/AE10/AE30/0 °C	2 days	–	45 °C	5 days
König et al. [205][a]	$n = 14$ ovaries (mares)	$-45\ °C$	2 mm	Acetone/$-25\ °C$	–	–	–	–	–	–	–
Steinke et al. [200][j]	$n = 2$ (1 pelvis and 1 foot)	Pre-cooled at 3 °C/frozen at -85 °C in 85% acetone	2 mm	Acetone/-25 °C/100%	4–6 weeks	–	Biodur® E12/E1		Sandwich	RT	5 days
Thorpe Lowis et al. [136]	$n = 16$ spines (7 males, 9 females)	$-80\ °C$ for 5 days	2.5 mm	Acetone/$-30\ °C$	3 weeks	Acetone/RT (22–24 °C)/2 weeks	Biodur® E12/E6/E600/0 °C	2 days	–	45 °C	5 days
Xu et al. [206]	$n = 6$ (2 females, 4 male cadavers)	$-80\ °C$ for 7 days	2.5 mm	Acetone/$-30\ °C$	4 weeks	Acetone/RT (22–24 °C)/3 weeks	Biodur® E12/E1/AE10/AE30/0 °C	2 days	–	45 °C	5 days
Basa et al. [77]	$n = 6$ limbs (3 felines)	20% gelatin/-80 °C for 24 h	1.5 ± 0.1 mm	Acetone/$-25\ °C$	7 days	Acetone/RT/5 days	Biodur® E12/AE30/AE10/E1 (95:5:20:26 pbw)/RT [3]	24 h	Sandwich	40 °C	–

(continued)

Table 6.7 (continued)

Authors	Specimens (fixation/vascular injection)	Blocks (freeze/cutting)		Dehydration		Defatting	Forced impregnation		Curing		
				Bath	Time		Mixture	Time	Flat chamber/ sandwich	Oven (t °C)	Time
Liang et al. [207][p]	n = 22 head sides (11 cadavers, 2 females, 9 males)	Cutting after impregnation and curing/epoxy block	250 μm (450–650 μm)	–	–	–	Biodur® E12/E6/E600	–	–	–	–
Vargas et al. [11][q]	n = 20 humeral joints (10 female Sprague-Dawley rats)	Cutting after impregnation and curing/epoxy block	230 μm	Acetone/−25 °C/100%	10 days	Methylene chloride/RT/4 days	Biodur® E12/E6/E600/ (100:50:0.2)/30–65 °C	5 days	–	–	–
Bond et al. [198]	n = 12 heads (7 males, 5 females)	−80 °C for 7 days	2.5 mm	Acetone/−30 °C/100%	4 weeks	Acetone/RT (22–24 °C)/3 weeks	Biodur® E12/E1/ AE20/0 °C	2 days	–	45–50 °C	5 days
Starchik et al. [25]	n = 14 human hearts	Cutting after impregnation and curing/epoxy block	2–4 mm and 0.5–2 mm/ diamond wire saw	Acetone/−25 °C	3–4 weeks	Acetone/RT/2 weeks	Epoxy resin (YD-128) + Catalyst (C −403) (20:1)	24 h	–	45 °C	–
Xu et al. [208][p]	n = 16 lumbar spines	−80 °C for 7 days	2.5 mm	Acetone/−30 °C	4 weeks	Acetone/RT (22–24 °C)/3 weeks	Biodur® E12/E1/AE10/ AE30/0 °C	24 h	–	45 °C	5 days
Xu et al. [196][r]	n = 8 cadavers	−80 °C for 7 days	2.5 mm (diamond band saw)	Acetone/−30 °C	4 weeks	Acetone/RT/3 weeks	Biodur® E12/E1/AE10/ AE30	–	–	45 °C	5 days
Zwirner et al. [197]	n = 2 calcaneal tendon	−80 °C for 7 days	2.5 mm	Acetone/−30 °C	4 weeks	Acetone/RT (22–24 °C)/3 weeks	Biodur® E12/E1/AE10/ AE30/0 °C	2 days	–	45 °C	5 days
Pieroh et al. [201][s]	n = 3 pelvis including proximal femoral region	Pre-cooled at 3 °C/Frozen at −85 °C in 85% acetone	1–2 mm	Acetone/−25 °C/100%	4–6 weeks	–	Biodur® E12/E1	–	Sandwich	–	–

Singh et al. [76]	$n = 18$ft	25% gelatin/pre-cooled at 4 °C/frozen at −20 °C and 80 °C for 5 days	2.5 mm (sagittal)	Acetone/−30 °C	3 weeks	Acetone/RT (22–24 °C)/2 weeks	Biodur® E12/E6/E600/0 °C	2 days	–	45 °C	5 days
Wu et al. [191]	$n = 16$ human heads	Cutting after impregnation and curing/epoxy block	250 µm (450–650 µm)	–	–	–	Biodur® E12/E6/E600	–	–	–	–
Texeira et al. [190]	$n = 5$ orbits (4 females, 1 male)	Cutting after impregnation and curing/epoxy block	0.3 mm	Acetone/−20 °C/100%	6 days	Acetone/RT/24 h	Biodur® E12/E6/E600 (100:50:0.2, pbw)/30–60 °C	2 days	Epoxy block	65 °C	6 days
Shi et al. [192]	$n = 9$ heads[w]/12 heads[x]	[w]Frozen and cutting, then plastination [x]Cutting after impregnation and curing/epoxy block	3 mm[y] 200–300 µm[z]	Acetone/−25 °C	–	–	[y]Biodur® E12/AE20/E1 [z]Biodur® E12/E6/E600	–	–	45 °C[y]	
Ma et al. [193]	$n = 18$ heads (16 males, 2 females)	Cutting after impregnation and curing/epoxy block	200–250 µm	Acetone/−25 °C	4 weeks	Acetone/RT/6 weeks	Biodur® E12/E6/E600	–	–	60 °C	7–10 days
Noriego et al. [202]	$n = 10$ adult upper limbs (5 right, 5 left)	−80 °C (elbow at 20° of flexion, forearm in neutral position)	1.5 mm (transepicondylar coronal serial slices)	Acetone/−25 °C	–	Dichloromethane/RT	Biodur® E12/E1	–	–	45 °C	72 h

References: *NFB* neutral-buffered formalin, ^A increased concentrations of acetone, *pbw* parts by weight, *h* human, *a* animal, *MeCl* methylene chloride

[a] von Hagens et al. [2]; Latorre et al. [177]

[b] von Hagens [1]

[c] von Hagens et al. [2]; Eckel et al. [6]

(continued)

Table 6.7 (continued)

[d] The block is impregnated during 5 days. The first 4 days at 30 °C and the fifth day at 60 °C. The resin-mix fluid viscosity must be controlled

[e] Sora and Cook [3]

[f] Sora et al. [33]

[g] Nash et al. [73]

[h] Rath et al. [151, 152]

[i] Fritsch [36]

[j] The samples were previously decalcified in 10% formic acid for 72 h

[k] Plastinated ultra-thin sections were subjected to histological staining: methylene blue and parafuchsin for connective tissue and acetone-diluted alizarin for bone

[l] The authors applied the plastination protocol of Soal et al. [43]

[m] Plastinated ultra-thin sections were subjected to histological staining: hematoxylin and eosin

[n] They applied the plastination protocol of Lane [5] and Sora and Cook [3]

[o] They applied the Sudan III stain on the sections of the samples, prior to the start of the dehydration stage, and applied the PAS reaction after dehydration

[p] Plastinated sections were stained with Stevenel blue and alizarin red S

[q] Prior to plastination, the samples were injected with red epoxy resin (E20 + E1, 100:40, Biodur) through the thoracic aorta

[r] In this work, from plastination, it was shown that the collagen, elastin, myofilaments, and neurofilaments of the samples are endogenously autofluorescent under 488 nm excitation

[s] After dehydration, pelvic axial sections were stained with methylene blue [159]. Coronal pelvic sections were stained with periodic acid-Schiff (PAS) [159]

[t] The sections were stained with Gomori's trichromic, using Wiegert's hematoxylin in the process

Conclusions

Plastination is a method of tissue preservation that has proven its applicability in different fields of teaching and research in morphological sciences and normal and pathological anatomy. Due to its superior quality in the preservation of sectioned tissues, plastination is a promising research tool. In addition, the dry and odorless specimens, which can be handled and maintained with ease, become useful tools to teach anatomy from a macro- and microscopic point of view. In relation to this last point, epoxy sheet plastination can be used as an alternative to the standard histological techniques [186], being able to apply histological stains to sections of macroscopic samples. According to Marks et al. [32], the transparency introduced by impregnation with epoxy is an advantage for microscopic analysis, and the rigidity can help stabilize the microstructure like it does in the samples impregnated with epoxy for ultrastructural electron microscopy.

In addition, sheet plastination is an anatomical technique that enables preservation in situ, without modification of the arrangement of the anatomical structures and permitting highly detailed visualization of the macro- and microscopic morphological characteristics. Therefore, specimens of high value are obtained, not only for undergraduate teaching but also for their application in programs for advanced training in sectional topography, training specialists in computed tomography and magnetic resonance imaging [3, 39], as well as facilitating surgical planning by enabling an adequate view of human and animal anatomy.

References

1. von Hagens G, editor. Heidelberg plastination folder. Collection of technical leaflets of plastination. 2nd English edn. Heidelberg: Biodur Products GmbH; 1986.
2. von Hagens G, Tiedemann K, Kriz W. The current potential of plastination. Anat Embryol (Berl). 1987;175(4):411–21.
3. Sora MC, Cook P. Epoxy plastination of biological tissue: E12 technique. J Int Soc Plastination. 2007;22:31–9.
4. Sora MC, Matusz P. General considerations regarding the thin slice plastination technique. Clin Anat. 2010;23(6):734–6; author reply 737–8.
5. Lane A. Sectional anatomy: standardized methodology. J Int Soc Plastination. 1990;4:16–22.
6. Eckel HE, Sittel C, Walger M, Sprinzl G, Koebke J. Plastination: a new approach to morphological research and instruction with excised larynges. Ann Otol Rhinol Laryngol. 1993;102(9):660–5.
7. Reidenbach MM, Schmidt HM. Prenatal development of the radial annular ligament. Ann Anat. 1993;175(5):459–67.
8. Ottone NE, Baptista CAC, Latorre R, Bianchi HF, Del Sol M, Fuentes R. E12 sheet plastination: techniques and applications. Clin Anat. 2018;31(5):742–56.
9. Ottone NE, Vargas CA, Veuthey C, del Sol M, Fuentes F. Epoxy sheet plastination on a rabbit head–new faster protocol with Biodur® E12/E1. Int J Morphol. 2018;36(2):441–6.
10. Ottone NE. Micro-plastination. Technique for obtaining slices below 250 µm for the visualization of microanatomy in morphological and pathological experimental protocols. Int J Morphol. 2020;38(2):389–91.
11. Vargas CA, Baptista CAC, Del Sol M, Sandoval C, Vásquez B, Veuthey C, Ottone NE. Development of an ultrathin sheet plastination technique in rat humeral joints with

osteoarthritis induced by monosodium iodoacetate for neovascularization study. Anat Sci Int. 2020;95(2):297–303. https://doi.org/10.1007/s12565-019-00500-7.

12. Ottone NE. Plastination: techniques fundamentals and implementation at Universidad de La Frontera. J Health Med Sci. 2018;4(4):293–302.
13. Bickley HC, von Hagens G, Townsend FM. An improved method for the preservation of teaching specimens. Arch Pathol Lab Med. 1981;105(12):674–6.
14. Ottone NE. Gunther von Hagens, creator of Plastination. Historical review and technical development. Rev Argent Anat Online. 2013;4(2):70–6. https://www.revista-anatomia.com.ar/archivos-parciales/2013-2-revista-argentina-de-anatomia-online-f.pdf.
15. Sora MC. Epoxy plastination of biological tissue: E12 ultra-thin technique. J Int Soc Plastination. 2007;2007(22):40–5.
16. Sora MC, von Horst C, López-Albors O, Latorre R. Ultra-thin sectioning and grinding of epoxy plastinated tissue. Anat Histol Embryol. 2019;48(6):564–71.
17. Latorre R, de Jong K, Sora MC, López-Albors O, Baptista C. E12 technique: conventional epoxy resin sheet plastination. Anat Histol Embryol. 2019;48(6):557–63.
18. Adds PJ, Al-Rekabi A. 3-D reconstruction of the ethmoidal arteries of the medial orbital wall using BiodurVR E12. J Plastination. 2014;26:5–10.
19. Ottone NE, del Sol M, Fuentes R. Report on a sheet plastination technique using commercial epoxy resin. Int J Morphol. 2016;34:1039–43.
20. Aufdemorte TB, Bickley HC, Krauskopf DR, Townsend FM. An epoxy resin and silicone impregnation technique for the preservation of oral pathology teaching specimens. Oral Surg Oral Med Oral Pathol. 1985;59(1):74–6.
21. Neusel E, Graf J, Kraft A, Niethard FU. Importance of the subchondral space for development of chondromalacia patellae—morphologic study of joint cartilage of the rabbit patella after experimental ischemia. Sportverletz Sportschaden. 1992;6(4):170–8.
22. Zhang M, An PC. Liliequist's membrane is a fold of the arachnoid mater: study using sheet plastination and scanning electron microscopy. Neurosurgery. 2000;47(4):902–8; discussion 908–9. 23, 33, 48, 56, 57, 174, 182.
23. Steinke H, Pfeiffer S, Spanel-Borowski K. A new plastination technique for head slices containing brain. Ann Anat. 2002;184(4):353–8.
24. Starchik DA. The methodological basis for the plastination of body sawcuts. Morfologiia. 2015;148(4):56–61.
25. Starchik D, Shishkevich A, Sora MC. Visualization of metal stents in coronary arteries with epoxy plastination. J Plastination. 2020;32(2):JP-20-07.
26. Xu Z, Chapuis PH, Bokey L, Zhang M. Denonvilliers' fascia in men: a sheet plastination and confocal microscopy study of the prerectal space and the presence of an optimal anterior plane when mobilizing the rectum for cancer. Colorectal Dis. 2017; https://doi.org/10.1111/codi.13906.
27. Scali F, Ohno A, Enix D, Hassan S. The posterior atlantooccipital membrane: the anchor for the myodural bridge and meningovertebral structures. Cureus. 2022;14(5):e25484.
28. Sora MC, Latorre R, Baptista C, López-Albors O. Plastination—a scientific method for teaching and research. Anat Histol Embryol. 2019;48(6):526–31.
29. Bickley HC, Townsend FM. Preserving biological material by plastination. Curator. 1984;27:65–73.
30. Shahar T, Pace C, Henry RW. Epoxy plastination of biological tissue: VisDocta EP73 technique. J Int Soc Plastination. 2007;22:46–9.
31. Minelli L, Yang HM, van der Lei B, Mendelson B. The surgical anatomy of the jowl and the mandibular ligament reassessed. Aesthetic Plast Surg. 2023;47(1):170–80. https://doi.org/10.1007/s00266-022-02996-3.
32. Marks DL, Chaney EJ, Boppart SA. Plastinated tissue samples as three-dimensional models for optical instrument characterization. Opt Express. 2008;16(20):16272–83.
33. Sora MC, Genser-Strobl B, Radu J, Lozanoff S. Three-dimensional reconstruction of the ankle by means of ultrathin slice plastination. Clin Anat. 2007;20(2):196–200.
34. Sora MC, Matusz P. Study of the vascular architecture of bones using the plastination technique. Clin Anat. 2012;25(2):258–9.

35. Fritsch H. Developmental changes in the retrorectal region of the human fetus. Anat Embryol (Berl). 1988;177(6):513–22.

36. Fritsch H. Staining of different tissues in thick epoxy resin-impregnated sections of human fetuses. Stain Technol. 1989;64(2):75–9.

37. Cook P, Al-Ali S. Submacroscopic interpretation of human sectional anatomy using plastinated E12 sections. J Int Soc Plastination. 1997;12:17–27.

38. Gruber H, Brenner E, Schmitt O, Fritsch H. The different growth zones of the fetal foot. Ann Anat. 2001;183(3):267–73.

39. Sora MC, Jilavu R, Matusz P. Computer aided three-dimensional reconstruction and modeling of the pelvis, by using plastinated cross sections, as a powerful tool for morphological investigations. Surg Radiol Anat. 2012;34(8):731–6.

40. Skalkos E, Williams G, Baptista CAC. The E12 technique as an accessory tool for the study of myocardial fiber structure analysis in MRI. J Int Soc Plastination. 1999;14:18–21.

41. Sora MC, Strobl B, Förster-Streffleur S, Staykov D. Aortic arch variation analyzed by using plastination. Clin Anat. 2002;15(6):379–82.

42. Sora MC, Strobl B, Staykov D, Traxler H. Optic nerve compression analyzed by using plastination. Surg Radiol Anat. 2002;24(3-4):205–8.

43. Soal S, Pollard M, Burland G, Lissaman R, Wafer M, Stringer MD. Rapid ultrathin slice plastination of embalmed specimens with minimal tissue loss. Clin Anat. 2010;23(5):539–44.

44. Kürtül I, Hammer N, Rabi S, Saito T, Böhme J, Steinke H. Oblique sectional planes of block plastinates eased by Sac Plastination. Ann Anat. 2012;194(4):404–6.

45. Fasel J, Mohler R, Lehmann B. A technical note for improvement of the E12 technique. J Int Soc Plastination. 1988;2:4–7.

46. Leaper M, Zhang M, Dawes PJ. An anatomical protrusion exists on the posterior hypopharyngeal wall in some elderly cadavers. Dysphagia. 2005;20(1):8–14.

47. Thorpe Lewis CG, Zhang M, Amin NF. Fine configuration of thoracic type II meningeal cysts: macro- and microscopic cadaveric study using epoxy sheet plastination. Spine (Phila Pa 1976). 2016;41(20):E1195–200.

48. Probst A, Kneissl S. Computed tomographic anatomy of the canine pancreas. Vet Radiol Ultrasound. 2001;42(3):226–30.

49. Henkel-Kopleck A, Schmidt HM. The architecture of the fibrous complex between the palmar aponeurosis and the flexor retinaculum. Handchir Mikrochir Plast Chir. 2001;33(5):294–8.

50. Ludwikowski B, Hayward IO, Fritsch H. Rectovaginal fascia: an important structure in pelvic visceral surgery? About its development, structure, and function. J Pediatr Surg. 2002;37(4):634–8.

51. Hermans JJ, Wentink N, Kleinrensink GJ, Beumer A. MR-plastination-arthrography: a new technique used to study the distal tibiofibular syndesmosis. Skeletal Radiol. 2009;38:697–701.

52. Andermahr J, Jubel A, Elsner A, Johann J, Prokop A, Rehm KE, Koebke J. Anatomy of the clavicle and the intramedullary nailing of midclavicular fractures. Clin Anat. 2007;20(1):48–56.

53. Hoch J, Stahlenbrecher A. Bottoming out in augmentation mammaplasty—correction and prevention. Handchir Mikrochir Plast Chir. 2006;38(4):233–9.

54. Greil GF, Wolf I, Kuettner A, Fenchel M, Miller S, Martirosian P, Schick F, Oppitz M, Meinzer HP, Sieverding L. Stereolithographic reproduction of complex cardiac morphology based on high spatial resolution imaging. Clin Res Cardiol. 2007;96(3):176–85.

55. Faymonville C, Andermahr J, Seidel U, Müller LP, Skouras E, Eysel P, Stein G. Compartments of the foot: topographic anatomy. Surg Radiol Anat. 2012;34(10):929–33.

56. Oppermann J, Franzen J, Spies C, Faymonville C, Knifka J, Stein G, Bredow J. The microvascular anatomy of the talus: a plastination study on the influence of total ankle replacement. Surg Radiol Anat. 2014;36(5):487–94.

57. Kostylenko YP, Starchenko II, Prylutsky OK. Structure of teeth in mature ovarian teratoma. Arkh Patol. 2015;77(2):28–31.

58. Perumal V. A sectional anatomy learning tool for medical students: development and user-usage analytics. Surg Radiol Anat. 2018;40(11):1293–300. https://doi.org/10.1007/s00276-018-2082-5.

59. Tote D, Tote S. Comparative study between various reagents of plastination in making museum specimen. Int J Curr Res Rev. 2020;12(22):126–8.
60. Elliott JM, Cornwall J, Kennedy E, Abbott R, Crawford RJ. Towards defining muscular regions of interest from axial magnetic resonance imaging with anatomical cross-reference: part II—cervical spine musculature. BMC Musculoskelet Disord. 2018;19(1):171.
61. Hryn VH, Kostylenko YP, Bilash VP, Ryabushko OB. Microscopic structure of albino rats' small intestine. Wiad Lek. 2019;72(5 cz 1):733–8.
62. Steinke H, Wiersbicki D, Völker A, Pieroh P, Kulow C, Wolf B, Osterhoff G. The fascial connections of the pectineal ligament. Clin Anat. 2019;32(7):961–9.
63. Didenko M, Starchik D, Natarova S, Pasenov G, Khubulava G. Morphometric characteristics of the right atrium and Bachmann's bundle region: implications for atrial pacing. P1525, ii263, Poster Session; 2013.
64. Párraga E, López-Albors O, Sánchez-Margallo F, Moyano-Cuevas JL, Latorre R. Effects of pneumoperitoneum and body position on the morphology of the caudal cava vein analyzed by MRI and plastinated sections. Surg Endosc. 2013;27:880–7.
65. Phan T, Lay J, Scali F. The alar fascia and danger space: a modern review. Cureus. 2022;14(12):e32871. https://doi.org/10.7759/cureus.32871.
66. Steinke H, Saito T, Kuehner J, Reibetanz U, Heyde CE, Itoh M, Voelker A. Sacroiliac innervation. Eur Spine J. 2022;31(11):2831–43. https://doi.org/10.1007/s00586-022-07353-1.
67. Hellmund C, Hepp P, Steinke H. The subpopliteal fat body. Ann Anat. 2023;245:151995. https://doi.org/10.1016/j.aanat.2022.151995.
68. Minelli L, van der Lei B, Mendelson BC. The deep fascia of the head and neck revisited: relationship with the facial nerve and implications for rhytidectomy. Plast Reconstr Surg. 2023; https://doi.org/10.1097/PRS.0000000000010556.
69. Minelli L, van der Lei B, Mendelson BC. The superficial musculo-aponeurotic system (SMAS): does it really exist as an anatomic entity? Plast Reconstr Surg. 2023; https://doi.org/10.1097/PRS.0000000000010557.
70. Minelli L, Brown CP, Warren RJ, van der Lei B, Mendelson BC, Little JW. Lifting the anterior midcheek and nasolabial fold: introduction to the melo fat pad anatomy and its role in longevity and recurrence. Aesthet Surg J. 2023;43(9):941–54. https://doi.org/10.1093/asj/sjad126.
71. Minelli L, Wilson JL, Bravo FG, Hodgkinson DJ, O'Daniel TG, van der Lei B, Mendelson BC. The functional anatomy and innervation of the platysma is segmental: implications for lower lip dysfunction, recurrent platysmal bands, and surgical rejuvenation. Aesthet Surg J. 2023;43(10):1091–105. https://doi.org/10.1093/asj/sjad148.
72. Nash L, Nicholson H, Lee AS, Johnson GM, Zhang M. Configuration of the connective tissue in the posterior atlanto-occipital interspace: a sheet plastination and confocal microscopy study. Spine (Phila Pa 1976). 2005;30(12):1359–66.
73. Nash L, Nicholson HD, Zhang M. Does the investing layer of the deep cervical fascia exist? Anesthesiology. 2005;103(5):962–8.
74. Liang L, Diao Y, Xu Q, Zhang M. Transcranial segment of the trigeminal nerve: macro-/microscopic anatomical study using sheet plastination. Acta Neurochir (Wien). 2014a;156(3):605–12.
75. Liang L, Gao F, Xu Q, Zhang M. Configuration of fibrous and adipose tissues in the cavernous sinus. PLoS One. 2014b;9(2):e89182.
76. Singh A, Zwirner J, Templer F, Kieser D, Klima S, Hammer N. On the morphological relations of the Achilles tendon and plantar fascia via the calcaneus: a cadaveric study. Sci Rep. 2021;11(1):5986.
77. Basa RM, Podadera JM, Burland G, Johnson KA. High field magnetic resonance imaging anatomy of feline carpal ligaments is comparable to plastinated specimen anatomy. Vet Radiol Ultrasound. 2018;59(5):597–606.
78. Stegmann T, Steinke H, Pieroh P, Dehghani F, Völker A, Groll MJ, Wolfskämpf T, Werner M, Kollan J, Hinz A, Leimert M. On the importance of the innervation of the human cervical longitudinal ligaments at vertebral level. Surg Radiol Anat. 2020;42(2):127–36.

79. Phillips MN, Nash LG, Barnett R, Nicholson H, Zhang M. The use of confocal microscopy for the examination of E12 sheet plastinated human tissue. J Int Soc Plastination. 2002;17:12–6.
80. Sora MC, Strobl B, Staykov D, Förster-Streffleur S. Evaluation of the ankle syndesmosis: a plastination slices study. Clin Anat. 2004a;17(6):513–7.
81. Sha Y, Zhang SX, Liu ZJ, Tan LW, Wu XY, Wan YS, Deng JH, Tang ZS. Computerized 3D-reconstructions of the ligaments of the lateral aspect of ankle and subtalar joints. Surg Radiol Anat. 2001;23(2):111–4.
82. Zhang M, Lee AS. The investing layer of the deep cervical fascia does not exist between the sternoclcidomastoid and trapezius muscles. Otolaryngol Head Neck Surg. 2002a;127(5):452–4.
83. Zhang WG, Zhang SX, Wu BH. A study on the sectional anatomy of the oculomotor nerve and its related blood vessels with plastination and MRI. Surg Radiol Anat. 2002b;24(5):277–84.
84. Qiu MG, Zhang SX, Liu ZJ, Tan LW, Wang YS, Deng JH, Tang ZS. Plastination and computerized 3D reconstruction of the temporal bone. Clin Anat. 2003;16(4):300–3.
85. Sebe P, Schwentner C, Oswald J, Radmayr C, Bartsch G, Fritsch H. Fetal development of striated and smooth muscle sphincters of the male urethra from a common primordium and modifications due to the development of the prostate: an anatomic and histologic study. Prostate. 2005;62(4):388–93.
86. Sebe P, Oswald J, Fritsch H, Aigner F, Bartsch G, Radmayr C. An embryological study of fetal development of the rectourethralis muscle—does it really exist? J Urol. 2005;173(2):583–6.
87. Sebe P, Fritsch H, Oswald J, Schwentner C, Lunacek A, Bartsch G, Radmayr C. Fetal development of the female external urinary sphincter complex: an anatomical and histological study. J Urol. 2005;173(5):1738–42; discussion 1742.
88. Lunacek A, Schwentner C, Fritsch H, Bartsch G, Strasser H. Anatomical radical retropubic prostatectomy: 'curtain dissection' of the neurovascular bundle. BJU Int. 2005;95(9):1226–31.
89. Fritsch H, Pinggera GM, Lienemann A, Mitterberger M, Bartsch G, Strasser H. What are the supportive structures of the female urethra? Neurourol Urodyn. 2006;25(2):128–34.
90. Konschake M, Fritsch H. Anatomical mapping of the nasal muscles and application to cosmetic surgery. Clin Anat. 2014;27(8):1178–84.
91. Reidenbach MM, Schmidt HM. Cartilage canals in the fetal elbow joint in man. Acta Anat (Basel). 1994;149(3):195–202.
92. Reidenbach MM, Schmidt HM. Vascularization of the fetal elbow joint. Ann Anat. 1994;176(4):303–10.
93. Reidenbach MM. Normal topography of the conus elasticus. Anatomical bases for the spread of laryngeal cancer. Surg Radiol Anat. 1995;17(2):107–11.
94. Reidenbach MM. The cricoarytenoid ligament: its morphology and possible implications for vocal cord movements. Surg Radiol Anat. 1995;17(4):307–10.
95. Schubert M, Bade H, Notermans HP, Knifka J, Koebke J. Functional gliding spaces of the dorsal side of the human finger. Surg Radiol Anat. 1996;18(1):17–22.
96. Reidenbach MM. The paraglottic space and transglottic cancer: anatomical considerations. Clin Anat. 1996;9(4):244–51.
97. Reidenbach MM. The attachments of the conus elasticus to the laryngeal skeleton: physiologic and clinical implications. Clin Anat. 1996;9(6):363–70.
98. Reidenbach MM. The cricothyroid space: topography and clinical implications. Acta Anat (Basel). 1996;157(4):330–8.
99. Reidenbach MM. The periepiglottic space: topographic relations and histological organisation. J Anat. 1996;188(Pt 1):173–82.
100. Reidenbach MM. Borders and topographic relationships of the paraglottic space. Eur Arch Otorhinolaryngol. 1997;254(4):193–5.
101. Matsumura A, Saito K. Distribution of muscle spindles in the extensor digitorum and hallucis brevis muscles of the macaque as determined by plastination. Acta Anat (Basel). 1997;158(1):59–67.
102. Reidenbach MM. Borders and topographic relations of the cricoid area. Eur Arch Otorhinolaryngol. 1997;254(7):323–5.

103. Reidenbach MM. Anatomical considerations of closure of the laryngeal vestibule during swallowing. Eur Arch Otorhinolaryngol. 1997;254(9-10):410–2.
104. Reidenbach MM. Subglottic region: normal topography and possible clinical implications. Clin Anat. 1998;11(1):9–21.
105. Reidenbach MM. Topographical anatomy and oncologic implications of the anterolateral surface of the arytenoid cartilage. Eur Arch Otorhinolaryngol. 1998;255(3):140–2.
106. Reidenbach MM. Aryepiglottic fold: normal topography and clinical implications. Clin Anat. 1998;11(4):223–35.
107. Reidenbach MM. Anatomical bases of glottic widening surgery related to arytenoidectomy. Clin Anat. 1999;12(2):94–102.
108. Sprinzl GM, Menzler A, Eckel HE, Sittel C, Koebke J, Thumfart WF. Bone density measurements of the paranasal sinuses on plastinated whole-organ sections: anatomic data to prevent complications in endoscopic sinus surgery. Laryngoscope. 1999;109(3):400–6.
109. Hoch J, Fritsch H, Frenz C. Does osseous extensor tendon avulsion or rupture really exist? Histologic plastination studies of insertion of the extensor aponeurosis and significance for operative therapy. Chirurg. 1999;70(6):705–12.
110. Frenz C, Fritsch H, Hoch J. Plastination histologic investigations on the inserting pars terminalis aponeurosis dorsalis of three-sectioned fingers. Ann Anat. 2000;182(1):69–73.
111. Eckel HE, Sprinzl GM, Sittel C, Koebke J, Damm M, Stennert E. Anatomy of the glottis and subglottis in the pediatric larynx. HNO. 2000;48(7):501–7.
112. Brenner E, Gruber H, Fritsch H. Fetal development of the first metatarsophalangeal joint complex with special reference to the intersesamoidal ridge. Ann Anat. 2002;184(5):481–7.
113. Gardetto A, Dabernig J, Rainer C, Piegger J, Piza-Katzer H, Fritsch H. Does a superficial musculoaponeurotic system exist in the face and neck? An anatomical study by the tissue plastination technique. Plast Reconstr Surg. 2003;111(2):664–72; discussion 673–5.
114. Oswald J, Brenner E, Deibl M, Fritsch H, Bartsch G, Radmayr C. Longitudinal and thickness measurement of the normal distal and intravesical ureter in human fetuses. J Urol. 2003;169(4):1501–4.
115. Thomas M, Steinke H, Schulz T. A direct comparison of MR images and thin-layer plastination of the shoulder in the apprehension-test position. Surg Radiol Anat. 2004;26(2):110–7.
116. Hoch J, Fritsch H, Lewejohann S. Congenital or acquired disposition of the separate compartment of the extensor pollicis brevis tendon associated with stenosing tendovaginitis (de Quervain's disease)? Macroanatomical and fetal-plastinationhistological studies of the first compartment of the wrist. Ann Anat. 2004;186(4):305–10.
117. Hoch J, Fritsch H, Lewejohann S. Plastination-histological investigations on the inserting extensor pollicis brevis tendon on the proximal phalanx of the thumb. Ann Anat. 2004;186(4):311–5.
118. Landes CA, Weichert F, Geis P, Wernstedt K, Wilde A, Fritsch H, Wagner M. Tissue-plastinated vs. celloidin-embedded large serial sections in video, analog and digital photographic on-screen reproduction: a preliminary step to exact virtual 3D modelling, exemplified in the normal midface and cleft-lip and palate. J Anat. 2005;207(2):175–91.
119. Dargel J, Schmidt-Wiethoff R, Schmidt J, Koebke J. Histomorphology and microradiography of quadriceps tendon-patellar bone grafts in press-fit anterior cruciate ligament reconstruction. J Orthop Res. 2005;23(5):1206–10.
120. Elsner A, Schiffer G, Jubel A, Koebke J, Andermahr J. The venous pump of the first metatarsophalangeal joint: clinical implications. Foot Ankle Int. 2007;28(8):902–9.
121. Steinke H, Hammer N, Slowik V, Stadler J, Josten C, Böhme J, Spanel-Borowski K. Novel insights into the sacroiliac joint ligaments. Spine (Phila Pa 1976). 2010;35(3):257–63.
122. Diao Y, Liang L, Yu C, Zhang M. Is there an identifiable intact medial wall of the cavernous sinus? Macro- and microscopic anatomical study using sheet plastination. Neurosurgery. 2013;73(1 Suppl Operative):ons106-9; discussion ons110.
123. Wiersbicki D, Völker A, Heyde CE, Steinke H. Ligamental compartments and their relation to the passing spinal nerves are detectable with MRI inside the lumbar neural foramina. Eur Spine J. 2019;28(8):1811–20.

124. Tomlinson J, Klima S, Poilliot A, Zwirner J, Hammer N. Fat is consistently present within the plantar muscular space of the human foot-an anatomical study. Medicina (Kaunas). 2022;58(2):154.
125. Arredondo J, López-Albors O, Recillas S, Victoria M, Castelán O, González-Ronquillo M, Becerril S, Latorre R. Three-dimensional virtual model of the elbow joint of the dog by means of ultrathin plastinated slices. Int J Morphol. 2016;34:1253–8.
126. Fritsch H. Sectional anatomy of connective tissue structures in the hindfoot of the newborn child and the adult. Anat Rec. 1996;246(1):147–54.
127. Fritsch H, Schmitt O, Eggers R. The ossification centre of the talus. Ann Anat. 1996;178(5):455–9.
128. Sora MC, Genser-Strobl B. The sectional anatomy of the carpal tunnel and its related neurovascular structures studied by using plastination. Eur J Neurol. 2005;12(5):380–4.
129. Sora MC, Jilavu R, Grübl A, Genser-Strobl B, Staykov D, Seicean A. The posteromedial neurovascular bundle of the ankle: an anatomic study using plastinated cross sections. Arthroscopy. 2008;24(3):258–263.e1.
130. Al-Ali S, Blyth P, Beatty S, Duang A, Parry B, Bissett IP. Correlation between gross anatomical topography, sectional sheet plastination, microscopic anatomy and endoanal sonography of the anal sphincter complex in human males. J Anat. 2009;215(2):212–20.
131. Elnady F, Sora MC. Anatomical exploration of a dicephalous goat kid using sheet plastination (E12). Congenit Anom (Kyoto). 2009;49(2):66–70.
132. Arredondo J, Agut A, Rodríguez MJ, Sarriá R, Latorre R. Anatomy of the temporomandibular joint in the cat: a study by microdissection, cryosection and vascular injection. J Feline Med Surg. 2013;15(2):111–6.
133. Chen S, Wang H, Fong AH, Zhang M. Micro-CT visualization of the cricothyroid joint cavity in cadavers. Laryngoscope. 2012;122(3):614–21.
134. Liu M, Chen S, Liang L, Xu W, Zhang M. Microcomputed tomography visualization of the cricoarytenoid joint cavity in cadavers. J Voice. 2013;27(6):778–85.
135. Bernal-Mañas CM, González-Sequeros O, Moreno-Cascales M, Sarria-Cabrera R, Latorre-Reviriego RM. New anatomo-radiological findings of the lateral pterygoid muscle. Surg Radiol Anat. 2016;38(9):1033–43.
136. Thorpe Lowis CG, Xu Z, Zhang M. Visualization of facet joint recesses of the cadaveric spine: a micro-CT and sheet plastination study. BMJ Open Sport Exerc Med. 2018;4(1):e000338.
137. Fritsch H, Hötzinger H. Tomographical anatomy of the pelvis, visceral pelvic connective tissue, and its compartments. Clin Anat. 1995;8(1):17–24.
138. Simank HG, Graf J, Schneider U, Fromm B, Niethard FU. Demonstration of the blood supply of human cruciate ligaments using the plastination method. Z Orthop Ihre Grenzgeb. 1995;133(1):39–42.
139. Fröhlich B, Hötzinger H, Fritsch H. Tomographical anatomy of the pelvis, pelvic floor, and related structures. Clin Anat. 1997;10(4):223–30.
140. Schiltenwolf M, Jakob DS, Graf J. Sheet plastination of the vascularity of the lunate bone—a morphological study. Acta Anat (Basel). 1997;158(1):68–73.
141. De Caro R, Aragona F, Herms A, Guidolin D, Brizzi E, Pagano F. Morphometric analysis of the fibroadipose tissue of the female pelvis. J Urol. 1998;160(3 Pt 1):707–13.
142. Andermahr J, Helling HJ, Landwehr P, Fischbach R, Koebke J, Rehm KE. The lateral calcaneal artery. Surg Radiol Anat. 1998;20(6):419–23.
143. Koebke J, Schäfer W, Aust T. Carpal tunnel topography during endoscopic decompression. J Hand Surg Br. 1999;24(1):3–5.
144. Andermahr J, Helling HJ, Tsironis K, Rehm KE, Koebke J. Compartment syndrome of the foot. Clin Anat. 2001;14(3):184–9.
145. Pavkov ML, Koebke J, Notermans HP, Brökelmann J. Quantitative evaluation of the utero-ovarian venous pattern in the adult human female cadaver with plastination. World J Surg. 2004;28(2):201–5.
146. Nash LG, Phillips MN, Nicholson H, Barnett R, Zhang M. Skin ligaments: regional distribution and variation in morphology. Clin Anat. 2004;17(4):287–93.

147. Prymka M, Wu L, Hahne HJ, Koebke J, Hassenpflug J. The dimensional accuracy for preparation of the femoral cavity in HIP arthroplasty. A comparison between manual- and robot-assisted implantation of hip endoprosthesis stems in cadaver femurs. Arch Orthop Trauma Surg. 2006;126(1):36–44.
148. Kneissl S, Probst A. Comparison of computed tomographic images of normal cranial and upper cervical lymph nodes with corresponding E12 plastinated-embedded sections in the dog. Vet J. 2007;174(2):435–8.
149. Soler M, Murciano J, Latorre R, Belda E, Rodríguez MJ, Agut A. Ultrasonographic, computed tomographic and magnetic resonance imaging anatomy of the normal canine stifle joint. Vet J. 2007;174(2):351–61.
150. Kneissl S, Probst A. Magnetic resonance imaging features of presumed normal head and neck lymph nodes in dogs. Vet Radiol Ultrasound. 2006;47(6):538–41.
151. Rath B, Notermans HP, Franzen J, Knifka J, Walpert J, Frank D, Koebke J. The microvascular anatomy of the metatarsal bones: a plastination study. Surg Radiol Anat. 2009;31(4):271–7.
152. Rath B, Notermans HP, Frank D, Walpert J, Deschner J, Luering CM, Koeck FX, Koebke J. Arterial anatomy of the hallucal sesamoids. Clin Anat. 2009;22(6):755–60.
153. Kaulhausen T, Siewe J, Eysel P, Knifka J, Notermans HP, Koebke J, Sobottke R. The role of the inter-/supraspinous ligament complex in stand-alone interspinous process devices: a biomechanical and anatomic study. J Neurol Surg A Cent Eur Neurosurg. 2012;73(2):65–72.
154. Kaulhausen T, Zarghooni K, Stein G, Knifka J, Eysel P, Koebke J, Sobottke R. The interspinous spacer: a clinicoanatomical investigation using plastination. Minim Invasive Surg. 2012;2012:538697.
155. Villamonte-Chevalier AA, Soler M, Sarria R, Agut A, Gielen I, Latorre R. Ultrasonographic and anatomic study of the canine elbow joint. Vet Surg. 2015;44(4):485–93.
156. Johnson G, Zhang M, Barnett R. A comparison between epoxy resin slices and histology sections in the study of spinal connective tissue structure. J Int Soc Plastination. 2000;15(1):10–3.
157. Steinke H. Plastinated body slices for verification of magnetic resonance tomography images. Ann Anat. 2001;183(3):275–81.
158. Porzionato A, Macchi V, Gardi M, Parenti A, De Caro R. Histotopographic study of the rectourethralis muscle. Clin Anat. 2005;18(7):510–7.
159. Macchi V, Porzionato A, Stecco C, Vigato E, Parenti A, De Caro R. Histo-topographic study of the longitudinal anal muscle. Clin Anat. 2008;21(5):447–52.
160. Scali F, Nash LG, Pontell ME. Defining the morphology and distribution of the alar fascia: a sheet plastination investigation. Ann Otol Rhinol Laryngol. 2015a;124(10):814–9.
161. Scali F, Pontell ME, Nash LG, Enix DE. Investigation of meningomyovertebral structures within the upper cervical epidural space: a sheet plastination study with clinical implications. Spine J. 2015b;15(11):2417–24.
162. Steinke H, Lingslebe U, Böhme J, Slowik V, Shim V, Hädrich C, Hammer N. Deformation behavior of the iliotibial tract under different states of fixation. Med Eng Phys. 2012;34(9):1221–7.
163. Alston M, Janick L, Wade RS, Weber W, Henry RW. A shark band saw blade enhances the quality of cut in preparation of specimens for plastination. J Int Soc Plastination. 1997;12:23–6.
164. Johnson GM, Zhang M. Regional differences within the human supraspinous and interspinous ligaments: a sheet plastination study. Eur Spine J. 2002;11(4):382–8.
165. Koslowsky TC, Berger V, Hopf JC, Müller LP. Presentation of the vascular supply of the proximal ulna using a sequential plastination technique. Surg Radiol Anat. 2015;37(7):749–55.
166. Koslowsky TC, Schliwa S, Koebke J. Presentation of the microscopic vascular architecture of the radial head using a sequential plastination technique. Clin Anat. 2011;24:721–32.
167. Wegmann K, Burkhart KJ, Buhl J, Gausepohl T, Koebke J, Müller LP. Impact of posterior tibial nail malpositioning on iatrogenic injuries by distal medio-lateral interlocking screws. A cadaveric study on plastinated specimens. Acta Orthop Belg. 2012;78:786–9.
168. Henry RW, Antinoff N, Janick L, Orosz S. E12 technique: an aid to study sinuses of psittacine birds. Acta Anat (Basel). 1997;158(1):54–8.

169. Aigner F, Zbar AP, Ludwikowski B, Kreczy A, Kovacs P, Fritsch H. The recto-genital septum: morphology, function, and clinical relevance. Dis Colon Rectum. 2004;47(2):131–40.
170. Sittel C, Eckel HE, Sprinzl GM, Stennert E. Plastination of the larynx for whole-organ sectioning. Eur Arch Otorhinolaryngol. 1997a;254(Suppl 1):S93–6.
171. Sittel C, Eckel HE, Ricke S, Stennert E. Sheet plastination of the larynx for whole-organ histology. Acta Anat (Basel). 1997b;158(1):74–80.
172. Qiu MG, Zhang SX, Liu ZJ, Tan LW, Wang YS, Deng JH, Tang ZS. Three-dimensional computational reconstruction of lateral skull base with plastinated slices. Anat Rec A Discov Mol Cell Evol Biol. 2004;278(1):437–42.
173. Sora MC, Brugger PC, Strobl B. Shrinkage during E12 plastination. J Int Soc Plastination. 2002;17:23–7.
174. Eckel HE, Koebke J, Sittel C, Sprinzl GM, Pototschnig C, Stennert E. Morphology of the human larynx during the first five years of life studied on whole organ serial sections. Ann Otol Rhinol Laryngol. 1999;108(3):232–8.
175. Sora MC, Dresenkamp J, Gabriel A, Matusz P, Wengert GJ, Bartl R. The relationship of neurovascular structures to the posterior medial aspect of the knee: an anatomic study using plastinated cross-sections. Rom J Morphol Embryol. 2015;56(3):1035–41.
176. Reed RB. Effects of reduced pressure on components of epoxy (E12) reaction mixture. J Int Soc Plastination. 2003;18:3–8.
177. Latorre RM, Reed RB, Gil F, Lopez-Albors O, Ayala MD, Martinez-Gomariz F, Henry RW. Epoxy impregnation without hardener: to decrease yellowing, to delay casting, and to aid bubble removal. J Int Soc Plastination. 2002;17:17–22.
178. Cook P. Sheet plastination as a clinically based teaching aid at the University of Auckland. Acta Anat (Basel). 1997;158:33–6.
179. Borzooeian Z, Enteshari A. Design of a silicone gasket with an iron core for polyester and epoxy sheet plastination. J Int Soc Plastination. 2006;21:17–20.
180. Johnson GM, Zhang M, Jones DG. The fine connective tissue architecture of the human ligamentum nuchae. Spine (Phila Pa 1976). 2000;25(1):5–9.
181. Panes-Villarroel C, Ponce N, Ottone NE. Modified Goldner Trichrome for Non-Decalcified Mineralized Tissue Embedded in Resin, XXV Congress of Anatomy of the Southern Cone. 2023;120.
182. Correa-Aravena J, Panes-Villarroel C, Prado A, Ponce N, Guzmán D, Vásquez B, Roa I, Veuthey C, Masuko T, Ottone NE. Development of a Micro-Plastination Protocol for Ultrathin Jaw Sections to Visualize the Dentogingival Junction, XXV Congress of Anatomy of the Southern Cone. 2023;38.
183. Ottone NE, del Sol M, Sánchez R, Zambrano F, Vásquez B. Ponce N. ANID FONDEQUIP EQM200228: "Adquisición de un TissueFAXS i PLUS para potenciar la investigación biomédica interdisciplinaria en la Universidad de La Frontera y la Región de La Araucanía - Acquisition of a TissueFAXS i PLUS to promote interdisciplinary biomedical research at Universidad de La Frontera and the La Araucanía Region", IX Concurso de Equipamiento Científico y Tecnológico mediano FONDEQUIP 2020 - Agencia Nacional de Investigación y Desarrollo (ANID), Ministerio de Ciencia, Tecnología, Conocimiento e Innovación, Gobierno de Chile, ANID-Fondequip-EQM200228. 2020.
184. Weber W, Henry RW. Sheet plastination of body slices—E12 technique, filling method. J Int Soc Plastination. 1993;7:16–22.
185. Windisch G, Weiglein AH. Anatomy of synovial sheaths in the talocrural region evaluated by sheet plastination. J Int Soc Plastination. 2001;16:19–22.
186. Beyersdorff D, Schiemann T, Taupitz M, Kooijman H, Hamm B, Nicolas V. Sectional depiction of the pelvic floor by CT, MR imaging and sheet plastination: computer-aided correlation and 3D model. Eur Radiol. 2001;11(4):659–64.
187. Sora MC, Strobl B, Radu J. High temperature E12 plastination to produce ultra-thin sheets. J Int Soc Plastination. 2004b;19:22–5.
188. Peschers UM, DeLancey JO, Fritsch H, Quint LE, Prince MR. Cross-sectional imaging anatomy of the anal sphincters. Obstet Gynecol. 1997;90(5):839–44.

189. Adds PJ, McCarthy P, Uddin J, Gore S. 3-D reconstruction of the retrobulbar orbital septa using Biodur E12®. J. Plastination. 2017;29:8–14.
190. Texeira DX, Cheung A, Naveed H, Adds PJ. Three-dimensional reconstruction of the orbital retrobulbar vasculature. Orbit. 2022;41(4):469–75.
191. Wu X, Ding H, Yang L, Chu X, Xie S, Bao Y, Wu J, Yang Y, Zhou L, Li M, Li SY, Tang B, Xiao L, Zhong C, Liang L, Hong T. Invasive corridor of clivus extension in pituitary adenoma: bony anatomic consideration, surgical outcome and technical nuances. Front Oncol. 2021;11:689943.
192. Shi K, Li Z, Wu X, Ma C, Zhu X, Xu L, Sun Z, Xu S, Liang L. The medial wall and medial compartment of the cavernous sinus: an anatomic study using plastinated histological sections. Neurosurg Rev. 2022;45(5):3381–91.
193. Ma C, Zhu X, Chu X, Xu L, Zhang W, Xu S, Liang L. Formation and fixation of the annulus of Zinn and relation with extraocular muscles: a plastinated histologic study and its clinical significance. Invest Ophthalmol Vis Sci. 2022;63(12):16.
194. Xu Z, Chapuis PH, Bokey L, Zhang M. Nature and architecture of the puboprostatic ligament: a macro- and microscopic cadaveric study using epoxy sheet plastination. Urology. 2017;110:263.e1–8.
195. Liugan M, Xu Z, Zhang M. Reduced free communication of the subarachnoid space within the optic canal in the human. Am J Ophthalmol. 2017;179:25–31.
196. Xu Z, Mei B, Liu M, Tu L, Zhang H, Zhang M. Fibrous configuration of the fascia iliaca compartment: an epoxy sheet plastination and confocal microscopy study. Sci Rep. 2020;10(1):1548.
197. Zwirner J, Zhang M, Ondruschka B, Akita K, Hammer N. An ossifying bridge—on the structural continuity between the Achilles tendon and the plantar fascia. Sci Rep. 2020;10(1):14523.
198. Bond JD, Xu Z, Zhang M. Fine configuration of the dural fibrous network and the extradural neural axis compartment in the jugular foramen: an epoxy sheet plastination and confocal microscopy study. J Neurosurg. 2020;135(1):136–46.
199. Villamonte-Chevalier A, Soler M, Sarria R, Agut A, Latorre R. Anatomical study of fibrous structures of the medial aspect of the canine elbow joint. Vet Rec. 2012;171:596.
200. Steinke H, Wiersbicki D, Speckert ML, Merkwitz C, Wolfskämpf T, Wolf B. Periodic acid-Schiff (PAS) reaction and plastination in whole body slices. A novel technique to identify fascial tissue structures. Ann Anat. 2018;216:29–35.
201. Pieroh P, Li ZL, Kawata S, Ogawa Y, Josten C, Steinke H, Dehghani F, Itoh M. The topography and morphometrics of the pubic ligaments. Ann Anat. 2021;236:151698.
202. Noriego D, Carrera A, Tubbs RS, Guibernau J, San Millán M, Iwanaga J, Cateura A, Sañudo J, Reina F. The lateral ulnar collateral ligament: anatomical and structural study for clinical application in the diagnosis and treatment of elbow lateral ligament injuries. Clin Anat. 2023;36(6):866–74. https://doi.org/10.1002/ca.23991.
203. Hammer N, Hirschfeld U, Strunz H, Werner M, Wolfskämpf T, Löffler S. Can the diagnostics of triangular fibrocartilage complex lesions be improved by MRI-based soft-tissue reconstruction? An imaging-based workup and case presentation. Biomed Res Int. 2017;2017:5870875.
204. Hedderwick M, Stringer MD, McRedmond L, Meikle GR, Woodley SJ. The oblique popliteal ligament: an anatomic and MRI investigation. Surg Radiol Anat. 2017;39(9):1017–27.
205. König HE, Sora MC, Seeger J, Donoso S. Anatomy of the mare ovary using plastinated sections by the method E12. Chil J Agric Anim Sci. 2017;33(1):59–63.
206. Xu Z, Tu L, Zheng Y, Ma X, Zhang H, Zhang M. Fine architecture of the fascial planes around the lateral femoral cutaneous nerve at its pelvic exit: an epoxy sheet plastination and confocal microscopy study. J Neurosurg. 2018;131(6):1860–8.
207. Liang L, Qu L, Chu X, Liu Q, Lin G, Wang F, Xu S. Meningeal architecture of the jugular foramen: an anatomic study using plastinated histologic sections. World Neurosurg. 2019;127:e809–17.
208. Xu Z, Lin G, Zhang H, Xu S, Zhang M. Three-dimensional architecture of the neurovascular and adipose zones of the upper and lower lumbar intervertebral foramina: an epoxy sheet plastination study. J Neurosurg Spine. 2020;1–11.

Chapter 7
Polyester Sheet Plastination Technique

General Description

Sheet plastination with polyester resin was designed by Gunther von Hagens to permanently preserve tissue slices, initially brain, within a rigid and durable sheet of cured resin [1–16]. Subsequently, this technique also began to be used to plastinate other body regions [4, 5, 17–29].

Toward the end of the 1980s, Prof. Gunther von Hagens developed the P35 polyester resin [3], which allowed the production of opaque brain slices, ensuring excellent differentiation between the gray and white matter, being identified as the gold standard in this type of sheet plastination technique [1–3, 19, 20, 30–36]. However, this technique presented certain difficulties and challenges in its implementation [20]. For this reason, Prof. von Hagens continued to investigate this type of polyester resin and managed to develop the P40 polyester resin [3], which was not only applied to brain sections, ensuring the differentiation between gray matter and white matter, but which was also identified by Latorre et al. in 2004 [37] as "an alternative for all tissues," and in this way it was also applied for the preservation of sections of body regions beyond the brain, ensuring the production of thin sections between 2 and 3 mm [3]. Subsequently, the authors Gao, Liu, Yu, and Sui, in 2006 [38] in the *Journal of the International Society for Plastination*, introduced a new polyester resin, defined by them as P45.

The steps of this technique are similar to those of the E12 technique, with variations particularly in the impregnation stage, in which the polyester resin can be used without a catalyst, and also in the curing stage. In this stage, it is necessary to create "flat chambers," also between two glass plates, but separated by a silicone tube and joined by paper fasteners. In this way, a curing chamber is made inside which the obtained slices, already impregnated, are placed and the chamber is filled with more polyester resin. Subsequently, the chamber is closed, and they are subjected to ultraviolet light (UV), whether from UV light tubes or they could also be subjected to

© The Author(s), under exclusive license to Springer Nature
Switzerland AG 2023
N. E. Ottone, *Advances in Plastination Techniques*,
https://doi.org/10.1007/978-3-031-45701-2_7

Table 7.1 Equipment, instruments, accessories, and chemicals used in sheet plastination with polyester

Steps [a]	Equipments and accessories	Chemicals
Fixation	Cannulas Basic dissection instruments Containers Injection pump	Formalin
Sectioning	Deli slicer Band saw Deep freezer (−80 °C)	Liquid nitrogen Gelatin Polyurethane foam
Dehydration	Freezer (−25 °C) Containers (HDPE plastic or stainless steel) Acetonometer/s Grids (plastic or metal)	Acetone
Defatting	Containers (HDPE plastic or stainless steel) Grids (plastic or metal)	Acetone Dichloromethane Extraction hood
Forced impregnation	Vacuum chamber (RT) Vacuum pump Vacuometer Valves (vacuum control) Grids (plastic or metal)	Polyester
Curing	Tempered glass (4–5 mm thickness) Silicone gasket Sealant Clamps or paper clips Steel ball or tiny metal Magnets UV lights	–

[a] The samples can be fresh, without fixation (except for brains, always fixated). *RT* room temperature

sunlight, but in the shade, not directly to sunlight, thus ensuring the proper temperature (which should not exceed 30 degrees) and thus allowing the final hardening of the samples. Table 7.1 lists the equipment, instruments, accessories, and chemical products used in sheet plastination with polyester. Figure 7.1 shows the basic steps of the classical technique of sheet plastination with polyester.

Fixation of the Samples

The samples, in the first place the brains, to assure an adequate consistency that facilitates the slicing and to avoid an excessive shrinkage during the plastination process, require to be fixed in 20% formalin, during 3–6 months. In this sense, it is best to start with concentrations of 5, 10, and 15% formalin, with initial successive changes of 1 week, reaching a concentration of 20% formalin in the fourth week, leaving the sample in this concentration for 30 more days. Prior to fixation, the dura mater and all blood vessels must be removed from the brains to facilitate the penetration of formalin into the nervous tissue. Fixation must be performed by

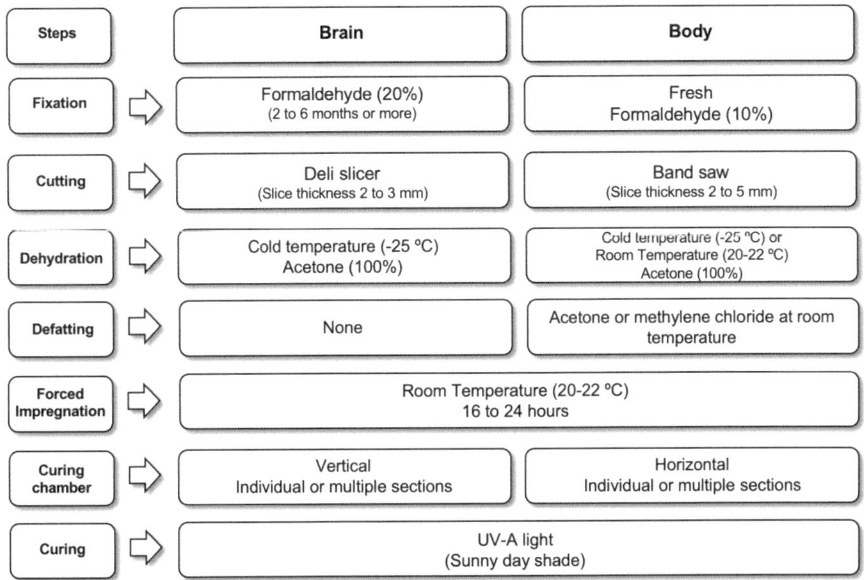

Fig. 7.1 Basic steps of the classical technique of sheet plastination with polyester

immersion. For fixation of body regions other than the brain, a 10% formalin fixation formulation can be used, ideally starting with concentrations of 2.5%, 5%, and 7.5%, until reaching the final 10%. The preservation of the samples, except for the brain, prior to plastination, must be done in refrigerated rooms at a temperature of 4–5 °C.

Washing of the Samples

Prior to sectioning, the samples must be washed in running water for a minimum of 1 week, to eliminate the excess formalin present in the tissues prior to sectioning. If fixation has been performed with other types of fixation solutions, for example, containing glycerin or other alcohols, before washing with running water, the samples must be washed in a mixture of 50% ethanol, 3% hydrogen peroxide, and the rest running water.

Positioning of Samples for Sectioning

In case the samples are sliced with a band saw, the sample must be immersed in a rectangular container of adequate size, in a warm solution of 20% gelatin, positioning the sample properly so that the block solidifies when it cools down. When the gelatin solidifies, the formed block is extracted from the container, and in this way the block of gelatin that will contain the brain in its interior is extracted, in order to proceed to its section. It is also possible to use polyurethane foam for the generation of the cutting block. Afterward, the gelatin/polyurethane foam block with the sample inside must be cooled in two ways. One option is to place the sample for 1 week in a regular freezer at −25 °C and then 1 day in an ultra-freezer at −80 °C; another option, if an ultra-freezer is not available, is to keep it in a regular freezer at −25 °C for 2 weeks. After cooling, it can be sliced with a circular saw. In the case of brains, if they are going to be sliced using a "deli slicer" saw, neither this step of inclusion in gelatin nor the cooling in a regular freezer and/or ultra-freezer will be necessary.

Sectioning of Samples Using a Circular Saw

The sample contained in the gelatin block will be extracted from the cold to proceed with the sectioning (see Chap. 6). It will be positioned in the way desired to make the section, facilitated by being a rectangular block, on the cutting plate of the circular saw. It is recommended to apply liquid nitrogen on the cutting guide, on which the block rests to subject it to the action of the saw blade, so that the heat of the saw does not thaw the sample and thus hinder the making of the slices, which may be between 2 and 5 mm thick. There are also circular saws that allow the adaptation of containers for the incorporation of alcohol and dry ice inside, thus allowing the cutting guides to be kept cold. Once the cut has been made, the solidified gelatin, located on the outside of the sample, can be removed manually. Immediately afterward, the slices obtained must be placed in cold acetone (−25 °C) at 100%.

Brain Section Using a "Deli Slicer" Saw

The brains will be sectioned into 2–3-mm-thick slices with a cutting machine with a stainless-steel saw (Fig. 7.2). The slices obtained will be placed between perforated grids, superimposing them one on top of the other and keeping the brain sections in between, keeping the slices separated, so that they can dehydrate properly. These grids must be knotted with thread, ideally fishing line, forming a package, to ensure that the slices do not move (Fig. 7.3). This package is then immersed in acetone to start the dehydration process, and without disassembling it, once the dehydration is finished, this package will be immersed in the polyester resin for the forced impregnation process. Immediately after cutting, the slices obtained should be placed in cold acetone (−25 °C) at 100%.

Fig. 7.2 Deli slicer saw

Fig. 7.3 Assembly of a package with grids that allow separating and preserving the anatomical integrity of the brain slices. This package will not be disassembled until the end of the plastination process, ensuring correct handling of the samples

Dehydration

The brain sections will be dehydrated in consecutive baths of cold acetone ($-25\ ^\circ$C) at 100%. The percentage of acetone must be controlled during the whole dehydration process. The duration of the acetone baths is variable and can be between 4 and 5 days. For this purpose, it is very important to control the acetone concentration daily with the acetometer/alcoholimeter, and prior to each extraction of acetone for measurement, it is advisable to shake the acetone to allow mixing of the acetone with the water and fixative liquid extracted from the samples, so that an adequate measurement of the concentration of acetone present in the respective bath can be made. As it is checked that the acetone concentration remains at the same value for 2 consecutive days, this is an indication that it is necessary to renew the acetone bath with 100% acetone, extracting the previous acetone and saving it in case it can be recycled. If after the dehydration process, it is found that the concentration of acetone is higher than 99.5%, the brain slices are considered completely dehydrated and ready to pass to the forced impregnation step.

Defatting

This step of defatting at room temperature should only be applied to slices of body regions, and not to brain sections. Related to brain sections, it goes directly from dehydration to forced impregnation. But in the case of body slices, it is necessary to carry out this defatting stage, for a variable period of 1–4 weeks, but depending on the visualization of the color of the acetone. Due to the fat, these will gradually turn yellowish in the initial stages of defatting and will remain transparent at the end of this stage, which will allow determining its completion. Consider that you should make weekly changes or as the acetone turns yellowish. Commonly, degreasing is done with acetone, and the last days, 2–5 days, the slices can be submerged in dichloromethane to enhance defatting (but this last step can be avoided, especially due to the toxicity and danger of using dichloromethane that requires an extraction hood for handling). Therefore, the entire defatting process can be done with acetone at room temperature.

Forced Impregnation

This step consists of removing the acetone and replacing it with polyester resin by generating a vacuum inside a vacuum chamber at room temperature (20–22 $^\circ$C) (Fig. 7.4). During the whole process of forced impregnation, the vacuum chamber must be kept covered, keeping the slices submerged in the polyester in darkness,

Fig. 7.4 The dehydrated samples are submerged in a container with polyester resin and placed in a vacuum chamber at room temperature. The figures show the start of the forced impregnation process with the presence of small acetone bubbles

since ultraviolet light, as will be seen in the curing stage, acts as a catalyst of the polyester resin and could harden very early, affecting the continuity of the plastination process. Pressure reduction will be performed for 24 h, from 760 mmHg to 10 mmHg. Vacuum must not be continued beyond 10 mmHg as there is a risk of starting to extract some components of the P40 [18, 20, 21]. The acetone extraction process will start at around 220 mmHg, pressure at which the acetone vaporizes and, by means of the vacuum generated and the existing pressure differences, this acetone in vapor form will be extracted from the samples, and the interstitial space left will be occupied by the polyester. The active-passive forced impregnation process can be applied [22], with two 8-h phases of active forced impregnation (16 h total with vacuum pump on) separated by an intermediate 8-h phase of passive forced impregnation (with vacuum pump off). Or perform forced impregnation continuously, with active extraction from 760 mmHg to the final 10 mmHg. The extraction of acetone and its replacement by polyester resin are visualized by the appearance of bubbles, which is the acetone being extracted from 220 mmHg (Fig. 7.5). Once the bubbling is over, the forced impregnation has ended, which occurs when 10 mmHg is obtained (Fig. 7.6).

Fig. 7.5 Intense bubbling corresponding to advanced phases of the forced impregnation stage at room temperature in the plastination of sections with polyester (the instruments that are displayed simply act as a weight to keep the package containing the sections submerged; these instruments must subsequently be placed in acetone for cleaning)

Assembly of Vertical Flat Curing Chambers

Once the forced impregnation is completed, the slices are ready to be incorporated into the vertical flat curing chambers (Fig. 7.7). This type of chamber is used because it allows the introduction of the impregnated brain slices through an upper opening, between two panes of glass. In this way, the brain and cerebellum slices will be extracted from the polyester resin and placed inside vertical flat chambers to be taken to the curing stage. These flat chambers consist of two 2–3-mm-thick glass plates, which face each other separated by a silicone gasket, and are held together using clamps or classic paper clips, which are placed on all sides of the glasses as

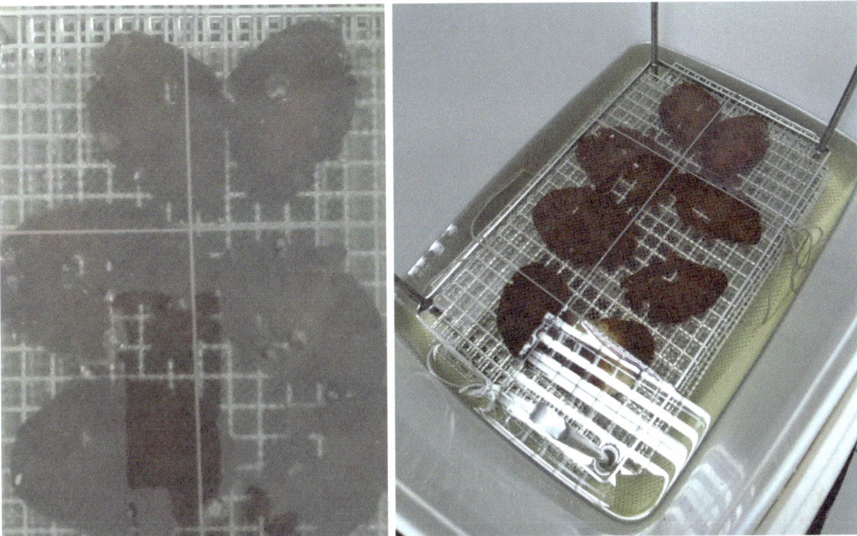

Fig. 7.6 Completion of the forced impregnation step, with a clear visualization of the color change of the samples, due to their impregnation

Fig. 7.7 Necessary elements to carry out the assembly of the curing chambers that will allow the curing/polymerization process of the sections impregnated with polyester resin to be carried out (see Table 7.1)

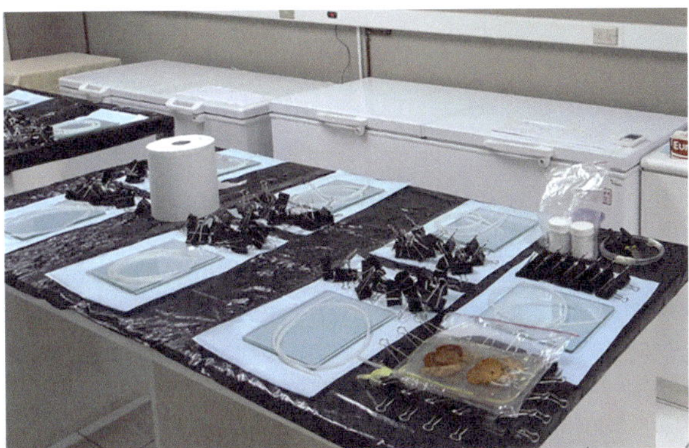

follows: glass + silicone gasket + glass, and always closing the clamp or paper clip above the level of the silicone gasket, where adequate pressure is required to prevent separation and opening of the flat chamber. In this way, the chamber is assembled with an upper opening, through which the polyester is incorporated first, until half of the vertical chamber is filled, and then the corresponding brain and cerebellum slices are introduced (Fig. 7.8). Depending on their size and species and the size of the chamber, one or several slices can be inserted. Before closing the curing chamber, a steel ball must be inserted, which has the function, once the curing chamber is closed and when a magnet is applied on it, of moving the sample slices to allow their correct positioning inside the curing chamber (Figs. 7.9 and 7.10). Subsequently,

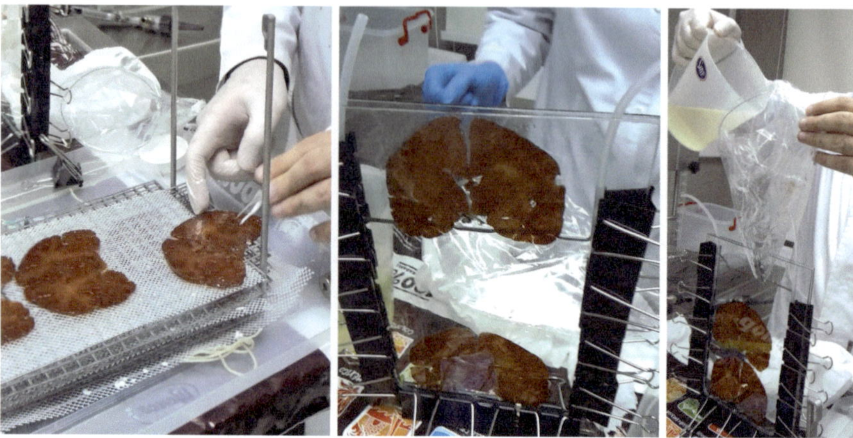

Fig. 7.8 Placement of the impregnated slices in the curing chambers, which must finally be filled until they cover all the cuts. The assembly of the curing chambers can be seen between two glass plates separated by a silicone gasket and joined together by paper clips

the chamber must be closed by placing a sealing putty in the joint between the ends of the silicone gasket.

Assembling of Horizontal Flat Curing Chambers

This is a second flat curing chamber option, but it is horizontal, in which larger glasses can be used, especially for large body region slices or for the final preservation of multiple brain and/or cerebellum slices on the same plate. In this type of chambers, tempered glass of a greater thickness than that of vertical flat chambers must be used, and the glass can be identified as lower (on which the work will be done) and upper (which will serve as a cover). Therefore, a "P" type silicone gasket must initially be placed on the lower pane, at its periphery, which will function as a containment dike for the polyester that must be incorporated and which will fill this space between the glasses and on which the already impregnated slices must be placed. This silicone gasket can be glued to the lower glass using a silicone-based glue. Once the lower pane is in place together with the "P" gasket, a minimal amount of polyester can be placed to fill the space over which the impregnated slices can be placed. Once all the slices have been located and positioned, they must be completely covered with polyester, and finally the upper glass must be placed over the "P" gasket to close the horizontal flat curing chamber. It can also be glued to the glass with the same glue previously used, although clamps and/or paper clips can also be used, placing them on the perimeter of the chamber, placing their closing pressure on the silicone "P" gasket, thus sealing and closing the chamber properly.

Fig. 7.9 Before closing the curing chamber, a steel ball or a small piece of metal must be inserted, which will serve, through the guidance of a magnet, to properly position the slices inside the curing chamber

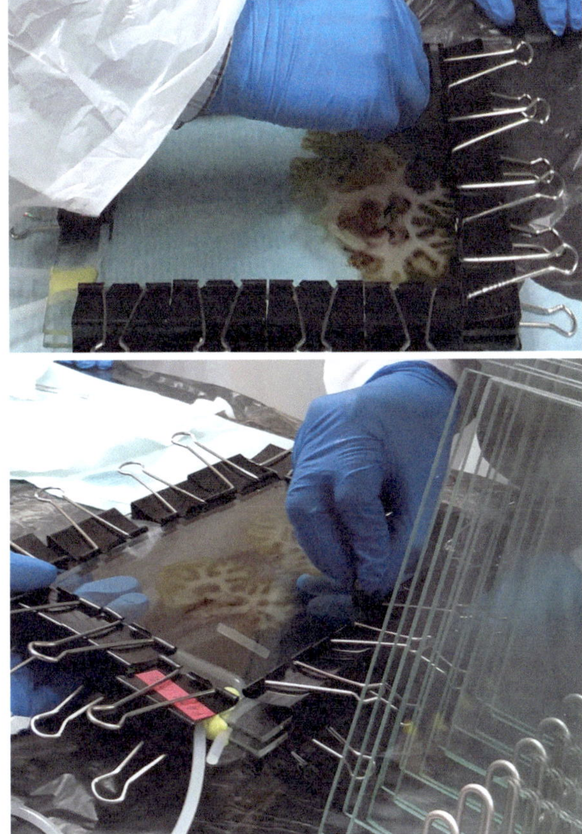

Subsequently, the chamber must be closed by placing a sealing putty in the joint between the ends of the "P" type silicone gasket.

Curing (Polymerization)

The curing chambers (horizontal and vertical) must be subjected to UV-A light (350–370 nm or 36–40 W), which polymerizes the polyester resin through an exothermic reaction, recommending the use of artificial UV lights, acting at the same time on both sides of the curing chambers, without interference between the light and the chamber, at a distance of approximately 30 cm from the light (Fig. 7.11). It is also important that the temperature does not exceed 40 °C, so it is recommended that the room where the curing takes place is air-conditioned at room temperature (20–22 °C), or a fan directed toward the UV-A lights can also be used. On the other hand, this curing process could also be performed by exposing the chambers to

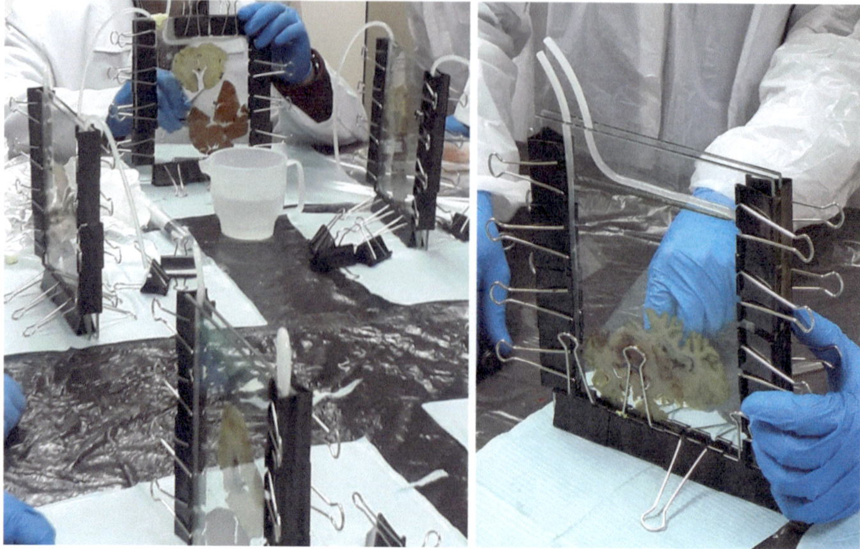

Fig. 7.10 Assembling the curing chambers and accommodation of the slices by using the magnet, before proceeding to close the chambers

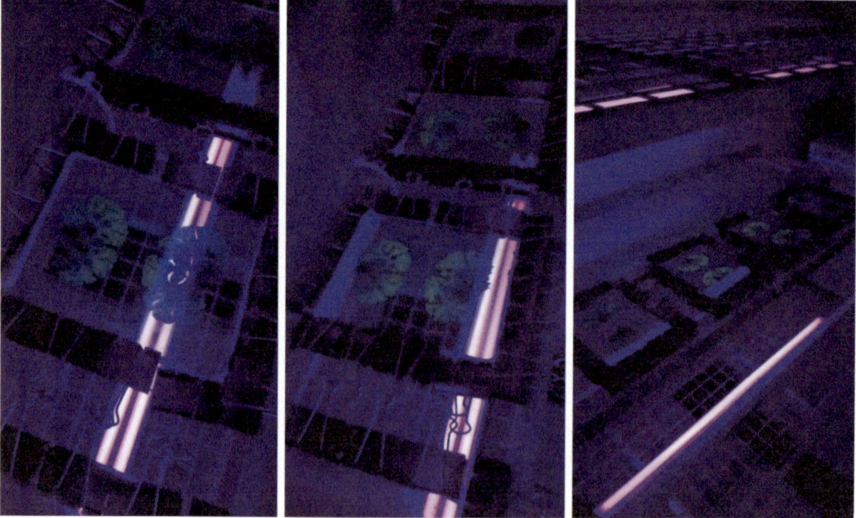

Fig. 7.11 Once the curing chambers are assembled, they are placed under the action of UV light, which will accelerate the polymerization of the polyester resin

sunlight, specifically in the shade of a sunny day, but always considering the afore-mentioned temperature limitation. The hardening of the polyester resin lasts between 12 and 24 h.

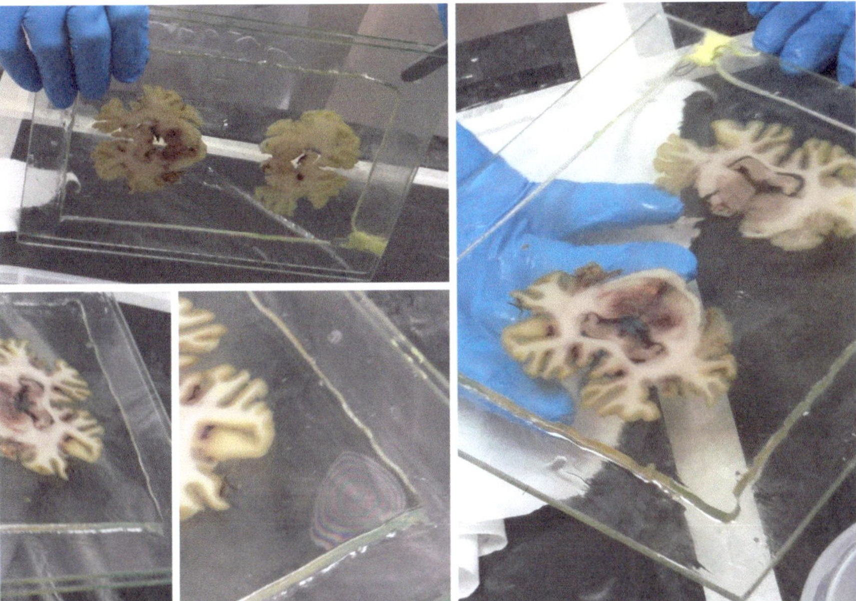

Fig. 7.12 Once the action of the UV light has finished, and the hardening of the polyester resin has been achieved, the curing chambers are disassembled. In the curing chambers, when disassembling is carried out, the characteristic effect can be seen at the level of the glass skimming

Disassembly of Curing Chambers

Once the curing process is finished, the curing chambers can be disassembled, with separation of the glass after removing the clamps or fasteners (Figs. 7.12 and 7.13). Immediately after separating the glass, the polyester plate with the slices inside must be covered with a transparent plastic wrap (Fig. 7.14). This will later allow cutting with a band saw the edges of the plate, where the uncured resin may remain and could damage the plates, in addition to protecting them from the dirt generated by the cut (Fig. 7.15). In addition, this cut will allow the polyester plates to be aligned for their subsequent presentation.

Discussion

The plastination protocols of slices with polyester resin, although initially intended for the preservation of brain slices due to its purpose of differentiation of gray matter and white matter, can also be extended to any anatomical region, taking into account that in comparison to the plastination of slices with epoxy resin, polyester determines a greater shrinkage of the tissues, without a greater transparency of the tissues [17]. However, excellent results are achieved. Moreover, in the long term,

Fig. 7.13 Disarming of curing chambers. (a) Curing chambers before disassembly. (b) Curing chambers after disassembly

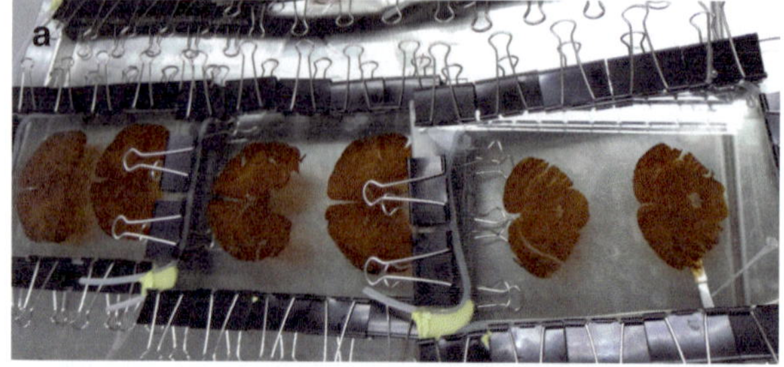

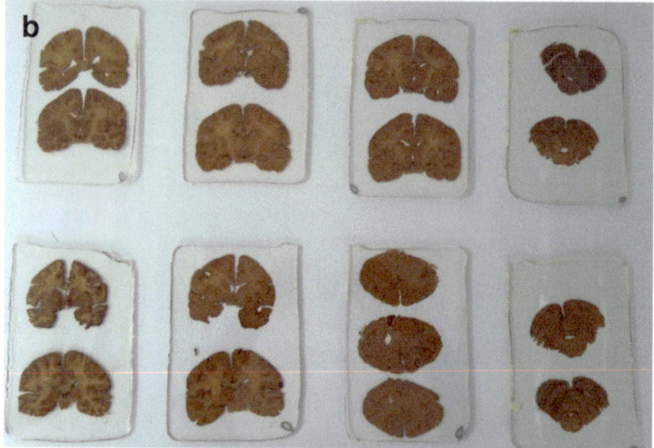

Fig. 7.14 Once the curing chambers have been disassembled, the polyester plates with the cuts inside must be wrapped in plastic wrap to avoid damaging the surface of the plates and to contribute to the drying of the edges, which can stabilize something sticky

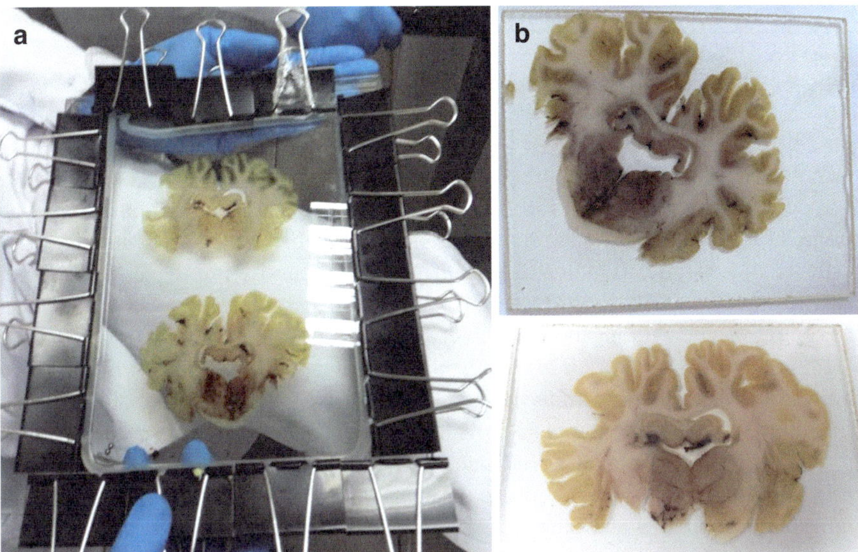

Fig. 7.15 Completion of the process of plastination of sections with polyester resin, with disassembly of the chambers and pairing by cutting the edges of the polyester plates at the level of the plastinated slices

slices plastinated with polyester resin do not turn yellow, as is the case with slices plastinated with epoxy resin [18, 20, 21].

Regarding the dehydration step, this must be performed at low temperature (-25 to -20 °C) with acetone concentrations above 90%, ideally 100%, in all acetone baths, where a concentration above 99% must finally be reached to finalize the dehydration process. The objective of dehydration at low temperatures is to reduce the shrinkage of brain tissue, which also occurs, but to a lesser extent due to the effect of acetone, which at room temperature also has a great capacity to eliminate fat, which would contribute to a greater shrinkage of brain tissue if left at room temperature. Sora and Brugger [39] conducted an interesting study in which they replace the use of acetone by methanol. These authors mention the difficulty in the use of acetone in laboratories due to biosafety conditions, especially the risks of explosion. However, in the Latin American region, there are several countries in which access to acetone is hindered by governmental measures, prohibiting its use for scientific research. In accordance with the characteristics of the components used in plastination, it is necessary that the ideal intermediate solvent for dehydration has a high vapor pressure and must be soluble in water, characteristics that acetone fulfills, which is why it is the one selected for this technique. However, Sora and Brugger [39], while searching for an alternative to acetone, found in methanol a water-soluble compound with high vapor pressure, similar to acetone. In this sense, methanol has a higher boiling point and lower vapor pressure, characteristics that make it less volatile than acetone. In addition, the flash point of acetone is significantly lower than the flash point of methanol, a fact that makes acetone more dangerous than methanol. However, its extraction during the forced impregnation stage is somewhat more complex, requiring precise handling and regulation of pressures, something that these authors only recommend for experienced plastinators [39].

With respect to forced impregnation, in the P40 technique, this is usually performed at room temperature (20 °C), and a pressure of at least 20 mmHg must be reached for the completion of impregnation within 24 h. Active impregnation can be achieved in less time, as in this case, with a period of passive impregnation and with the vacuum pump turned off. During the whole process, the vacuum chamber must be kept covered, without the ultraviolet light affecting the polyester, to avoid the beginning of polymerization of the resin and the darkening of the brain slices. The impregnation is performed only with polyester resin, without the use of a catalyst. This allows the reuse of the resin, which significantly reduces costs in the preparation of samples. In addition, as indicated by Baptista et al. [20], the use of polyester resin without a catalyst allows its use for many years, if it is kept away from a UV-A light source and if it is also stored cold (4–5 °C). They also indicate that before reusing that cold-stored polyester resin, it must be brought to room temperature [18, 20, 21, 40]. Furthermore, in the case of using a catalyst together with the polyester resin, incomplete polymerization can be caused if the UV-A light does not penetrate deeply into the tissues, and this can be seen not only in the case of the brain but also in the bones and the liver [20].

On the other hand, forced impregnation can be performed in a total of 16 h of active forced impregnation (8 h of activity in 2 consecutive days) with a period of 12 h of passive forced impregnation (vacuum pump off), differing from the classic techniques of plastination of slices with polyester resin, in which forced impregnation is continuous, without turning off the vacuum pump [2].

With regard to curing, there are protocols in which P40 polyester is combined with a catalyst. According to our experience, we do not consider it necessary to combine the polyester with a catalyst. However, in those protocols that do combine them, it is crucial to ensure a correct fixation of the brains, since a deficient fixation will determine the appearance of orange spots in the cerebral cortex; these are due to the presence of active peroxidase in the brain tissue, which reacts with the catalyst in the P40 polymer. Therefore, this peroxidase can be largely inactivated by the fixatives, and, for this reason, complete fixation (10% formalin) is vital for success with the P40 technique in combination with a catalyst [32].

Gao et al. [38] have published a plastination protocol with P45 polyester resin from the Chinese company Hoffen. This company, whose president is Prof. Dr. Sui, presents a new polyester resin. In this publication, they have presented the plastination of dolphin slices, presenting a novelty in their protocol in the curing step, where they replaced the use of UV light with a water bath. With this new curing method, they were able to reduce the space required to cure the samples by positioning them vertically, and the water bath ensured the same temperature dispersion in all the samples during curing, preventing temperatures from rising too high, keeping the temperatures even in all the slices.

Reed et al. [41] compared in their research different methods to accelerate polymerization and achieve curing of the samples. In this research, they concluded that ultraviolet light sources provided the fastest cure rates compared to sunlight and other artificial ultraviolet light sources. For example, they found that exposing the samples to the shade of sunlight doubled the curing time compared to direct exposure to other UV light sources. The same is true for direct exposure to mercury vapor lamps. In all cases, it is critical to control the temperature, since direct exposure to UV light sources, without temperature control, would cause the glass in the curing chambers to break [41]. It is important not to exceed 30 °C [4]; above this temperature, there is a risk of glass breakage. Reed et al. [41] also found that fluorescent lighting had no effect on the curing of P40 resin, with no observation of temperature rise nor exothermic reaction, as occurs in the curing of P40 resin. When brain slices are plastinated with polyester, they become hard and partially see-through. The reason for this partial transparency is that both the gray matter and the polyester resin have similar refractive indices. Consequently, the polyester-plastinated slices display a distinct gray-white differentiation [42] .

Shrinkage of Brains in the Technique of Slice Plastination with Polyester Resin

Brown et al. [43] conducted a study on the shrinkage levels caused by dehydration with acetone at room and cold temperature as well as with methanol. In relation to dehydration at room temperature with acetone, they found that the shrinkage was 20.2%, while, at cold temperature, the shrinkage was 14.5%. On the other hand, in the dehydration with methanol at room temperature, the shrinkage was higher, 22.6%. However, in relation to dehydration with acetone at room temperature, for Holladay [44] it was 35%. Meanwhile, Sora and Brugger [39] analyzed the shrinkage levels of the tissues during plastination with epoxy resin, finding a shrinkage of between 2 and 4% after dehydration and between 6 and 10% at the end of the whole plastination process with epoxy resin. According to Sora et al. [16] plastination is a method that is ideal for creating three-dimensional reconstructions and for taking precise morphometric measurements. So it is essential to consider the retraction that plastinated slices may experience after the technique is applied. The authors analyzed the retraction in the P40, E12 standard, and E12 ultra-fine techniques. The shrinkage rate for the P40 technique was 5.74%, for the standard E12 technique it was 6.54%, and for the E12 thin slices technique it was 6.23%.

Measurements of brain slices were taken before dehydration, after dehydration, and after forced impregnation. Data collection was recorded in a Microsoft Office Excel spreadsheet; a descriptive analysis of the data was performed: mean, median, variance, minimum, maximum, standard error of the mean, standard deviation, and coefficient of variation. The normal distribution of the data was established using the Shapiro-Wilk test. The student's t-test for related samples was used to establish the mean difference of the shrinkage, in the case of data following a normal distribution, and the Wilcoxon rank test for related samples, for data following a non-normal distribution. The alpha error was set at $p < 0.05$. For the determination of the shrinkage percentage, the mean and the values of its confidence interval were transformed into percentages, and the result was subtracted from 100: p. retraction = $(((\bar{\chi})$ of retraction$)/(\bar{\chi}$ before the process$) \times 100)-100$. The IBM SPSS Statistics V23 statistical program was used for data analysis. One-way ANOVA studies and the Kruskal-Wallis test for independent samples were used, and 95% confidence intervals were constructed for the respective variables. IBM SPSS Statistics (version 23.0) was used for data analysis. A value of $p < 0.05$ was chosen as the threshold for significance.

There was a statistically significant difference between the initial measurement values and after the first and second dehydration at both the latero-lateral and superior-inferior levels.

It was found that the shrinkage percentages between the initial values and dehydration are between 2.14% and 6.22%; and with respect to forced impregnation, they are between 7.02% and 10.62%, the latter percentages being the values corresponding to the shrinkage caused by the technique of plastination of slices with polyester resin in human brain sections of 3 mm thickness.

In this sense, our results are closer to those established by Sora and Brugger [39] and Sora et al. [16], especially when considering the performance of dehydration at cold temperature (−25 °C). Thus, according to von Hagens [1], it can be concluded that dehydration at low temperatures (−25 °C to 5 °C) will allow less tissue shrinkage.

Sheet Plastination with P45 Polyester Resin

In 2006, the authors Gao, Liu, Yu, and Sui published in the *Journal of the International Society for Plastination* a paper titled: "A New Polyester Technique for Sheet Plastination" [38]. In this work they presented a new polyester resin for sheet plastination technique, the P45 polyester resin (Dalian Hoffen Biotechnic Co. Ltd., Dalian, China), which was applied for the conservation of tissue slices from a cape dolphin (*Delphinus capensis*). In this sense, one of the objectives of this article, in addition to investigating the cape dolphin, was to present a new sheet plastination protocol with polyester resin that is easier to develop and more cost-effective than previous techniques [38].

The P45 sheet plastination technique consists of the following steps [38]:

Cutting of Samples

In order to cut the samples, as previously mentioned in this chapter, once the samples have been selected, they must be frozen for 2 weeks in a ultra-deep freezer at −70 °C and then placed inside a box, embedded in expanded polyurethane, to generate the cut block, which must later be frozen again at −70 °C, but for only 2 days. Once this is done, the blocks are ready to be cut on a high-speed band saw, achieving a thickness of between 2 and 3 mm.

Bleaching of Samples

In this technique, which could also be applied to the techniques described above, the samples must be washed with cold water overnight and then submerged in 5% hydrogen peroxide, also overnight, to achieve the desired bleaching. The type of bleaching achieved will depend on the percentage of hydrogen peroxide used, as well as the time spent for it.

Dehydration and Degreasing of the Samples

Once the desired bleaching of the samples has been achieved, the samples must be subjected to the dehydration process [38, 45]. To do this, they must first be pre-cooled to 5 °C to avoid ice crystal formation and to minimize shrinkage upon placement into cold acetone. Then, the samples were placed in 100% pure acetone, at −2 °C, for 7 days. Subsequently, the authors indicate that the samples should be passed to a second bath of 100% acetone concentration, but at −1 °C, for 10 days. The authors indicate that when the acetone concentration remains unchanged for 3 consecutive days, through measurements made with the acetonometer, the samples are transferred to a next dehydrating bath. Finally, the authors proceed to the degreasing process of the samples, at room temperature, in 100% acetone, for 1 week. This dehydration and defatting process is similar to that originally described for the plastination techniques described by von Hagens [1, 2]. However, in other publications, carried out by the same research group, the following dehydration process is described: the slices were pre-cooled [46, 47] and then were dehydrated in the following way: first acetone bath of 85% acetone, at −25 °C, for 5 days [46, 47]; second acetone bath of 90% [47] to 93% [46] acetone, at −15 °C for other 5 days; and defatting at room temperature in 100% acetone [46, 47].

Forced Impregnation of the Samples

To carry out the forced impregnation process, once the previous stage of dehydration and defatting has finished, the slices must be incorporated into a glass chamber, whose assembly is similar to that described earlier in this chapter ("Assembling of horizontal flat curing cameras"). Therefore, slices are placed in these chambers pre-filled with P45 polyester resin. Once the chambers are assembled, they are placed in a vertical position inside the vacuum chamber at room temperature to start the forced impregnation process. The authors indicate that the pressure reduction must be carried out until reaching pressures very close to 0 mmHg (20, 10, 5, and 0 mmHg) [38], according to the size of the bubbles. In this way, they described that in more than 8 h, the pressure reduction can be achieved until reaching 0 mmHg, and in the absence of visualization of bubbling, the forced impregnation can be finished.

Curing of the Samples

Once the forced impregnation is finished, the bubbles that could be kept inside the chambers must be removed. Subsequently, the samples are placed for 3 days in a hot water bath at 40 °C. After this period, the chambers can be disassembled, and the samples included in the polyester can be extracted, which must be covered with bonded plastic foil to protect the surface of the polyester slices. These slices can be cut to size or to straighten the edges of the resulting polyester plates to give the final result a better appearance.

The technique of plastination of sections with P45 polyester resin was applied in approximately 24 articles, from its first description in 2006 [38] to the present [19, 35, 48–68].

Comparative Details of the Sheet Plastination Technique with P35/P45 Polyester Resin, in Relation to the Sheet Plastination Technique with P40 Polyester Resin

In 1997, Weiglein [42] and Barnett et al. [69] described the techniques for sheet plastinating with P35 polyester resin, developed originally by Gunther von Hagens. This technique differs from the process already described for the sheet plastination technique with P40 polyester resin. In the impregnation of the P35 technique, a hardener is added along with the polyester resin. Additionally, the slices need to be immersed in this impregnation mixture for 24 hours at 5°C, followed by placing the sections in an immersion mixture for 24 hours under vacuum, and then at room temperature for another 24 hours. However, in the P40 technique, a single impregnation mixture is used. There is another difference in the processes at the time of curing. In P35, the slices not only need to be exposed to UV light for polymerization, but after this exposure, slices must be placed for 5 days in an oven at 45°C [70]. On the other hand, the main differences that are established between the P40 [3] and P45 [38] techniques occur at the level of the forced impregnation and curing stages. In this sense, in the P40 technique, the process of forced impregnation of the slices of the corresponding sample (whether from the brain or any other body region) follows the classical plastination model: the slices must be submerged, organized in separators perforated plastics, in a container with polyester resin, and in this way the forced impregnation is developed inside a vacuum chamber at room temperature. However, in the P45 technique, the sections are individually impregnated in flat glass chambers (similar to the flat chambers used for the curing process in the P40 technique), which are placed vertically inside a vacuum chamber, at room temperature. In relation to the differences in the curing stage, in the P40 technique, the fully impregnated slices must be extracted from the polyester resin inside the vacuum chamber, and they are introduced into the flat curing chambers, which are exposed to UV light, which acts by accelerating the polymerization and hardening of the polyester. Typically, in 24 h or less, hardening of the polyester occurs. For its part, in the P45 technique, the slices are already in flat chambers from the forced impregnation stage, and in this case, these chambers are placed in a hot water bath at 40 °C for 3 days. The stages of cutting, dehydration, and defatting are similar in both techniques, responding to what was originally described by Prof. Gunther von Hagens [1–3].

References

1. von Hagens G. Heidelberg plastination folder. Collection of technical leaflets of plastination. Heidelberg: Biodur Products GmbH; 1986.
2. von Hagens G, Tiedemann K, Kriz W. The current potential of plastination. Anat Embryol. 1987;175(4):411–21. https://doi.org/10.1007/BF00309677.
3. von Hagens G. Plastination of brain slices according to P40 procedure. A step-by-step description. In: Heidelberg plastination folder. Collection of technical leaflets of plastination. Heidelberg: Biodur Products GmbH; 1994. p. 1–23.
4. Ottone NE. Plastination: techniques fundamentals and implementation at Universidad de La Frontera. J Health Med Sci. 2018;4(4):293–302.
5. Ottone NE, Baptista CAC, Latorre R, Bianchi HF, Del Sol M, Fuentes R. E12 sheet plastination: techniques and applications. Clin Anat. 2018;31(5):742–56. https://doi.org/10.1002/ca.23008.
6. Weiglein AH. Plastinated brain-specimens in the anatomical curriculum at Graz University. J Int Soc Plast. 1993;7:3–7.
7. Sora MC, Brugger P, Traxler H. P40 plastination of human brain slices: comparison between different immersion and impregnation conditions. J Int Soc Plast. 1999;14(1):22–4.
8. Steinke H, Pfeiffer S, Spanel-Borowski K. A new plastination technique for head slices containing brain. Ann Anat. 2002;184(4):353–8. https://doi.org/10.1016/S0940-9602(02)80055-3.
9. Lozanoff S, Lozanoff BK, Sora MC, Rosenheimer J, Keep MF, Tregear J, Saland L, Jacobs J, Saiki S, Alverson D. Anatomy and the access grid: exploiting plastinated brain sections for use in distributed medical education. Anat Rec B New Anat. 2003;270(1):30–7. https://doi.org/10.1002/ar.b.10006.

10. Bravo H. Plastination and additional tool to teach anatomy. Int J Morphol. 2006;24(3):475–80. https://doi.org/10.4067/S0717-95022006000400029.

11. Borzooeian Z, Enteshari A. Design of a silicone gasket with an iron core for polyester and epoxy sheet plastination. J Int Soc Plast. 2006;21:17–20.

12. Henry RW, Latorre R. Polyester plastination of biological tissue: P40 technique for brain slices. J Int Soc Plast. 2007;22:59–68.

13. Pashaei S. A brief review on the history, methods and applications of plastination. Int J Morphol. 2010;28(4):1075–9. https://doi.org/10.4067/S0717-95022010000400014.

14. Rabiei AA, Asadi MH, Esfandiari E, Taghipour M, Bahadoran H, Setayesh M, Shamoosi A, Mardani M, Rashidi B, Fathollahpour A. Preparation of flexible plastinated sheets of human brain by P87 polyester. J Isfahan Med Sch. 2011;28(124):1961–5.

15. Wadood AA, Jabbar A, Da N. Plastination of whole brain specimen and brain slices. J Ayub Med Coll Abbottabad. 2011;13(1):11–3.

16. Mohamed SKA, El-Behery EI, Mahdy EAERA. Computed tomography scan and polyester resin 40 plastination technique: Teaching aids to illustrate anatomical structure of donkey brain. World Vet J. 2019;9(3):230–40.

17. Ottone NE, Guerrero M, Alarcón E, Navarro P. Statistical analysis of shrinkage levels of human brain slices preserved by sheet plastination technique with polyester resin. Int J Morphol. 2020;38(1):13–6. https://doi.org/10.4067/S0717-95022020000100013.

18. Latorre R, Henry RW. Polyester plastination of biological tissue: P40 technique for body slices. J Int Soc Plast. 2007;22:69–77.

19. Sui HJ, Henry RW. Hoffen P45: a modified polyester plastination technique for both brain and body slices. J Plast. 2015;27(2):4–8. https://doi.org/10.56507/ONMI1596.

20. Baptista CAC, DeJong K, Latorre R, Bittencourt AS. P40 polyester sheet plastination technique for brain and body slices: the vertical and horizontal flat chamber methods. Anat Histol Embryol. 2019;48(6):572–6. https://doi.org/10.1111/ahe.12486.

21. Guerrero M, Vargas C, Alarcón E, del Sol M, Ottone NE. Development of a sheet plastination protocol with polyester resin applied to human brain slices. Int J Morphol. 2019;37(4):1557–63. https://doi.org/10.4067/S0717-95022019000401557.

22. Ottone NE, Cirigliano V, Bianchi HF, Medan CD, Algieri RD, Borges Brum G, Fuentes R. New contributions to the development of a plastination technique at room temperature with silicone. Anat Sci Int. 2015;90(2):126–35. https://doi.org/10.1007/s12565-014-0258-6.

23. Latorre R, Arencibia A, Gil F, Rivero M, Ramirez G, Vaquez-Auton JM, Henry RW. P-40 and S10 plastinated slices: an aid to interpreting MR images of the equine tarsus. J Int Soc Plast. 2003;18:14–22.

24. Genser-Strobl B, Sora MC. Potential of P40 plastination for morphometric hip measurements. Surg Radiol Anat. 2005;27(2):147–51. https://doi.org/10.1007/s00276-004-0298-z.

25. Ali RA, Ebrahim E, Morteza H, Atefeh S, Mohammad M, Bahman R, Shahnaz R, Hamid S, Ali V. Comparing using of natural and chemical dyes in process of flexible 3-dimentional plastination in heart. J Isfahan Med Sch. 2010;28(109):385–92.

26. Cires DN, Sinescu C, Dodenciu D, Ardelean L, Negrutiu ML, Halga AD, Rominu M. Plastination in dentistry: methods and polymers. Mat Plast. 2012;49(3):196–7.

27. Kürtül I, Hammer N, Rabi S, Saito T, Böhme J, Steinke H. Oblique sectional planes of block plastinates eased by Sac plastination. Ann Anat. 2012;194(4):404–6. https://doi.org/10.1016/j.aanat.2011.11.006.

28. Valenzuela OM, Azocar SC, Werner FK, Vega PE, Valdés GF. Plastination experience in polyester resin (P-4). Int J Morphol. 2012;30(3):810–3. https://doi.org/10.4067/S0717-95022012000300006.

29. Sivagnanam S, Geetha R, Kannan TA, Jeyachandra K. Polyester resin plastination for light weight poultry specimens. Asian J Sci Technol. 2014;5(3):183–4.

30. Weber W. Sheet plastination of the brain, P35 technique, filling technique. J Int Soc Plast. 1992;6(1):6–7.

31. Weber W, Henry RW. Sheet plastination of the brain - P 35 technique, filling method. J Int Soc Plast. 1992;6:29–33.

32. de Boer-van RT, Cornelissin CJ, ten Donkelaar HJ. Sheet plastination of the human head. J Int Soc Plast. 1993;6(1):20–4.
33. Weiglein AH. Preparing and using S-10 and P-35 brain slices. J Int Soc Plast. 1996;10:22–5.
34. Weber W, Weiglein A, Latorre R, Henry RW. Polyester plastination of biological tissue: P35 technique. J Int Soc Plast. 2007;22:50–8.
35. Sui HJ, Henry RW. Polyester plastination of biological tissue: Hoffen P45 technique. J Int Soc Plast. 2007;22:78–81.
36. Riederer BM. Plastination and its importance in teaching anatomy. Critical points for long-term preservation of human tissue. J Anat. 2014;224(3):309–15. https://doi.org/10.1111/joa.12056.
37. Latorre R, Arencibia A, Gil F, Rivero M, Ramirez G, Vaquez-Auton JM, Henry RW. Sheet plastination with polyester: an alternative for all tissues. J Int Soc Plast. 2004;19:33–9.
38. Gao H, Liu J, Yu S, Sui H. A new polyester technique for sheet plastination. J Int Soc Plast. 2006;21:7–10.
39. Sora MC, Brugger P. P40 brain slices plastination using methanol for dehydration. J Int Soc Plast. 2000;15(1):22–4.
40. Barnett RJ. Plastination of coronal and horizontal brain slices using the P40 technique. J Int Soc Plast. 1997;12(1):33–6.
41. Reed RB, Helms L, Rowe JA. Curing times of P40 exposed to different light sources. J Int Soc Plast. 2008;28:25–9.
42. Weiglein AH. Plastination in the neurosciences. Acta Anat. 1997;158:6–9. https://doi.org/10.1159/000147902.
43. Brown MA, Reed RB, Henry RW. Effects of dehydration mediums and temperature on total dehydration time and tissue shrinkage. J Int Soc Plast. 2002;17:28–33.
44. Holladay SD. Experiments in dehydration techniques. J Int Soc Plast. 1988;2(2):17–20.
45. Zhang XH, Gong J, Song Y, Hack GD, Jiang SM, Yu SB, Song X, Zhang J, Yang H, Cheng J, Sui HJ, Zheng N. An anatomical study of the suboccipital cavernous sinus and its relationship with the myodural bridge complex. Clin Anat. 2023;36:726–36. https://doi.org/10.1002/ca.24048.
46. Jiang WB, Song TW, Sun SZ, Li C, Adds P, Xu Q, Liu C, Tang W, Chi YY, Chen W, Yu SB, Sui HJ. Bony structures at the attachment areas of the cruciate ligaments in human knee joints and their clinical significance-studied using the P45 plastination technique. Int J Morphol. 2022;40(6):1579–86. https://doi.org/10.4067/S0717-95022022000601579.
47. Jiang WB, Li C, Gilmore C, Yu SB, Sui HJ. Detailed anatomy of the corpora-glans ligament via the P45 plastination method. Int J Morphol. 2023;41(1):264–7. https://doi.org/10.4067/S0717-95022023000100264.
48. Schwab K, von Hagens G. Freeze substitution of macroscopic specimens for plastination. Acta Anat. 1981;111:139–40.
49. Zheng N, Yuan XY, Li YF, Chi YY, Gao HB, Zhao X, Yu SB, Sui HJ, Sharkey J. Definition of the to be named ligament and vertebrodural ligament and their possible effects on the circulation of CSF. PLoS One. 2014;9(8):e103451. https://doi.org/10.1371/journal.pone.0103451.
50. Chun P, Yu S, Qin H, Sui H. The use of P45 plastination technique to study the distribution of preseptal and preaponeurotic fat tissues in Asian eyelids. Cell Biochem Biophys. 2015;73(2):313–21. https://doi.org/10.1007/s12013-015-0581-0.
51. Yuan XY, Yu SB, Li YF, Chi YY, Zheng N, Gao HB, Luan BY, Zhang ZX, Sui HJ. Patterns of attachment of the myodural bridge by the rectus capitis posterior minor muscle. Anat Sci Int. 2016;91(2):175–9. https://doi.org/10.1007/s12565-015-0282-1.
52. Zhang JH, Tang W, Zhang ZX, Luan BY, Yu SB, Sui HJ. Connection of the posterior occipital muscle and dura mater of the Siamese crocodile. Anat Rec. 2016;299(10):1402–8. https://doi.org/10.1002/ar.23445.
53. Zheng N, Yuan XY, Chi YY, Liu P, Wang B, Sui JY, Han SH, Yu SB, Sui HJ. The universal existence of myodural bridge in mammals: an indication of a necessary function. Sci Rep. 2017;7(1):8248. https://doi.org/10.1038/s41598-017-06863-z.

54. Liu P, Li C, Zheng N, Xu Q, Yu SB, Sui HJ. The myodural bridge existing in the Neophocaena phocaenoides. PLoS One. 2017;12(3):e0173630. https://doi.org/10.1371/journal.pone.0173630.
55. Zhang JF, Du ML, Sui HJ, Yang Y, Zhou HY, Meng C, Qu MJ, Zhang Q, Du B, Fu YS. Investigation of the ischioanal fossa: application to abscess spread. Clin Anat. 2017;30(8):1029–33. https://doi.org/10.1002/ca.22901.
56. Liu P, Li C, Zheng N, Yuan X, Zhou Y, Chun P, Chi Y, Gilmore C, Yu S, Sui H. The myodural bridges' existence in the sperm whale. PLoS One. 2018;13(7):e0200260. https://doi.org/10.1371/journal.pone.0200260.
57. Okoye CS, Ya-Ru D, Sui HJ. Tissue shrinkage after P45 plastination. J Plast. 2019;31(2):26–30. https://doi.org/10.56507/DNIB6497.
58. Okoye CW, Sui HJ. Updated protocol for the Hoffen P45 sheet plastination technique. J Plast. 2019;31(2):22–5. https://doi.org/10.56507/ONMI1596.
59. Zheng N, Chung BS, Li YL, Liu TY, Zhang LX, Ge YY, Wang NX, Zhang ZH, Cai L, Chi YY, Zhang JF, Samuel OC, Yu SB, Sui HJ. The myodural bridge complex defined as a new functional structure. Surg Radiol Anat. 2020;42(2):143–53. https://doi.org/10.1007/s00276-019-02340-6.
60. Ma XD, Chun P, Zhang C, Li FF, Qin T, Qin HZ, Sui HJ. Investigation of retro-orbicularis oculi fat and associated orbital septum connective tissues in upper eyelid surgery. Ann Palliat Med. 2020;9(6):3899–908. https://doi.org/10.21037/apm-20-1822.
61. Sun SZ, Jiang WB, Song TW, Chi YY, Xu Q, Liu C, Tang W, Xu F, Zhou JX, Yu SB, Sui HJ. Architecture of the cancellous bone in human proximal tibia based on P45 sectional plastinated specimens. Surg Radiol Anat. 2021;43(12):2055–69. https://doi.org/10.1007/s00276-021-02826-2.
62. Jiang WB, Sun SZ, Li C, Adds P, Tang W, Chen W, Yu SB, Sui HJ. Anatomical basis of the support of fibula to tibial plateau and its clinical significance. J Orthop Surg Res. 2021;16(1):346. https://doi.org/10.1186/s13018-021-02500-8.
63. Zhang ZX, Gong J, Yu SB, Li C, Sun JX, Ding SW, Ma GJ, Sun SZ, Zhou L, Hack GD, Zheng N, Sui HJ. A specialized myodural bridge named occipital-dural muscle in the narrow-ridged finless porpoise (Neophocaena asiaeorientalis). Sci Rep. 2021;11(1):15485. https://doi.org/10.1038/s41598-021-95070-y.
64. Qin T, Chun P, Li FF, Yu SB, Hwang K, Sui HJ. Medial and lateral canthal ligaments shown in P45 sheet plastination and dissection. Indian J Ophthalmol. 2021;69(5):1150–4. https://doi.org/10.4103/ijo.IJO_2848_20.
65. Chi YY, Zhuang J, Jin G, Hack GD, Song TW, Sun SZ, Chen C, Yu SB, Sui HJ. Congenital atlanto-occipital fusion and its effect on the myodural bridge: a case report utilizing the P45 plastination technique. Int J Morphol. 2022;40(4):796–800. https://doi.org/10.4067/S0717-95022022000300796.
66. Zhuang J, Gong J, Hack GD, Chi YY, Song Y, Yu SB, Sui HJ. A new concept of the fiber composition of cervical spinal dura mater: an investigation utilizing the P45 sheet plastination technique. Surg Radiol Anat. 2022;44(6):877–82. https://doi.org/10.1007/s00276-022-02962-3.
67. Weber W. Sheet plastination of brain slices. J Int Soc Plast. 1994;8:23.
68. Hwang K, Sui HJ, Han SH, Kim H. The nasolabial area shown on histology and P45 sheet plastination. J Craniofac Surg. 2021;32(2):771–3. https://doi.org/10.1097/SCS.0000000000007032.
69. Barnett R, Burland G, Duxson M. Plastination of coronal slices of brains from cadavers using the P35 technique. J Int Soc Plast. 2005;20:16–9.
70. Üzel M, Weiglein AH. P35 Plastination: experiences with delayed impregnation. J Plast. 2013;25(1):9–11. https://doi.org/10.56507/PQUD7016.
71. Sora MC, Binder M, Matusz P, Ples H, Sas I. Slice plastination and shrinkage. Mat Plast. 2015;52(2):186–9.
72. Sora MC, Latorre R, Baptista C, López-Albors O. Plastination-A scientific method for teaching and research. Anat Histol Embryol. 2019;48(6):526–31. https://doi.org/10.1111/ahe.12493.

Chapter 8
Research Applications of Plastination

Introduction

Currently, more than 400 institutions located in countries around the world use Gunther von Hagens' invention to preserve anatomical specimens for the teaching of anatomy in all careers of health sciences. In the last 10 years, the plastination technique has taken center stage in morphological research, and scientific publications using this technique have increased remarkably. To take this topic further, from 1979 (the year in which von Hagens published the first article disclosing the plastination technique) to May 2023, in a PubMed search on plastination, we can find a total of 470 articles. In this sense, the articles present sheet plastination with epoxy resin (E12) as the main technique, a fundamental technique for research, thanks to the benefits it gives to the preserved material: transparency, practically zero shrinkage, and the possibility of visualizing paths of arteries, veins, nerves, muscles, ligaments, and all kinds of anatomical structures, without the need for decalcification. There are also many scientific articles aimed at demonstrating the usefulness of plastination in teaching anatomy, as well as the ethical implications of plastination. In this chapter, a large number of scientific publications that used plastination to investigate and generate new knowledge from the morphological point of view will be described.

Head and Neck

Sittel et al. [1] described two plastination techniques: block plastination and slice plastination. In the block plastination, they first plastinated the complete samples, without cutting them, and for this purpose fresh larynxes were destined for processing, carrying out dehydration in acetone at −23 °C for 1 month, and renewing the

© The Author(s), under exclusive license to Springer Nature Switzerland AG 2023
N. E. Ottone, *Advances in Plastination Techniques*,
https://doi.org/10.1007/978-3-031-45701-2_8

acetone weekly. Subsequently, for the degreasing stage, they used methylene chloride for 4 days. Next, the samples were placed in an impregnation mixture, made up of E12/E6/E600 (100:70:0.15%), to develop the forced impregnation. Then, hardening was achieved by placing the samples in an oven at 60 °C for 2 weeks. Finally, the resin blocks were cut, reaching sections of 0.8 mm. Subsequently, to improve transparency, the sections were polished and embedded xylol-based mounting medium. On the other hand, for the sheet plastination technique, they first made the slices of the sample and then plastinated them. In this case, they used larynxes frozen in liquid nitrogen, which were later cut with a traditional saw, obtaining 4-mm-thick slices. Next, the laryngeal sections were fixed in 15% formalin for several hours, and then 10% vol. methanol was added to prevent the formation of ice crystals. Dehydration was then carried out for 48 h and with four changes of acetone. After this, defatting was carried out in methylene chloride for 2 h at room temperature. Subsequently, the samples were immersed in an impregnating mixture, whose components were E50/E7/AE15/E700 (100:80:25 ppw and 0.1% of the total volume, Biodur). For the next step of curing (polymerization) of the slices, they described the assembly of a curing chamber, made up of two tempered glass plates, covered by sheets of polyethylene, and separated from each other by a flexible elastic joint. In this way, the plastinated sections are introduced into this curing chamber, which is filled with the same impregnating mixture. The curing chambers are placed inside the vacuum chamber, and the bubbles are removed by vacuum for 24 h, to reach the final cure in 3 days. Subsequently, once the plastination is finished, the 4-mm-thick samples will be cut, to further reduce their thickness, using a saw with a diamond blade and a polisher. In this way, Sittel et al. [1] investigated the human larynx and applied histological stains, without the need for decalcification, in results obtained with both plastination techniques, without explaining the histological procedure in detail.

Johnson et al. [2] investigated the arrangement of connective tissue at the level of the human nuchal ligament, analyzing the muscular and ligamentous structures of the posterior region of the cervical spine. Although they had a total of nine cadavers (two male and seven female), they applied the E12 plastination technique on only one cadaver, and the rest of the cadavers were destined for anatomical dissection. There was no further explanation about the E12 plastination protocol development, only the mention of its implementation, and the obtaining of 44 cross sections of 2.5 mm thickness, using a saw blade of 1.6 mm thickness for cutting. Once the plastinated sections were obtained, they were visualized by means of low and high magnifications using a Leica MZ8 stereoscopic microscope with a magnification range of X0.68-5. From the morphological point of view, they were able to identify the close relationship between the aponeurotic fascias of the bilaminar muscles and the arrangement of the connective tissue in the conformation of the nuchal ligament, thus promoting the importance of the muscles in the stability of the cervical spine. From these conclusions, and from the comparison between the possibilities offered by plastinated sections in relation to histological sections, it was possible to identify the advantages of plastination in relation to the inclusion in paraffin of histological techniques, in relation to the maintenance of the anatomical structures and their surrounding relationships in large body regions, avoiding the disruption of the tissues in the plastinated samples and allowing without inconveniences, despite being larger slices, and the correct identification and distinction of the connective tissue from the connective tissue in relation to the adipose tissue.

Zhang and An [3] described a technique of plastination of sections with epoxy resin to obtain anatomical sections that were studied by means of electron microscopy, with the objective of visualizing the subarachnoid space and, in particular, Liliequist's membrane, wall trabecular double arachnoid fold. For this research, the sample consisted of 38 adult human cadavers, of which 3 were destined for the plastination process [4]. Prior to the start, they stained the arachnoids with Gill's hematoxylin, through perforations made in the skull. Likewise, through the carotid arteries, they were injected with a red gelatin solution (eosin 6%), and through the venous system, a blue gelatin solution (methyl blue 1.2%) was injected. Subsequently, the heads of the bodies were separated at the level of the second cervical vertebra, and they were submerged in a 20% gelatin solution and frozen at $-30\ °C$, to generate a block of gelatin with the head inside. The gelatin block was frozen at $-80\ °C$ for 24 h and then cut, obtaining 2.5-mm-thick sections. Subsequently, the sections were dehydrated in acetone (95–100%) at $-30\ °C$ for 6 weeks, and then they were defatted at room temperature (22–24 °C) for 2 weeks. Next, the samples were immersed in the impregnation mixture, to develop the forced impregnation stage, in a formulation consisting of E12/E1/AE10/AE30 (100:28:20:5, Biodur), at 0 °C for 24 h. Finally, the sections were placed inside an oven, at 32 °C and 45 °C, for a week at each temperature. In this way, it was possible to obtain epoxy resin sections, which presented the subarachnoid space transparent, the transcisternal arteries colored red and the veins blue, in addition to achieving the identification of the purple-blue arachnoid trabeculae and the cranial nerves. They were also able to visualize very well preserved, in a natural state, not collapsed, and well differentiated by the incorporated dyes.

Leaper et al. [5] investigated the causes of dysphagia in the elderly, characterized mainly by dysfunction of the cricopharyngeal muscle, identifying the characteristics of the posterior hypopharyngeal wall in 31 adult human cadavers (15 female and 16 male, aged 60–97 years, average age 77 years), especially associated with the anatomical causes of the protrusion of this region. For this, they developed a technique of sheet plastination with epoxy resin (E12 Biodur) in two cadavers (both male), identifying the fundamental steps of the technique, which also allowed them to study the relationships between the structures with the surrounding tissues, such as the retropharyngeal spaces, prevertebral fascia, and vertebral fascia. In relation to the plastination technique with epoxy resin (E12 Biodur), firstly they fixed the corpses by means of the classic procedure of 10% formalin, to then freeze the corpses at $-80\ °C$ for 24 h, to then proceed to cutting, and to obtain sections 2.5 mm thick. Next, these sections were dehydrated in increasing concentrations of acetone (95–100%) at $-30\ °C$ for 6 weeks, with a subsequent defatting process at room temperature (22–24 °C) for 2 weeks. Particularly, and without detailed explanation, the sections were then immersed in a polymer mixture (Biodur E12/E1/AE10/AE30, 100:28:20:5) at 0 °C for 24 h. This same mixture was used for the forced impregnation, without indicating the temperature at which it was developed nor the time used. Finally, they indicate that the slices were placed between plastic sheets, and the slices were cured (polymerized) for a week at temperatures between 32 °C and 45 °C. Finally, the sections were visualized using a stereomicroscope and an Olympus BHB optical microscope.

Sora et al. [6] developed an investigation oriented to the application of plastination of sections with epoxy resin (E12) to demonstrate the possibility that this technique contributes to carry out morphological measurements and also the possibility of examining the obtained sections histologically. To do this, they applied the proposed technique in a human sample that presented compression of the optic nerve due to hypertrophy and degeneration of the extraocular muscles near the apex of the orbit. The sample corresponded to a fresh, unfixed human orbit, corresponding to a 71-year-old male cadaver. The sample was frozen at −80 °C for 1 week and then cut into ten sections of approximately 3.8 ± 0.46 mm. Sections were stored at −25 °C and scanned to assess shrinkage. Sections were dehydrated in cold acetone (−25 °C) and then defatted with methylene chloride for 1 week. Forced impregnation was carried out at +5 °C, using the following formula for the impregnation mixture: E12/E1/AE10 (100:30:20, Biodur), using the same proportions suggested by Gunther von Hagen in his handbook on plastination of 1985. Subsequently, the sections were placed in curing chambers made up of two glass plates separated by a 4 mm flexible joint that acted as a separator. In this way, they were filled with the impregnating mixture, the sections were placed inside, and the curing chambers were left tilted at an angle of 15° at room temperature for 1 day, to then place the curing chambers in an oven at 45 °C for 4 days, to achieve final polymerization. The sections, in turn, were cut even thinner with a saw with a diamond blade in sections of 300 um and then polished and thinned even more, until reaching 150 um. Finally, each slice was scanned for visualization. In a subsequent step, each section was stained with hematoxylin and eosin, and the histology of the samples was analyzed at up to ×40 magnification; however, the protocol developed to achieve this histological staining is not described. Sora et al. [6] indicate that the E12 plastination technique allowed them to obtain high-quality anatomical sections, with perfect coloration and transparency, without evidence of tissue retraction. Thus, they also indicated that plastination allows the topographic study of all the anatomical structures of the orbit in a non-collapsed or distorted state, keeping the orbit intact and without the need to resort to bone decalcification for the development of the technique. Finally, they indicate that three-dimensional reconstructions could be made from these plastinated sections.

Zhang and Lee [7] identified the great difficulties of routine dissection to allow the identification and distinction to distinguish from the membranous (or fibrous) portion of the subcutaneous tissue, the deep fascia, the epimysium, and the epitendineum, indicating that histological techniques allow this differentiation; however in histology there is the limitation of the size of the sample. Zhang and Lee [7] therefore consider that the slice plastination technique provides a new approach to illustrate the detailed structural structure of connective tissue at the macro- and microscopic levels. An important finding of this investigation consisted in the demonstration of the absence of aggregation of fibrous connective tissue that makes a connection between the sternocleidomastoid and trapezius muscles, this space being completely occupied by fatty tissue and being indistinguishable from subcutaneous cell tissue. Regarding the protocol for plastination of sections with epoxy resin, they applied the technique in five cadavers, developing coronal slices in one cadaver, transversal slices in two cadavers, and sagittal slices in the two remaining cadavers. The samples were fixed in 10% formalin, to be subsequently frozen at −80 °C for at least 24 h. The samples were sectioned and 2.5-mm-thick slices were obtained with a 2 mm saw blade. Next, these anatomical sections were placed in dehydration, in increasing concentrations of acetone (95–100%) at a temperature of −30 °C for 6 weeks, and subsequently placed at room temperature (22–24 °C) for 2 weeks to favor the defatting of the samples. Subsequently, the samples were immersed in the resin mixture, in a container at 0 °C for 24 h. The impregnating mixture consisted of E12/E1/AE10/AE30 (100:28:20:5, parts by weight, Biodur). The authors do not indicate whether the impregnation continued at 0 °C and only indicate that the sections were subsequently placed between plastic sheets and cured at 32 °C and 45 °C for 1 week at each temperature.

Zhang et al. [8] investigated the course of the oculomotor nerve and the corresponding blood vessels in the region, studying the anatomical correlation between the plastinated sections obtained from three cadaveric samples and the images obtained from the analysis of 140 living subjects from magnetic resonance studies. Regarding the plastination technique performed, it was a technique of sheet plastination with epoxy resin, indicating only the components of the impregnation mixture used: E12/E6/E600 (Biodur). No proportions of the components of the mixture were indicated nor were further details of the remaining steps of the plastination. The samples were used to make transverse, sagittal, and coronal plastinated slices, with a thickness of 1 mm, with a loss of 0.3 mm due to the thickness of the saw blade. Among the conclusions associated with the utility of the sheet plastination technique in this type of research, Zhang et al. [8] concluded that the use of the plastination technique can result in a low rate of tissue retraction, maintaining the original shape and spatial relationships between the different anatomical structures, with very low tissue loss due to slices, and obtaining transparent anatomical sections that are as thin as those obtained by magnetic resonance. In addition, they identified that working at low temperature (without indicating which or at what point in the plastination process) allowed them to improve the protocol, thereby significantly reducing tissue shrinkage. In addition, they also indicated that arteries and nerves can be accurately visualized due to the thinness of the sections. It was possible to confirm what was seen in the plastinated sections, in relation to the anatomical relationships of the analyzed structures (oculomotor nerve, basilar artery, posterior cerebral artery, superior cerebellar artery, and posterior communicating artery, among other anatomical structures) with what was obtained from magnetic resonance images. Obviously, they indicate that plastination cannot allow the visualization of anatomical variations or compression of blood vessels by nervous structures, particularly due to the limited number of samples that could be subjected to the plastination technique, thus not being a problem of the plastination technique itself, and rather this being a limitation of the research itself, since if an anatomical variation is found, it could be correctly visualized in a plastinated sample (as evidenced by a large number of publications).

Qiu et al. [9] carried out a work aimed at investigating the lateral base of the skull and its relationship with adjacent structures. In this sense, they used three fresh human heads, which were fixed with formalin and analyzed by helical computed tomography, obtaining 1.0 mm sections. Subsequently, they extracted blocks of 80 mm per side, which contained the lateral region of the skull base on both sides, and these blocks are the ones that proceeded to plastinate. First, the blocks were cooled at -4 °C for 24 h, and then they were dehydrated in 100% acetone and at a temperature of -25 °C, for a total of 35 days. Next, the forced impregnation of the blocks was carried out, in an impregnation mixture composed as follows: epoxy resin E12, hardener E6, and accelerator E500 (1:0.5:0.1, W/W/V, Biodur), without indicating the forced impregnation time. Finally, the samples were subjected to the curing process (polymerization) in an oven at 50 °C for 2 weeks, to later be cut on a saw with a diamond blade (Ebner, Mannheim, Germany), obtaining ultrafine slices of 700 um. In this way, the samples were cut with a resolution down to the microscopic level and scanned to achieve three-dimensional reconstructions, and then they were compared with the computed tomography images, demonstrating the plastinated slices a superior differentiation to the images of computed tomography in relation to the identification of the different anatomical structures. In this way, they confirm that the possibility of plastinating at levels of ultrafine slices with the subsequent possibility of performing three-dimensional reconstructions ensures the possibility of generating 3D anatomical models for their application in surgical procedures, in space, for preoperative analysis, in this case, in surgeries oriented toward otorhinolaryngology and neurosurgery.

Nash et al. [10] carried out an interesting investigation on the configuration of connective tissue at the level of the posterior atlanto-occipital interspace, indicating that as a result of the plastination process, connective tissue, especially collagen, has the ability to be endogenously autofluorescent upon excitation of 488 nm [10–12]. In this sense, the authors applied the technique of plastination of sections with epoxy resin in seven human samples, intended for sagittal (2), coronal (2), and transverse (3) sections. The samples were placed in 20% gelatin to generate a block, which was subsequently frozen at -80 °C for 24 h. This frozen gelatin block was cut with a traditional saw, achieving 2.5 mm slices. Subsequently, the sections obtained were dehydrated in acetone (86.5–100%) at -25 °C for 8 weeks, to then be degreased for 4 weeks at room temperature (18–24 °C). Once the degreasing was finished, the sections were subjected to the forced impregnation process, using an impregnation mixture composed of E12/E1/AE10/AE30 (100: 28:20:5) (Biodur). The impregnation temperature was 0 °C, and it was applied for 24 h. Finally, the sections were placed between 50-um-thick plastic sheets, to carry out the curing process (polymerization), in an oven at a variable temperature of 30–40 °C, for 3 days. In this way, the plastinated sections were obtained, which were then visualized in a Leica MZ8 stereoscopic microscope, with a magnification of 1.25–5×. Nash et al. [11] developed these early studies of the deep cervical fascia in human samples. Additionally, Nash et al. [12] determined that tissue can be visualized using confocal microscopy, and there is also the possibility of reconstructing three-dimensional images of soft tissue structures. Likewise, in a later work, Nash et al. [11] also conclude the relevant importance of the anatomical knowledge obtained from the plastination of sections with epoxy resin in relation to its use by anesthetists who perform cervical plexus blocks, as well as health professionals, such as surgeons, which allows them to develop pre-surgical planning of endoscopic approaches, in this case, of the regions of the anterior and lateral cervical triangle, thyroid gland, and lymph nodes of the same region.

Chen et al. [13] indicated in their research that they applied the same protocol for plastination of sections with epoxy resin developed by Nash et al. [10] which allowed them to carry out the first investigations through plastination, in four cadavers, on the fibrous configuration of the capsule of the cricothyroid joint, also establishing the correlation of the joint cavity at the macro- and microscopic level. Likewise, they complemented their research with the development of anatomical dissections, in 18 human cadavers; performing micro-CT scans and three-dimensional reconstructions in 9 of the 18 cadavers subjected to the anatomical dissection process; and application of histological techniques in 3 of the samples destined for micro-CT, which were first decalcified to then continue with the histological staining process, applying hematoxylin-eosin and van Gieson stains; finally, they also subjected two specimens to magnetic resonance scanning. For their part, Liu et al. [14], also applying the epoxy resin sheet plastination technique developed by Nash et al. [10], investigated the cricoarytenoid joint and its correlation with the capsule of the joint itself. Both the works developed by Chen et al. [13] and that of Liu et al. [14] studied, respectively, the biomechanics of the cricothyroid and cricoarytenoid joints and their corresponding rotational movements, especially the movements of the vocal fold and the surgical correlation in cases of joint dysfunction [14]. Also in the research by Liu et al. [14], complementary analyses of the samples were performed through micro-CT, histological techniques, and three-dimensional reconstructions from the plastinated sections.

Diao et al. [15] applied the technique of plastination of sections with epoxy resin developed by Zhang and An [3] for the investigation of the medial wall of the cavernous sinus, whose anatomical knowledge is fundamental in determining the direction of growth of pituitary adenomas, with which, in turn, it will be possible to identify a better planning of pituitary surgery. Diao et al. [15], based on the results obtained from the plastinated sections developed, were able to define the existence of a wall between the pituitary gland and the cavernous sinus, made up of both the meningeal dura mater and a loose fibrous network in the form of a net, which are located, respectively, in the anterosuperior and posteroinferior aspects of the cavernous sinus.

Arnts et al. [16] investigated the connections between the white matter and the cerebellar nuclei, specifically oriented to the existing interest in recent years on the knowledge of the anatomy of the human white matter. In order to develop an applied study of the white matter, and an adequate clinical interpretation of the images, the authors combined fiber dissection with plastination and three-dimensional reconstruction, seeking to better demonstrate the existing anatomical interrelationship and improve the understanding of the morphological characteristics of the white fibers and the imaging tratographies. For the development of this research, Arnts et al. [16] used four human brains (two male and two female) from births older than 60 years, without neurological diseases. The brains were fixed in 10% formaldehyde and then washed in water overnight, after which the arachnoid membrane, pia mater, and vasculature were removed. The brains were then placed back in 10% formaldehyde and stored cold (-10 to $-15\,^{\circ}$C) for 8–10 days. Subsequently, the brains were again washed in running water for 1 day and stored again in the refrigerator for another 8–10 days, as indicated by De Castro et al. [17]. Finally, the brains were thawed in running water for 1 day to then start the dissection process. During the dissection, the brains were stored in 10% formaldehyde. After the fiber dissection tasks carried out on the brains, the finished preparations were sent to the cold silicone plastination process [18]. In this way, they first washed the brains in running water and then cooled them (at 4 $^{\circ}$C) before starting the cold dehydration stage ($-20\,^{\circ}$C) for 3 weeks. In this type of sample, cold is essential to avoid defatting the samples and also to avoid excessive retraction of the brains. They finished the dehydration process when they reached a final acetone percentage higher than 99%. Subsequently, the samples were immersed in an impregnation mixture made up of silicone (PR10, Dow Corning) and catalyst (Cr20, Dow Corning), in a ratio of 100:5, respectively, to start the forced impregnation process. The forced impregnation stage ended at 23 mbar and upon detecting the absence of bubbling. The samples were then subjected to the curing process, placing the brains on absorbent paper to initially remove excess silicone for 8–15 days. Once this superficial drying stage was finished, the samples were brushed superficially with a mixture of Ct32 catalyst (Dow Corning) and acetone (1:5), without wrapping them in film paper to avoid, as the authors indicate, leaving marks on the preparations. In a total of 3 weeks, in which the samples were dried with absorbent paper and brushed with the Ct32:acetone mixture, the samples finally reached the definitive curing (polymerized).

Liang et al. [19] carried out a very interesting investigation associated with the possibilities of mechanical compression of the trigeminal nerve and its association with the development of trigeminal neuralgia. They applied a technique of sheet plastination with epoxy resin developed by Zhang and An [3], obtaining anatomical sections of between 2.5 and 3.00 mm. In addition, as it had also been previously indicated by Nash et al. [10, 11], the plastination process allowed collagenous fibers and neurofilaments to become endogenously autofluorescent at 488 nm excitation. In this way, they were able to visualize the trigeminal nerve together with the surrounding anatomical structures, achieving a great structural differentiation thanks to the autofluorescence of the nerve fibers, being able to make identifications from morphology, fluorescence intensity, and anatomical distribution. From the results obtained, they were able to identify the nature, architecture, and location of the fibrous trabeculae of the cavernous sinus, the dural trabeculae, and the intracavernous cranial nerves, as well as the identification of two types of network-shaped fibrous networks, which form a true skeletal framework at the level of the lateral dural wall of the cavernous sinus, acting as sleeves or covers for each of the intra-cavernous cranial nerves. In turn, they identified a fine network of trabecular or adipose tissue, which formed the matrix of the cavernous sinus and was distributed along the medial aspect of the intracavernous cranial nerves, forming a dumbbell-shaped fatty zone [19].

Scali et al. [20, 21] developed in 2015 two investigations oriented to the study of anatomical spaces delimited by connective tissues, reporting, on the one hand, the first analysis, through the sheet plastination with epoxy resin, of the alar fascia in situ and its association with potential adjacent spaces, product of possible patho-logical processes such as deep neck infections, allowing the development of a more precise pre-surgical planning of retropharyngeal lymphadenectomy. Thus, through this work they were able to define the inability to adequately differentiate the alar fascia by previous studies, by not developing sheet plastination protocols, and in this way there would be studies that would be giving an erroneous interpretation in the number of potential infection spaces existing, definitively affecting the clinical interpretation of deep neck pathologies. However, the second work was oriented to the investigation of the intervertebral and epidural spaces at the C3 level, being able to observe it through a Leica MZ8 stereoscopic microscope [2, 13–15, 19–22], the upper cervical region without compromising the morphological arrangement. In this sense, the plastination of sections with epoxy resin allowed to increase and clarify the information related to the intricate layers that form the upper cervical epidural space, the posterior atlanto-occipital membrane, identifying thanks to this work the participation of soft tissue components from multiple sources. In these works, they were able to identify a limitation of the epoxy resin sheet plastination technique, associated with the possibility of causing retraction of soft tissues such as muscles. But on the other hand, and as is currently known, they confirmed the possibility that this technique provides for an adequate separation of the fascial planes, allowing a clearer identification and delimitation of the dangerous preverte-bral space, something that traditional conservation methods and above of anatomi-cal dissection do not allow identification. Regarding the developed protocol, the sample consisted of 32 cadavers (15 female, 17 male, between 67 and 89 years of age). Three of these bodies were destined for the technique of sheet plastination with epoxy resin, proceeding to freeze the samples at −85 °C for 48 h to later pro-ceed to cut them. Sections with thickness of 2 mm were obtained, and these were placed in dehydration at −25 °C, for a period of 20–22 weeks, including defatting. Subsequently, forced impregnation was carried out under cryogenic conditions (between −8 °C and 0 °C) for a period of 2 days. They do not indicate pressure ranges or the impregnating mixture used, in addition to not specifying the brand of the resins used. However, they end by stating that the anatomical sections were placed in a warm water bath followed by placing the anatomical sections in a 35 °C oven for 24 h to achieve resin solidification.

Bernal-Mañas et al. [23] studied the arrangement of the muscle fibers of the human lateral pterygoid muscle, by making slices 2–3 mm thick, through the appli-cation of a sheet plastination technique with E12 epoxy resin (Biodur). In this sense, four human temporomandibular joint blocks, previously studied by MRI, were frozen at $-20\,°C$ for 48 h and then at $-80\,°C$ for another 7 days. Next, they were cut in four planes (axial, oblique-coronal, oblique-sagittal perpendicular to the main axis of the mandibular condyle, and oblique-sagittal parallel to the fibers of the superior fascicle of the lateral pterygoid muscle), correlating with the MRI images. Subsequently, the slices obtained were photographed and subjected to the process of sheet plastination with epoxy resin (the details of which are not provided in the manuscript). Based on the results obtained, they were able to make a correct topographic identification of the anatomical structures of the temporomandibular joint, particularly those related to the insertion fascicles of the lateral pterygoid muscle, the venous drainage of the joint, and other structures, demonstrating the great utility of the technique of sheet plastination with epoxy resin, in the microanatomical analysis of this type of anatomical regions of difficult access to routine dissection.

Thorpe Lowis et al. [24] investigated, for the first time in situ, the arrangement of type II thoracic meningeal cysts, seeking to demonstrate the important relationships it establishes with the surrounding anatomical structures. They carried out the processing of the samples through the application of the protocol for plastination of sections with epoxy resin established by Nash et al. [10, 11]. And as specific details of the plastination protocol, they indicated that they froze the samples at $-80\,°C$ for 5 days and then made the respective slices, obtaining 2.5-mm-thick cross sections. Dehydration was developed at $-30\,°C$, for 3 weeks (without indicating concentration levels of acetone), followed by 2 weeks of defatting at $22–24\,°C$ temperature. Subsequently, they indicated that the forced impregnation was carried out at $0\,°C$ for 2 days, implementing an impregnating mixture that consisted of the following components: E12/E6/E600 (Biodur), without indicating the specific proportions. Finally, they carried out the curing process at $45\,°C$ for 5 days. The obtained plastinated sections were visualized in a Leica MZ8 stereoscopic microscope, and some sections were also scanned with an EPSON Perfection V700 scanner, obtaining 1200 dpi images. Regarding the objective of the manuscript, Thorpe Lowis et al. [24] raised the special difficulty of studying meningeal cysts due to routine dissection activities, in addition to the limitation provided by histology, due to the size of the samples, for which they determined that plastination was the best protocol for obtaining precise anatomical analysis information in this type of research, also finding the possibility of better demonstrating the anatomical relationship of the cysts with the meninges, roots and spinal nerve ganglia, vertebral pedicle, epidural space, and intervertebral foramen in the thoracic spine.

Liang et al. [25] carried out a histological study, with three-dimensional reproduction, on the jugular foramen for the precise identification of its meningeal structures, with the aim of providing anatomical information necessary for the safe surgical approach of the cranial nerves of the region. For this, they used 22 posterior skull base blocks, corresponding to 11 human heads, between 57 and 90 years of age. These blocks were subjected to the ultrafine sheet plastination technique with epoxy resin, using the following formulation for the forced impregnation mixture: epoxy resin (E12), hardener (E6), and accelerator (E600) (Biodur). Once the epoxy resin blocks were obtained with the samples inside, they were cut into sections 250 microns thick (450–650 microns). Subsequently, the sections were stained with Stevenel blue and Alizarin red S, after mounting and polishing. Finally, they were visualized and photographed using a Leica DM6 B microscope and reconstructed using the free-use 3D Slicer software (http://www.slicer.org).

Wu et al. [26] investigated the anatomical bases of the pituitary adenoma, in relation to the possibility of invading the clivus corridor, analyzing the anatomical details of the considered anatomical region, as well as its application in surgery. They developed the technique of plastination of ultra-thin sections with epoxy resin described by Liang et al. [25], on 16 human heads, therefore having 32 samples. These blocks, without decalcifying, were impregnated in a mixture made up of the following components: epoxy resin (E12), hardener (E6), and accelerator (E600) (Biodur). And once the epoxy resin blocks were obtained, they were cut ultrafine and stained with Stevenel blue and Alizarin red S, for subsequent visualization by microscopy.

Thorax

Didenko et al. [27] communicate in a poster the performance of a plastination technique of sections with epoxy resin in 84 human hearts, in which they developed the study of the anatomical arrangement of the triangle of the atrioventricular node (Koch's triangle, trigonum nodi atrioventricularis) and its application in the clinic.

Starchik et al. [28] studied 14 human hearts by plastination of sections with epoxy resin, with the particularity that their coronary arteries had metallic stents. The objective of this work was to visualize the metal stent and its relationship with the wall of the coronary artery. For this, a block with the stented coronary artery was extracted from the samples, and these blocks were first subjected to dehydration in cold acetone (-25 °C) for 3–4 weeks, replacing it every 7 days. Then the blocks were degreased with pure acetone or methylene chloride at room temperature for 2 weeks, changing the solvent on the seventh day. Once this was done, the samples were impregnated (for 24 h and until a pressure of less than 5 mmHg was reached), with an impregnation mixture made up of epoxy resin (YD-128) and a catalyst (C-403) in a ratio of 20:1. Subsequently, the samples were removed from the impregnating mixture and placed in a polyethylene container to cure (polymerize) the samples at room temperature. After 2–3 days, the samples were placed in an oven at 45 °C to achieve the final polymerization of the samples. The mixture used for the curing process is not indicated in detail. Nor if any type of accelerator was used, they only indicate the use of epoxy resin and catalyst. Subsequently, the blocks were cut into sections of 2–4 mm, obtaining in some cases thinner sections of 0.5–2 mm, using a high-speed saw with a diamond blade. The sections were then re-impregnated using the same impregnating mix components used previously, but in a 10:1 ratio. This re-impregnation step was carried out for 3–5 h, reaching final pressures between 5 and 10 mmHg. These re-impregnated sections were placed in a plexiglass chamber, which was filled with a new mixture of re-impregnation resin, and these chambers were placed in an oven at 45 °C for 7 days, until the complete curing of the sections was achieved. Once this was achieved, the cameras were dismantled, and the plastinated sections were scanned to achieve adequate visualization of the coronary metal stents.

Skalkos et al. [29] carried out an advanced work for the time, in which the disposition of the myocardial fibers was investigated, through the application of the technique of sheet plastination with epoxy resin and its comparison with magnetic resonance images, performed previously on the same hearts that were later subjected to the plastination technique. For this, the hearts, after being scanned by MRI, were destined for plastination. They were submerged in a mixture of gelatin and polyethylene glycol to generate a "plastic block" in order to cut the hearts as precisely as possible, using a circular meat slicer, obtaining 3-mm-thick slices, parallel to the axis length of the left ventricle. Subsequently, the sections were placed between plastic grids and submerged in cold acetone (-25 °C) to proceed with their dehydration. Then, once the dehydration was finished, the sections were defatted in methylene chloride for 2 weeks. Subsequently, the forced impregnation process was developed for 24 h, placing the samples inside the vacuum chamber submerged in an impregnation mixture, made up of E12/AT30/AT10/E1 (95:5:20:26 pbw, Biodur). In this case, as in others in which degreasing is carried out with methylene chloride, the high boiling speed with which the methylene chloride is extracted from the interior of the samples causes a significant drop in temperature, which contributed to controlling the exothermic reaction, as indicated by Weber and Henry [30]. They managed to visualize in three-dimensional form the details of the heart with the sheet plastination. Once the forced impregnation was completed, the heart slices were subjected to the curing process (polymerization), placing them between two tempered glass plates, separated by a flexible joint. In this case, the following mixture of resins was introduced between the glass plates to achieve the final hardening of the core sections: E12/AT30/E1 (95:5:26 pbw, Biodur). The curing chambers were placed for 45 min to 1 h inside the vacuum chamber so that the small bubbles could be eliminated through the vacuum, eliminating the larger ones manually. Subsequently, the chambers were placed inside an oven at 45 °C, for 2–3 days, until the complete hardening of the mixture was detected. According to Skalkos et al. [29], the hardening ended when the "Newton" rings appeared in the curing chambers. Finally, the curing chambers were opened, removing the glass plates. Skalkos et al. [29] concluded that sections of the heart plastinated with E12 epoxy resin allowed correct three-dimensional visualization of the heart, vessels, and myocardial bundles, achieving comparison with magnetic resonance images.

Pelvis

Porzionato et al. [31] carried out an investigation aimed at the morphological description of the rectourethral muscle, which, despite its importance in perineal surgery, as in the case of radical prostatectomies, had not been adequately identified from the point of view of its topography and morphology. In this way, although they used 16 bodies as a sample, only 4 pelvis blocks were used to develop a technique of sheet plastination with epoxy resin. In this sense, firstly, they froze the pelvic blocks at −20 °C to achieve 2–3-mm-thick slices, these being transversal (in two blocks) and sagittal (in the remaining two blocks). Subsequently, the sections were dehydrated in acetone at low temperatures (−25 °C) for 2 weeks and then defatted at room temperature for 1 more week. Once these dehydration and defatting stages were completed, the sections were subjected to the forced impregnation stage, in a mixture of E12 epoxy resin (Biodur). However, they did not identify other components or the proportions used in this impregnating mixture, identifying only the epoxy resin. In addition, they did not indicate the time that the forced impregnation took nor the temperature at which it was carried out. Finally, they carried out the curing process (polymerization) by placing the slices under ultraviolet light and heat (50 °C). Also in this investigation, they developed anatomical dissection tasks, for the identification of the rectourethral muscle, in 12 specimens, which were fixed with 10% formalin. To do this, after fixation, the samples were extracted en bloc by means of a transverse circumferential cut through the parietal peritoneum using the pelvic margin as a reference point. Likewise, they applied histology and immuno-histochemistry techniques, in samples taken from the specimens that were destined for anatomical dissection, applying hematoxylin-eosin, Azan-Mallory, and Weigert staining for elastic fibers. In the histological sections, they evaluated the morphology of the rectourethral muscle, with respect to its shape and relationships with the rectum, urethra, and musculoaponeurotic structures of the perineum. In the sections that were subjected to immunohistochemical techniques, they identified the distribution of smooth and striated muscle fibers within the rectourethral muscle, in addition to the relationships with the levator ani muscle. Among some of the conclusions reached in this work, Porzionato et al. [31] were able to identify that the rectourethral muscle fibers join the longitudinal muscle layer of the rectum and anal canal, in addition to being able to identify the presence of two muscle components, the rectoperineal muscle and the anoperineal muscle. They also demonstrated, at the level of the anterior portion of the muscle, a clear separation of the smooth muscle fibers from the prostatic and membranous membrane and their direct fixation to the perineal body.

In the same year, Sebe et al. [32] also investigated the rectourethral muscle, debating its real existence, and for this they based themselves on the development of the histological plastination technique developed by Fritsch in 1988 [33]. In this way, 15 male human fetuses were fixed in formalin at 4% and immersed in the same solution for at least 3 months. Subsequently, they applied the epoxy resin section plastination technique developed by Fritsch [33] to the pelvic blocks of the previously indicated samples, based on a plastination technique aimed at visualizing histology and therefore the development of histological staining techniques on plastinated sections. In this way, the pelvic blocks were subjected to the process and subsequently cut in the transverse, coronal, and sagittal planes, obtaining ultra-thin sections of 200–600 microns. Finally, the sections were mounted and polished to be stained with methylene blue and basic fuchsin stains. On the other hand, from the same samples, they bled the membranous urethra, prostate, and distal part of the anorectal canal, in which they applied standard histological and immunohistochemical staining techniques. Both the plastinated sections and the histological and immunohistochemical sections were visualized under a bright field microscope. From the results obtained, they were able to determine that the name of the recto-urethral muscle is incorrect because they were able to define the non-existence of a developing muscular structure that connects the rectum with the membranous urethra. In addition, they were able to identify the adult rectoperineal muscle in early fetal development, formed as a longitudinal muscle composed of striated and smooth fibers, with a deep location, anchored to the perineal body, and located independently between the caudal rhabdosphincter and the external muscularis sheath of the anorectal canal. In a second work, Sebe et al. [34] investigated the female external urinary sphincter and the striated sphincter in fetal development, with the particular aim of identifying the growth and organization of muscle fibers around the urethra and morphological changes due to vaginal development. They developed the research in 28 human fetuses, implementing the technique of plastination of sections with epoxy resin and development of Fritsch's [35] histological stains. They applied the same histological stains as in their previous work [32].

Lunacek et al. [36] carried out an investigation associated with radical retropubic prostatectomy, one of the treatments indicated for prostate cancer, but with high morbidity, associated with the possibility of damaging the cavernous nerves, with a notable impact on quality of life afterward surgery, by affecting, for example, urinary control and sexual function. For this reason, through the plastination technique, they analyzed the cavernous nerves together with the surrounding anatomical structures (seminal vesicles, prostate, and urethra), with the aim of improving the surgical technique associated with radical prostatectomy and achieving fundamental preservation of the nerves and especially its functioning. For this, they carried out the technique of sheet plastination with epoxy resin with the intention of obtaining ultrafine slices of 300–700 microns. They began the process by fixing the samples (20 male fetuses, 9–37 weeks gestation) in 4% formaldehyde for approximately 12 weeks. Subsequently, they developed dehydration in acetone at −25 °C for approximately 6 weeks, followed by 1–2 weeks of defatting in methylene chloride at room temperature. Then, after dehydration and defatting, the samples were subjected to the forced impregnation process for approximately 2 weeks, in an impregnation mixture that consisted of epoxy resin, catalyst, and accelerator. However, they did not indicate the brand of the products or the proportions used. Finally, they made the slices of the samples in a saw with a diamond blade, achieving plastinated sections of 300–700 microns, which were later polished prior to the application of histological stains (methylene blue/azure II), to later be visualized, and the stained sections were photographed under a binocular microscope at ×4–400.

Fritsch et al. [37] carried out an investigation on the existing connective tissue structures around the female urethra, aimed at analyzing those anatomical structures that provide structural support to the female urethra. In this sense, it should be noted that Fritsch established in 1988 [33] and 1989 [35] an interesting concept of "histology plastination," describing in detail the possibility of making ultrafine slices in blocks impregnated with epoxy resin, to then apply different histological stains, in order to visualize through macroscopic sections the histology of the anatomical region studied. This concept was applied in this 2006 article, in which he developed the research on 30 female fetuses, between 9 and 37 weeks of gestation, and on 6 female male pelvises. In 29 fetuses, they applied the technique of plastination of sections with epoxy resin, and for this, the samples were first fixed in 4% formaldehyde for 3 weeks. Next, they started the plastination process, subjecting the samples to washing in tap water for 24 h and then placing the samples in acetone at −25 °C for 5–6 weeks for the dehydration process, followed by 1–2 weeks degreasing at room temperature in methylene chloride. Once these dehydration and degreasing stages were completed, the samples were placed in a mixture of epoxy resin, hardener, and accelerator (E12, E1, E600, Biodur, respectively) for 2 weeks, citing her works from 1988 [33] and 1989 [35], from which we can deduce that the proportions used were the following: E12 (2 parts), E1 (1 part), and E600 (0.2 ml per 100 ml of mixture). In this way, they developed the vacuum forced impregnation step. Once the forced impregnation was completed, the samples were polymerized and cut with a saw with a diamond blade. They managed to obtain thin sections between 300 and 700 microns, which were then polished to be able to apply histological staining techniques, using methylene blue stains in alkaline solution. These sections were visualized under a microscope with a magnification of 4–80× [33, 35]. On the other hand, the adult pelvis samples were subjected to the technique of sheet plastination with epoxy resin of thin slices, 3–5 mm thick. For them they applied the mixture of epoxy resin E12 (Biodur) described by von Hagens [38]. They compared the results with magnetic resonance images of 41 volunteers participating in this research. Among the general conclusions reached, associated with this type of plastination technique, they considered that these anatomical investigations based on transparent plastinated slices allow a sectional study of the anatomy, without finding alterations in the topography similar to the living state of the subject. They also consider that anatomical dissection or conventional histological studies may present distortions that alter the morphology of the studied regions, and also the thin and ultra-thin sections ensure a real interpretation of the spatial perspective of the anatomical regions under study. In relation to the morphological conclusions of the study, in general, they identified that there is no direct ligamentous fixation of the female urethra to the pubic bone. Likewise, they visualized that the fascia of the levator ani muscle seems to give support to the urethra and the rhabdosphincter, the most important being the support corresponding to the dorsal fixation of the connective tissue in the ventral wall of the vagina. This information is crucial for its implementation in the surgical treatment of female urinary incontinence.

For their part, Macchi et al. [39] investigated the longitudinal anal muscle, at the level of the identification of its histological structure, but also of its adhesions and topography, associated with the role in continence and defecation, as well as in relation to the application of anatomy surgery in the region. For this, they considered 16 samples, corresponding to 8 male cadavers and 8 female cadavers (aged between 52 and 72 years), with the absence of pelvic pathology. From each sample (24 h after death), they extracted en bloc, and prior to embalming, the region of the pelvis. Of the total samples, four pelvis blocks were subjected to the technique of sheet plastination with epoxy resin. In this way, the blocks were frozen at $-20\ °C$ to proceed with their cutting with a saw, allowing the generation of thin slices of 2–3 mm. Subsequently, the sections were subjected to the dehydration process, in acetone at $-25\ °C$ for 2 weeks, followed by a week of defatting, also in acetone, at room temperature. Subsequently, the samples were placed in epoxy resin (E12, Biodur) to carry out forced vacuum impregnation. Finally, the samples were polymerized by being exposed to ultraviolet light and then heat (50 °C). Histological and immunohistochemical staining techniques were applied to the rest of the samples. In relation to the results obtained from the plastinated slices, they were able to identify the longitudinal anal muscle located between the circular layers of the internal and external anal sphincters, with a helical arrangement of smooth and striated muscle fibers, being also identifiable on the lateral faces of the anal canal, contributing to the narrowing and shortening of the anal canal during contraction of the anal sphincter.

Al-Ali et al. [40] carried out an investigation on the anal sphincter, through the technique of sheet plastination combined with conventional histological techniques and its visualization by microscopy. They used 18 male cadavers, of people between 62 and 82 years of age, specifically, the pelvic regions. These bodies were embalmed vascularly (through the carotid and femoral arteries) to then carry out the plastination process in two pelvises without dissecting. The samples were then frozen for 4 days at a temperature of $-80\ °C$. One of the pelvises was cut coronally, while the other was cut axially, into thin sections 2.5 mm thick, using a high-speed saw. The sections obtained were subsequently dehydrated in acetone by freezing and defatted in methylene chloride (it is estimated at room temperature, although it was not declared). Next, forced impregnation of the sections was carried out, using E12 epoxy resin (Biodur), but without indicating the use of other compounds. Finally, the samples were polymerized through the process indicated by Cook and Al-Ali [41]. In this investigation, the transillumination achieved by the technique of sheet plastination with epoxy resin stands out, allowing to reveal the presence of smooth muscle fibers in the upper muscular portion of the sphincter and of fibroelastic fascial layers in the lower muscular portion of the anal sphincter, the latter extending toward the interior and exterior of the anal sphincter, also reaching the perianal dermis. Likewise, in the perianal region, thanks to plastination, fatty bags with the appearance of a honeycomb were identified at the level of the fibroelastic fascial layers.

Kaulhausen et al. [42] analyzed the supraspinous and interspinous ligaments, associated with one of the most common degenerative pathologies in elderly people, intermittent claudication resulting from lumbar spinal stenosis. In this sense, they used a device for decompression of the interspinous process, in particular to analyze the development of surgical techniques and the association with the damage caused by surgical implantation. In a fresh cadaver, four interspinous spacers were implanted percutaneously, to later analyze these implants through the technique of sheet plastination with epoxy resin. Therefore, they subsequently fixed the sample with a 4% formaldehyde solution and then implemented the steps of dehydration, degreasing, and forced impregnation with an impregnation mixture made up of Biodur resins (E12, E6, E600), without indicating proportions or impregnation times. The entire process took 12 weeks. Next, and after taking radiographic images, the plastinated block was cut with a diamond blade saw into 4-mm-thick sections. In addition, to achieve greater transparency of the slices, the sections were included in a curing chamber with a mixture made up of E12 epoxy resin, E1 hardener, and AE30 accelerator (Biodur), and then the final polymerization of the transparent slices was carried out. Thanks to the development of these plastinated slices, the macroanatomical position of the implants in relation to the osteoligamentous structures was analyzed.

Xu et al. [43] carried out an investigation aimed at providing greater clarity in relation to the suspensory system of the urinary tract, in direct relation to urinary incontinence, aimed at providing precise anatomical information when developing surgical approaches in the region, with the aim of achieving preservation of the puboprostatic ligament and pubococcygeal fibers during radical retropubic prostatectomy. In this way, they carried out a study based on the technique of plastination of thin sections with epoxy resin and visualized these sections through confocal microscopy in addition to performing a histological analysis of the samples obtained. This plastination technique allows studying the anatomy in situ, in the natural state of the anatomical structures, without requiring decalcification of the samples and allowing the macro- and microscopic study of the corresponding region. For this, they used six male cadavers (from 46 to 87 years of age), without prostate pathologies. The pelvises were obtained from these cadavers, performing a block extraction of the same from the fourth lumbar vertebra to the upper part of it. This block was frozen at −80 °C for 4 days and then cut, obtaining 2.5-mm-thick anatomical sections. Next, the sections were dehydrated for 4 weeks at −30 °C, to then be defatted at room temperature (22–24 °C) for another 3 weeks. After dehydration and defatting, the samples were subjected to the forced impregnation phase, in an impregnation mixture made up of E12/E1/AE10/AE30 (Biodur), at 0 °C for 2 days. Once the forced impregnation was completed, the samples were subjected to the curing phase (polymerization) at 45 °C for 5 days. Finally, the 2.5 mm plastinated sections were studied and visualized by confocal laser microscopy (Nikon AIR, Tokyo, Japan).

Xu et al. [44] carried out a research work in search of the identification of the facial planes corresponding to the lateral femoral cutaneous nerve, associated with its mechanical understanding and the appearance of meralgia paresthetica, characterized by tingling, numbness, and burning pain in the lateral part of the thigh. This anatomical detail had not yet been fully identified, and therefore, Xu et al. applied the technique of sheet plastination with epoxy resin combined with the visualization and analysis of the anatomical slices obtained by means of confocal microscopy. In this way, out of a total of 30 corpses destined for this research, 6 were used to develop the technique of sheet plastination with epoxy resin [3], using 2 female bodies and 4 male bodies (between 46 and 87 years old). Among the details provided about the developed technique, the samples were frozen at −80 °C for 7 days and then the slices were made, with a thickness of 2.5 mm. Then, firstly, the dehydration stage was developed, placing the sections in cold acetone (−30 °C), for 4 weeks, and secondly, the defatting stage was developed, at room temperature (22–24 °C) for 3 weeks. After these two stages, the samples were placed in an impregnation mixture made up of E12/E1/AE10/AE30 (Biodur), for the forced impregnation process, at a temperature of 0 °C, for 2 days. Finally, the sections were cured (polymerized) at 45 °C, for a total of 5 days. The finalized sections were visualized and studied using a Leica MZ8 stereoscopic microscope. The visualization by confocal microscopy (Nikon laser scanner) allowed to confirm the endogenous autofluorescence of the plastinated samples, at 488 nm of excitation, allowing the identification of collagen, elastin, myofilaments, and neurofilaments. Likewise, they managed to develop three-dimensional reconstructions of the thigh region, from the slices plastinated with epoxy resin.

Upper Limb

Sora and Genser-Strobl [45] analyzed 12 carpal regions of fresh cadavers, without fixations, in which they in turn investigated the morphological characteristics of the carpal tunnel and neurovascular relationships using the technique of sheet plastination with epoxy resin. For this, all the samples were frozen at −80 °C, previously positioning the palmar in maximum abduction and extension and then making slices with an average thickness of 1.7 ± 0.26 mm, obtaining a total of 20 slices (with a loss of 0.4 mm between each cut, due to the thickness of the saw blade). Frozen sections were scanned and then subjected to the plastination process. Sections were stored overnight at −25 °C, to then begin dehydration in acetone at −25 °C and defatting for 1 week in methylene chloride. Subsequently, forced impregnation was continued at +5 °C, submerging the sections in an impregnation mixture made up of E12/E1/AE10 (Biodur) [38]. Once the impregnation was finished, the sections were placed between glass plates separated by a flexible joint, creating a 2 mm gap. These curing chambers were placed in a vacuum chamber to extract the bubbles and then left at room temperature for 24 h, before being placed in an oven at +45 °C for 4 days. This last process is polymerization, which ensures the hardening and drying of the resin. Once the plastinated sections were obtained, they were scanned again, with which they were able to identify the retraction between the fresh and plastinated sections (but this was not reported in the publication). In this way, they were able to identify the anatomical characteristics of the structures of the hand, establishing morphometric measurements between each one of them. It is possible to clearly identify the topographic relationships between the different neurovascular, muscular, and tendon anatomical elements, essential for the recognition of precise surgical approaches when defining and planning the surgical act in the region.

Koslowsky et al. [46] investigated the anatomical basis associated with surgical approaches to nonunion fractures or olecranon osteotomies, as well as muscle insufficiencies following total elbow arthroplasty. To do this, they applied the technique of sheet plastination with epoxy resin to analyze the periosseous and intraosseous vascular distribution of the radial head, as well as the distal vascularization of the humerus, and its relationship with the implants. In this sense, they used 11 fresh upper limbs, from cadavers between 71 and 85 years of age, with no history of elbow injuries. All samples were isolated from the carcasses and first subjected to a saline wash. Subsequently, the samples were fixed with 4% formaldehyde at 1–3 °C for 24 h, incorporating the fixing solution via the arterial route. After this, the samples were washed with acetone 10 times, seeking to restore the natural color, to then inject red epoxy resin also through the arterial route. The specimens were subjected to X-rays prior to the start of the plastination process. The dehydration of the samples was carried out with 99% acetone, at −25 °C, for 4 months, followed by several weeks of degreasing with methylene chloride, to favor the degreasing of the samples. Next, the specimens were submerged in epoxy resin for 3 weeks to force-impregnate them, by applying a vacuum. Once the hardening of the resin blocks was achieved, with the samples inside, they were sectioned into 4-mm-thick slices. Finally, the sections were placed in curing chambers and filled with a mixture of epoxy resin and hardener, to achieve transparency of the plastinated sections. With the vascular injection, before plastination, of colored epoxy resin, the transparency of the sections resulting from the process of plastination of sections with epoxy resin ensures perfect visualization of the vascular distribution of the radial head, which ensures remarkable identification of periosseous and intraosseous vascularization of the region under study, thus allowing the definition of a new vascular distribution of the radial head, a subject little investigated up to that time.

Xu et al. [47] carried out a study through plastination of sections with epoxy resin and confocal microscopy of the rectoprostatic fascia (Denonvilliers' fascia) in men, with the aim of providing anatomical information to the study of the prerectal space and the generation of better mobilizations of the rectum in the surgical approach of the region. For this research, they developed the technique of sheet plastination with epoxy resin described by Zhang and An [3], in six male cadavers (from 46 to 87 years of age), which were initially fixed 24 h after the death of the patient donor, with an embalming mixture consisting of 2% formaldehyde, 60% ethanol, 15% glycerin, and 7% phenoxetol (all diluted in water). Among the data provided in the article, the authors indicate that the pelvic blocks were frozen at −80 °C to be cut into 2.5-mm-thick sections after 7 days. Subsequently, the sections were dehydrated in acetone at −30 °C for 4 weeks, followed by defatting at room temperature (22–24 °C) for 3 weeks. Next, the sections were subjected to the forced impregnation stage, in an impregnation mixture formed by E12/E1/AE10/AE30 (Biodur), at 0 °C, for 2 days. Curing (polymerization) was performed at 45 °C for 5 days. The plastinated sections were visualized under a Leica MZ8 stereoscopic microscope and scanned for imaging at 1200–6400 dpi. The authors indicate that this plastination process allows collagen, elastin, myofilaments, and neurofilaments to become endogenously autofluorescent at 488 nm excitation.

Lower Limb

Fritsch [48] investigated the sectional anatomy of the connective tissue structures of the foot in samples from newborns and adults, in order to apply this anatomical knowledge to deforming pathologies of the foot. In this way, he developed a technique of plastination of sections with epoxy resin, combined with histology, defined as "plastination histology" [33, 38]. The sample destined to obtain ultra-thin sections consisted of 7 feet of newborn children, which were initially dehydrated in acetone at −25 °C and defatted in methylene chloride at room temperature. Subsequently, the samples were subjected to the forced impregnation process using E12 epoxy resin (Biodur). Once the impregnation was finished and the plastinated block was obtained with the foot inside it, these blocks were cut with a saw with a diamond blade, in transversal, coronal, and sagittal planes. Ultrafine sections between 300 and 500 microns were obtained. Finally, they applied histological stains to the plastinated sections, after they had been mounted and polished, using methylene blue staining. To obtain thin plastinated sections, they applied the classic epoxy resin sheet plastination technique [38] to the feet of seven adult cadavers, between 59 and 81 years of age. They began by freezing the samples at −80 °C, and these were then cut with a saw in transverse, coronal, and sagittal planes. The thickness of the slices obtained was 3–4 mm. Then, these sections were placed in acetone at −25 °C for 5 weeks, for the dehydration step, and later they were placed at room temperature in methylene chloride, for 2 more weeks. After dehydration and defatting were complete, the samples were placed in E12 epoxy resin (Biodur) [38], where forced impregnation was performed. Once this process was completed, the sections were placed between glass plates, in curing chambers, to develop the curing or polymerization stage in an oven at 50 °C. Among some of the conclusions reached in this investigation, the topographic relationships between the bone, cartilage, and connective tissue structures of the foot were defined, which remain undisturbed in the plastinated sections, in addition to maintaining the characteristics that in the living state and its total correlate with the corresponding results of magnetic resonance images.

Sora et al. [49] investigated the human ankle, specifically developing a computerized model through the application of anatomical sections plastinated by the technique of sheet plastination with epoxy resin. In particular, they developed a plastination technique for ultra-thin slices, through which they obtained slices with a thickness of 1 mm. In this work they used an unfixed ankle corresponding to the cadaver of a man, between the foot and the distal third of the tibia. The block corresponding to the ankle was frozen at −80 °C for 1 week. Subsequently, the block was placed in a refrigerator at −25 °C for 2 days and then submerged in 25 liters of cold acetone (−25 °C) to develop the dehydration process (reaching acetone concentrations of 92% after 4 weeks, 97% after 3 more weeks, and finally 99% after 2 more weeks). In total, dehydration lasted 9 weeks. Next, the sample was placed in methylene chloride at room temperature (+15 °C), to favor the defatting of the sample, for 4 weeks. Once the dehydration and degreasing stages were completed, the sample was immersed in an impregnating mixture, made up as follows: epoxy resin (E12), hardener (E6), and accelerator (E600) (100/50/0.2, Biodur) [50], to proceed with the forced impregnation of the sample, the fundamental step in the plastination technique. This step was carried out inside an oven with a vacuum chamber at a temperature of +30 °C. The process began 24 h after placing the sample in the impregnating mixture and began with a vacuum, reducing the pressure to 80 mmHg daily, until reaching a final pressure of 2 mmHg after 5 days, varying the temperature on last day, in which it rose to +60 °C. After the forced impregnation process, the sample was extracted from the oven with a vacuum chamber and placed in a mold, inside an oven at 65 °C for 4 days, immersing the sample in a mixture similar to that used in the forced impregnation, to achieve hardening

of the polymer. Once the polymerization of the block was achieved, it was cut with a saw with a diamond blade, achieving 1-mm-thick sections. These ultra-thin sections were then scanned to proceed with the assembly of the three-dimensional reconstruction of the ankle. In this way, the authors defined that the plastination of sections with epoxy resin, and especially the method of obtaining ultra-thin sections, allows obtaining exact, precise, and transparent sections, achieving remarkable visual clarity from macroscopic to the submicroscopic level, also allowing, through the three-dimensional reconstruction of said ultra-thin sections, improve teaching in medical residency, allowing the use of said models for surgical planning and the development of new ways of approaching the anatomical planes during surgeries.

Sora et al. [51] developed an investigation applying the sheet plastination with epoxy resin to analyze the topographic distribution of the posteromedial neurovascular elements of the ankle, with the aim of providing details on the morphological arrangement of these anatomical structures when planning the development of minimally invasive surgery. In this way, they used 12 lower limbs, without pathologies, from individuals between 65 and 84 years of age. All the samples were frozen at −80 °C and then cut, obtaining sections of 1.5 ± 0.26 mm, obtaining a total of 20 slices from each sample. These sections were immediately scanned prior to plastination. Next, the sections were stored at −25 °C overnight, to begin dehydration the next day in cold acetone (−25 °C) in several sessions, followed by defatting with methylene chloride for a week. Once dehydration and degreasing were completed, the samples were subjected to the forced impregnation process at +5 °C with E12 epoxy resin (Biodur). Then, the impregnated sections were placed inside curing chambers, made up of two glass plates separated by a 2 mm flexible joint. These chambers were placed in a vacuum chamber, to extract the bubbles, and after being placed for 1 day and at room temperature in a 15° horizontal orientation, the chambers are placed in an oven at 45 °C for 4 days to finish with the curing process (polymerization). Once the plastination is finished, the slices are scanned again, and the three-dimensional reconstruction of the slices was developed for the generation of 3D models of the posteromedial neurovascular bundle of the ankle.

Rath et al. [52, 53] carried out investigations using the technique of plastination of sections with epoxy resin, but with prior arterial injection of colored epoxy resin in order to study the microvascularization of sesamoid bones [52] and the metatarsal bones [53], fundamental anatomical information when analyzing arterial lesions that could arise from foot trauma or also after performing surgical procedures in these foot regions. In both investigations, 22 feet of adults between 49 and 77 years of age were used, in which the technique of sheet plastination with epoxy resin was applied, based on previously published works [38, 50].

Faymonville et al. [54] analyzed the topographic anatomy of the compartments of the foot by means of vascular injection techniques and then by plastination of sections with epoxy resin, with the aim of reviewing the anatomy of this region, associated with the importance of applying this anatomical knowledge in resolution of compartment syndromes. For the vascular injection technique, the authors used 12 feet from fresh human cadavers, which were injected vascularly with different colored epoxy resins (Biodur), of very low viscosity, to then cut the injected samples to be subjected to the process of sheet plastination with epoxy resin.

Sora et al. [55] investigated, for the first time, through the technique of sheet plastination with epoxy resin, the distal tibiofibular joint, the ankle syndesmosis, thus ensuring an unaltered study of this anatomical region, whose approach through dissection routine anatomical movement would move the anatomical structures of the region, causing an opening of the joint and, therefore, an alteration of the anatomical relationships of the ankle syndesmosis. To do this, these researchers applied the technique of sheet plastination with epoxy resin to study in particular the space between the bone structures of the joint, and for this plastination was ideal for analyzing the region without moving the anatomical structures involved and through the generation of sections with transparency, ensuring an exact visualization of the connective tissue. For this, the sample consisted of 20 lower right limbs, from male cadavers, between 65 and 91 years of age. Samples were frozen at $-80\ °C$ and then cut into 1.6+ sections of 0.26 mm thick, achieving 20 slices for each sample. These sections were initially scanned and then subjected to the plastination process. In this way, these sections were refrigerated at $-25\ °C$ overnight, to start the dehydration process in cold acetone $(-25\ °C)$ the next day, in several changes of acetone, to then leave them for a week in methylene chloride for the degreasing process. Once these dehydration and defatting processes were completed, the sections were submerged in E12 epoxy resin (Biodur), at a temperature of $+5\ °C$, to proceed to the forced impregnation stage, using the following impregnation mixture: E12/E1/AE10 [38]. Once the impregnation of the sections was completed, they were incorporated into the curing chambers, made up of two glass plates separated by a 3-mm-thick flexible joint. The curing chambers, once closed, were placed in a vacuum chamber to extract the bubbles, and after being placed for 1 day and at room temperature in an orientation of 15° horizontally, the chambers were placed in an oven at 45 °C for 4 days to finish the curing process (polymerization). Next, the sections were scanned again, developing three-dimensional reconstructions and generating three-dimensional models of the ankle syndesmosis. This research provided the first data on the ankle syndesmosis, at the level of its length, width, and inclination of the fibular notch, data that will allow us to analyze, for example, the association with the decrease in the inclination of the fibular notch, which is associated with lesions of this joint.

Opperman et al. [56] carried out an investigation to identify the intraosseous microvascularization of the talus, also associated with the relationship established between the implants necessary to solve fractures at the level of the talus and its microvascularization. In this way, they developed a technique of plastination of sections with epoxy resin, obtaining thin sections between 2.6 and 3 mm thick, after vascular injection of the samples with epoxy resin to identify the microvascular distribution of the talus. In this work, 8 fresh lower limbs and 20 macerated tali were used. As a general conclusion, the authors were able to demonstrate that the greatest quantity and density of nutrient foramina were located in the sulcus tali, while through the plastination technique, when analyzing a type implant in particular, the detriment of this type of implant was demonstrated in relation to the vascular distribution of the talus at the level of the deep end of the incorporated implant, with the possibility of generating avascular necrosis of the talus as a result of a search for a surgical solution.

Villamonte-Chevalier et al. [57] carried out an investigation associated with veterinary anatomy, in which they analyzed the elbow joint in the dog, through transparent slices plastinated with epoxy resin, carrying out a comparison with ultrasonographic images, identifying an adequate correlation when analyzing the soft tissues and bony structures visualized in ultrasonographic studies and plastinated sections. For the anatomical study, they used ten elbows that were injected with epoxy resin through the arterial line and then frozen at −20 °C, developing blocks that were then cut with a saw, after freezing the blocks for a week at −70 °C. Sections 2 mm thick were obtained, in planes similar to those obtained through ultrasonographic studies. These sections were subsequently subjected to the previously reported technique of plastination of sections with epoxy resin [38, 58].

Xu et al. [59] developed an investigation in which they applied, in addition to cadaveric dissection, the technique of sheet plastination with epoxy resin on the fibrous configuration of the fascia iliaca compartment, oriented in particular to its application in anesthesia and analgesia techniques in knee and hip surgeries, related to the appropriate site for needle placement at the time of performing the indicated procedures. For this research, the authors used a total of 46 human cadavers (from 38 to 97 years of age), of which 9 were destined for the development of the technique of sheet plastination with epoxy resin, as indicated in a publication by our research group [60]. In relation to this, the fundamental steps of the developed technique consisted, firstly, in freezing the specimens at −80 °C for 7 days and then cutting them with a saw with a diamond blade, obtaining slices of 2.5 mm thick. Tissue loss was only 0.9 mm, corresponding to the thickness of the cutting blade. Next, the sections were dehydrated at −30 °C, in acetone, for 4 weeks and defatted, also in acetone but at room temperature (22–24 °C), for 3 weeks. Subsequently, after defatting, the samples were subjected to the forced impregnation phase, using an impregnation mixture made up of E12/E1/AE20/AE30 (Biodur), to finally polymerize the sections at 45 °C for 5 days. As explained in previous articles [10–12, 19, 43, 44, 47], this plastination process causes collagen, elastin, neurofilament, and myofilaments to become endogenously autofluorescent at 488 nm excitation and thus can be visualized by confocal laser microscopy (Nikon, Tokyo, Japan). In this way, it was also possible to perform three-dimensional reconstructions of the anatomical region, from the plastinated sections, using the Thermo Scientific Amira software.

Studies Oriented to the Diagnosis by Imaging of Plastinated Models

The development of imaging studies from radiographs to MRI requires a new vision in the anatomical discipline, constituting plastination a fundamental tool that can be applied to enhance anatomical knowledge in advanced training programs in sectional topography, allowing a better interpretation and diagnostic analysis of images, as well as for their application in improving surgical planning, with their corresponding correlate in CT and MRI [23, 26, 41, 61–75].

The 3D reconstructed model can also be used for residency education, for performing unusual surgery, and for developing new surgical approaches [25, 44, 59, 69, 74, 76–81].

Plastination of sections with epoxy resin provides better image quality than two-dimensional ultrasound and would be very useful to complement this technique for teaching human and veterinary anatomy [14, 63, 72, 73, 82].

Hermans et al. [83] studied six frozen legs which were injected with gadolinium and subjected to magnetic resonance arthrography. Subsequently, they applied the plastination technique to the entire legs, to then submit them again to magnetic resonance imaging. Finally they made 3 mm slices and finished the plastination process. These researchers demonstrated, for the first time, the possibility of mixing a polymer with the gadolinium solution, making it possible to obtain an adequate signal intensity to achieve effective scanning of plastinated pieces. Adding a polymer to the gadolinium solution causes a decrease in signal intensity over time because curing of the polymer results in increased solidification of the mixture. The mixture that allowed the highest signal intensity for the longest period of time was a slow curing polymer with a 1:50 dilute gadolinium solution.

Plastinated slices of an anatomical specimen can be directly compared to previously obtained MRIs, and the same structures can then be identified [63].

Beyersdorff et al. [62] applied the slice plastination technique to visualize pelvic floor muscle structures and their relationship to adjacent structures down to the microscopic level. These authors indicate, on the one hand, that the best diagnostic visualization of the pelvic floor was achieved by magnetic resonance, with soft tissue contrast that allowed differentiation of the venous plexus, fascia, and fatty tissue, and, on the other hand, indicate that the computed tomography, despite using a spatial resolution similar to that of the magnetic resonance, represented the structures of the pelvic floor with lower quality due to a lower contrast of the soft tissues.

The best optical quality for slices can only be achieved using fresh samples [84].

Studies Oriented to the Three-Dimensional Reconstruction of Plastinated Models

Shah et al. [76] carried out three-dimensional reconstructions of the ligaments of the lateral aspect of the ankle and the subtalar joint from plastinated sections obtained by the technique of plastination of sections with epoxy resin. They used four samples from two embalmed bodies (28 and 35 years old). The four samples were cooled at −4 °C for 24 h, and then dehydration began, in four successive baths, in 100% acetone, at −25 °C. Subsequently, the samples were subjected to the forced impregnation stage, in an impregnation mixture made up of epoxy resin (E12, Biodur), hardener (E6, Biodur), and accelerator (E500, Biodur) (W/W/V, 1/0.5/0.1). Once the impregnation was completed, the samples were subjected to the curing stage (polymerization) at 50 °C for 2 weeks. Next, the generated blocks were cut, using a saw with a diamond blade, to obtain 1.2-mm-thick slices, each sample in a different plane: coronal, sagittal, axial, and oblique axial. Among some of the comments made in this publication, the authors indicate that the ligaments cannot be visualized in X-rays and are not easy to identify with MRI, which is an advantage with these three-dimensional reconstructions, since each of the structures can be reconstructed individually viewable. In addition, they allow obtaining, in this type of reconstruction, all kinds of information on parameters of the articular surface.

Qiu et al. [77] developed the three-dimensional reconstruction of the temporal bone and the surrounding anatomical structures, to contribute to the development of anatomical knowledge associated with the interpretation of fundamental anatomical relationships for the development of surgical approaches in this intracranial region. In this way, they defined three stages in the development of the three-dimensional reconstruction of the temporal bone: plastination, acquisition of the slices, segmentation, and generation of the three-dimensional model. For this research, the authors used three formalin-fixed human heads, which in turn were sectioned in such a way that blocks containing the region of interest were generated. The blocks of the samples were dehydrated in cold (−25 °C) 100% acetone, for 4 weeks, to then carry out the forced impregnation in an impregnation mixture made up of epoxy resin (E12, Biodur), hardener (E6, Biodur), and accelerator (E500, Biodur), in the following proportions (1/0.5/0.1; W/W/V). Subsequently, the samples were placed at +50 °C, for 2 weeks, to cause their polymerization. The blocks were cut with a saw with a diamond blade, in six sections of 1.00 mm thickness (in coronal, sagittal, and axial planes). Once the sections were obtained, they were scanned to continue with the process of generating three-dimensional reconstructions. Qiu et al. [77] highlight some advantages of performing the plastination technique to later develop the three-dimensional reconstruction, indicating that from plastination body regions and complete organs can be investigated without the need to carry out a prior decalcification of the samples, which would cause a great retraction of the tissues, and in this case, the morphometric measurements would not be precise and much less reliable. Thus, plastination becomes a unique method in anatomical techniques for the preservation and conservation of biological materials. Furthermore, in this case, the presence of transparency in the sections plastinated with epoxy resin allows for better levels of photography in terms of resolution compared to frozen sections; therefore this type of technique is very useful in the development of radiological studies. For its part, plastination ensures the generation of anatomical sections with fully intact tissues that allow them to be reconstructed three-dimensionally in a better way, ensuring the possibility of selecting objectives and performing animation as well as a wide variety of manipulation modes, which allow visualizing all the anatomical structures of the region together with the anatomical relationships established with the surrounding elements. In addition, the generation of three-dimensional models is very useful for the development of surgical planning and intraoperative navigation, especially in this type of anatomical regions that are difficult to access.

Steinke et al. [68] carried out a research work in which they sought to demonstrate a novel view of the ligaments of the human sacroiliac joint, by generating three-dimensional reconstructions of plastinated sections, in addition to performing magnetic resonance studies on the corresponding specimens. They developed the technique of plastination of sections with epoxy resin, according to the protocols of Fritsch [35] and Steinke [63], in two fresh pelvises (a pelvis from a 71-year-old male donor, and a pelvis from an 85-year-old female donor), which previously underwent CT and MR studies. Once the samples were plastinated, the male pelvis was cut with a standard saw, obtaining 65 thin sections of 2.5 mm thickness, while the female pelvis was cut with a saw with a diamond blade, obtaining 127 ultra-thin slices, 0.7 mm thick. In turn, the sections were stained (Giemsa stain) following the protocol established by Steinke et al., to then be scanned for the three-dimensional reconstruction process. Steinke et al. [68] generally concluded that the possibility of combining high-resolution magnetic resonance imaging and plastination of thin and ultra-thin sections makes it possible to develop highly accurate three-dimensional reconstructions, allowing in situ anatomical visualization and morphometric analysis, in this case, of the bones and ligaments of the sacroiliac joint.

Sora et al. [69] presented the three-dimensional reconstruction of the human female pelvis, an anatomical region of high morphological complexity, allowing the development of a three-dimensional model that allows stereoscopic visualization of the anatomical relationships between the constituent anatomical structures of the corresponding region. In this way, the authors developed a plastination process using epoxy resin slices in a human female pelvis from a fresh, unfixed cadaver. The anatomical block was initially scanned using a high-resolution magnetic resonator, allowing the generation of high-contrast images that allowed it to be correlated with the plastinated sections. The ultra-thin sheet plastination technique with epoxy resin was applied after obtaining the MR images, and the pelvis block was frozen at −80 °C for a week, to then begin the cold dehydration process (−25 °C), in 100% acetone. Subsequently, the sample was defatted with methylene chloride. Next, the already dehydrated and degreased sample was placed in an impregnation mixture made up of epoxy resin (E12), hardener (E6), and accelerator (E600) (Biodur) [49], to develop the forced impregnation of the sample. The block is then placed inside a mold, submerged in a mixture of resins similar to that of the forced impregnation process, and placed in an oven for 4 days at +65 °C. In this way, a suitable block was generated to be cut with a saw with a diamond blade, obtaining ultrafine slices of 1.6 ± 0.26 mm thick. These ultra-thin sections were later scanned to develop the three-dimensional reconstruction. Among some of the comments made by the authors, it is found that the generation of three-dimensional models from real cadaveric samples, and plastinated ones, ensures the generation of models that provide greater realism, which is essential when using 3D virtual representations for teaching anatomy, also ensuring accurate visualization of complex anatomical regions. In turn, these three-dimensional reconstructions become fundamental tools for the interactive development of training in technical skills of medical residents, being very useful for planning complex surgeries, based on the development of new surgical approaches. In this case, a variety of disciplines (gynecologists, radiologists, surgeons, urologists, and physiotherapists, among others) would benefit from the generation of this virtual model from ultra-thin plastinated sections, in the search for generating a model that favors the approach of the diffusions of the female pelvic floor.

Adds and Al-Rekabi [85] studied the ethmoid arteries of the medial wall of the orbit from the three-dimensional reconstruction of slices obtained from the ultra-thin slice plastination technique with epoxy resin. In this sense, the researchers, prior to starting the plastination process, injected low-viscosity red silicone resin (Biodur KEM 06) through the carotid artery to reach the ophthalmic artery. The head was then submerged in 4% formaldehyde for 1 week prior to the development of the anatomical dissection. Subsequently, the sample was washed, and the region of the orbit was extracted, keeping the lateral wall, together with the eyeball, the lateral ocular adnexa, and the periorbital fat, in order to expose the extraocular muscles and the ophthalmic artery together to its branches. The samples were then cold dehydrated (−20 °C) in 100% acetone, replacing the acetone every 3 days and renewing it with new 100% acetone. Subsequently, the samples were placed at room temperature, also in a new acetone, to favor the defatting of the samples. Next, forced impregnation was carried out, immersing the samples in an impregnating mixture composed of epoxy resin (E12, Biodur), hardener (E6, Biodur), and accelerator (E600, Biodur), in the following proportions (100/50/0.2). The forced impregnation process was carried out inside an oven with a vacuum chamber, starting the process at +30 °C, finally reaching +60 °C during the pressure reduction, finally reaching 2 cmHg, being able to develop this process in just 24 h. Subsequently, the samples were placed in molds, inside an oven at 65 °C, submerged in a resin mixture similar to the impregnation mixture, for 5 days, until the block hardened. Once this was achieved, the block was taken to a low-speed saw with a diamond blade, to be cut into ultrafine slices 0.3 mm thick. Finally, the sections were stained with Miller's stain for elastin (see Chapter E12). In this investigation, from the plastinated ultra-thin sections, it was possible to improve the visibility of the vascular structures, especially small anatomical regions such as the one studied, without the need to decalcify the sample, highlighting the arterial walls thanks to the incorporated Miller's staining, thus facilitating the three-dimensional reconstruction process.

Studies Related to Histology Analysis from Plastinated Models

Helga Fritsch, in her detailed publications in 1988 [33] and 1989 [35], described the "plastination histology" method, which consisted of a plastination technique of ultra-thin sections with epoxy resin, subsequently applying histological stains, in order to visualize the histology of anatomical regions studied from macroscopic sections. In 1988 [33], he developed this novel protocol by studying the retrorectal region in human fetuses, while in 1989 [35] in the pelvis of human fetuses. In this sense, in 1989 he described in great detail the histological staining protocol. He indicates in this publication that von Hagens [50] proposed the staining technique of Richardson et al. [86]. However, Fritsch indicated that this technique resulted in a monochromatic staining with difficulties in achieving contrast and adequate differentiation between the different tissues studied in the corresponding sample. In this way, from the work published in 1988 [33], Fritsch tested a modification of the staining of Laczkó and Lévai [87], based on methylene blue/azur II and basic fuchsin, demonstrating an improved quality of staining in thick fetal tissue sections preserved with epoxy resin (E12, Biodur) through the technique of plastination of sections. In Fritsch's 1988 article [33], the technique of plastination of sections with epoxy resin is described in detail, while in 1989 [35], the histological staining technique applied in said plastination technique is described in detail. Therefore, in relation to the plastination protocol, the samples initially fixed with formaldehyde by immersion were subsequently measured and classified (22 human fetuses). Prior to dehydration, three samples were treated with red stain for muscles (Biodur CB) [50], a process that lasted 6–8 days, reaching a light brown coloration in the fetal tissue. Next, all the samples were washed in tap water for 24 h to remove excess formaldehyde. Subsequently, the samples were cold dehydrated in acetone at −25 °C for 3–4 weeks, renewing the acetone weekly. At the end of the dehydration, defatting was carried out using methylene chloride at room temperature, for 1–2 days. Then, the samples were placed in an impregnation mixture consisting of epoxy resin (E12, Biodur, 2 parts), hardener (E6, Biodur, 1 part), and accelerator (E600, Biodur, 0.2 ml per 100 ml mixing) [50]. In this way, the forced impregnation process begins, a fundamental step in the plastination technique, for 6–10 days. During this stage, Fritsch describes a gradual reduction in pressure, until reaching 5 mbar to achieve the complete entry of the resin into the samples [88]. The forced impregnation was terminated when no bubbles corresponding to the intermediate solvent were detected. Next, the polymerization of the samples was achieved by placing them in an oven at 50 °C for a week, obtaining epoxy resin blocks with the fetuses inside. These epoxy resin blocks were cut with a diamond blade saw [89], in transverse, sagittal, and coronal planes, obtaining slices between 400 and 800 microns. Among some of the details found by Helga Fritsch when analyzing the slices obtained, she detected the presence of undulations on their surface, as well as opacity, all produced by the diamond blade of the saw. In order to solve these problems and obtain the transparency of the sections to be able to be visualized by transillumination, the sections must be polished, and for them it is necessary that they be mounted on a slide, using a mixture of epoxy resin with hardener (E12, Biodur, 10 parts + E1, Biodur, 3 parts + benzyl benzoate, 4 parts). On the other side of the cut, it was covered with an adhesive plastic sheet to protect it from the mounting medium. These mounted sections were placed inside an oven at 50 °C, for 24 h, to achieve the gluing of the section to the slide, also managing to detach the protective plastic sheet. Next, the final polishing of the slices was carried out, in order to obtain smooth and transparent sections, by using 4-degree wet sandpaper (600, 800, 1000, and 1200), to then be manually burnished with polishing paste. In this way, with the polishing applied, it was possible to reduce the thickness of the plastinated sections to an approximate thickness of between 80 and 100 microns. Finally, the sections were stained using the process proposed by Laszkó and Lévai [87], modified by Fritsch [33] based on methylene blue/azur II and basic fuchsin, and with which the collagen fibers could be visualized in a blue-violet coloration and elastic fibers red. This process was described in detail by Fritsch, in his 1989 publication [35], detailing the staining of Laszkó and Lévai [87] as follows, consisting of three solutions:

- Solution (1): Methylene blue and azure II, each 0.25% in a 0.5% solution of sodium carbonate in distilled water
- Solution (2): 0.5% basic fuchsin, in distilled water
- Solution (3): 0.5% sodium carbonate in distilled water

The sections are first immersed in solution (1) (at 90 °C) for 3–10 min, depending on the desired color intensity, to then wash the sections in solution (3) (at room temperature). The counterstaining was carried out with solution (2), for 2–5 min (at 90 °C). These solutions can be used up to 8 times if kept at room temperature. Once the staining is complete, the sections are washed in running water, and if there are stain precipitates, they can be removed with acetone-soaked paper. To achieve visualization of the slices, immersion oil can be placed directly on them. Fritsch [48] demonstrated that the retraction of plastinated sections presents less contraction than conventional histological sections [38], offering this a great advantage when investigating fetal tissues that tend to contract due to high water content thereof.

From the development of this concept of "plastination histology," a large number of publications can be seen in which Helga Fritsch participated and in which they managed to apply this method of combining plastination with histological staining techniques, some of which were described in this chapter [37, 48]. In addition, the following investigations can also be highlighted: Fritsch and Hötzinger [90], studies of the tomographic anatomy of the pelvis, the viscera, and its connective tissue, in relation to the identification of its compartments; Fritsch [48], research on the central ossification of the talus, in human fetuses; as well as the article by Konschake and Fritsch [91], in which they investigated the human nasal muscles; among others.

Eckel et al. [91] plastinated the human larynx for the first time, from the application of a plastination technique of ultra-thin sections with epoxy resin. The samples used for the development of this research work consisted of 11 portions of the trachea (first tracheal rings) of pigs. First, the samples were fixed with 15% formaldehyde, with 10% methanol added to the fixation formula (added to prevent the formation of ice crystals), storing these fixed samples at +5 °C for 5 days. Next, the plastination process began, through dehydration, carried out in cold acetone (-25 °C) for 3 weeks, making between 4 and 5 changes of acetone, corresponding to the last change to methylene chloride, carrying out this last change, the degreasing process. Subsequently, once the dehydration and degreasing were completed, the forced impregnation of the samples continued, in an impregnation mixture made up of epoxy resin (E12, Biodur, 100 parts), hardener (E6, Biodur, 70 parts), and accelerator (E600, Biodur, 0.15% of the total volume). The duration of forced impregnation was 3–4 days. Once the forced impregnation was finished, the samples were left in the vacuum chamber for 4 or 4 more days, to contribute to the hardening of the mixture and thus obtain a block of resin with the sample inside. Final hardening (curing) was achieved by placing the samples in an oven at 50 °C for 1 week. Once the hard blocks of epoxy resin were obtained, they were cut using a saw with a diamond blade, obtaining 25–30 slices of 0.8 mm thickness from each sample. Finally, these thin sections were polished, smoothing their surfaces, and transparency was achieved by mounting the sections in a toluene-based liquid medium and between two slides, hardening after a few hours. In this way, the sections could be visualized under an optical microscope.

Also a fundamental conclusion that can be reached from the development of histological staining techniques in plastinated samples is the one indicated by Sebe et al. [32, 34], who comment that plastination offers the advantage of analyzing tissue sections with a much greater thickness and size (50:1 ratio) compared to conventional staining techniques, thus being able to in situ anatomy and morphology of any organ or bodily region magnified.

Johnson et al. [92] made a comparison between sections preserved with epoxy resin and histological sections, in samples destined for the study of the structure of the spinal connective tissue, corresponding to samples corresponding to the cervical region, between the external occipital protuberance and the first thoracic vertebra. The slices obtained by the technique of sheet plastination with epoxy resin were 2.5 mm thick [30]. Among the advantages identified for the plastination technique, the authors indicate that there is no need to decalcify the samples, allowing an intact study of the plastinated samples, allowing the relationships of the neighboring anatomical structures to be preserved, ensuring a visualization of the microanatomy from macroscopic sections.

Deplastination for Histopathological Research and DNA Detection Methods in Plastinated Tissues

Frierson et al. [93] carried out an interesting investigation with the aim of demonstrating whether plastinated neoplastic pathological samples could be suitable for ploidy analysis, since the microscopic preservation of plastinated tissues had already been demonstrated. This is how they developed the deplastination process, from which the plastination process applied to anatomical samples is reversed and in this way multiple analyses can be carried out on the tissues, in this case aimed at the possibility of detecting genetic material. In other words, to demonstrate that plastination allows not only the preservation of the microscopic characteristics of biological tissues at the morphological level but also the properties of the genetic material with DNA preservation. In this sense, they used samples plastinated with silicone from human squamous cell carcinoma of the esophagus and adenocarcinoma of the colon. The samples were immersed in 5% sodium methoxide dissolved in methanol for 48 h to remove the silicone. The samples were then washed with fresh methanol. This process is what is defined as deplastination. Subsequently, the authors subjected half of the samples to DNA ploidy analysis, without further processing, while the remaining half were processed for histological studies by forming paraffin blocks. The authors identified that the deplastinized samples processed in paraffin were the ones that offered the best results. In addition, they managed to communicate a protocol that allows performing ploidy analysis on plastinated tissues.

Grondin et al. [94] developed a publication in which they used human samples, plastinated with silicone (S10, Biodur) corresponding to the spleen and pancreas, to perform the same ultrastructural analyses from their visualization by optical microscopy and electron microscopy. For this, they implemented the deplastination technique, similar to that applied by Frierson et al. [93]. The samples were then previously plastinated, using the silicone plastination process, beginning with the cooling of the samples at 4 °C for 12 h and then immersing them in pure acetone at −20 °C, to dehydrate them. The acetone changes were made every 5 days, renewing the acetone at each change. After dehydration, the samples were placed in an impregnation mixture, at −20 °C, and inside a vacuum chamber, made up of the following compounds: silicone (S10, Biodur) (100 pbw), catalyst (S3, Biodur) (1 pbw), and xylene 20%. The forced impregnation process was carried out for 3 weeks, reaching a final pressure value of 15 mmHg. Subsequently, the specimens were polymerized through a rapid curing process, inside a hermetic chamber, subjecting the samples to a catalyst (S6, Biodur). Finally, the samples were stored inside containers for a month to develop the secondary curing and final drying of the samples. In relation to the deplastination process, portions of the spleen and pancreas samples were extracted, which were submerged in a saturated solution of sodium methoxide in methanol, stirring gently, in order to extract the silicone from the samples (no percentages nor indications were given more information regarding the deplastination formulation, as previously indicated by Frierson et al. [93]). Based on these reported protocols, it was shown that plastinated samples can be used for the development of optical and electron microscopy studies. The authors also mentioned that proper formaldehyde or glutaraldehyde fixation of the samples tends to preserve the tissue structure during the plastination and deplastination process. In addition, they also indicate that retrospective studies could be carried out from pathological samples, although they did not carry out any type of analysis at the level of the genetic material.

Francis and Rabi [95] investigated the applications of deplastination in histopathology studies, in a human sample of liver plastination, through the silicone plastination technique [50]. After 3 months of applying plastination to the sample, they performed the deplastination process, which consisted of immersing the liver sample in a 5% sodium methoxide solution in 90% methanol, similar to those described by Frierson et al. [93]. Subsequently, they developed a histological processing of the deplastinized sample to obtain the final paraffin blocks, from which 5 micron sections were cut, which were stained with hematoxylin and eosin. The authors state that deplastination occurs when the Na^+ ion breaks down the crosslinking of silicon polymers. They further cited Ripani et al. [96], indicating that these authors suggested other components to perform deplastination, using alcohol, methylbenzene, methylene, and acetone bichloride, and although Ripani et al. [96] managed to develop the deplastination of the plastinated samples with these components, they indicated that they failed to obtain good results in the histopathological analyses performed on the deplastinated sections. In this way, and by comparing, Francis and Rabi [95] established that deplastination from the use of sodium methoxide and methanol is the best alternative for obtaining histopathology information from plastinated samples.

Ramos et al. [97] made a comparison between different deplastination methods, from samples plastinated with silicone, for the development of histological studies. The samples used corresponded to pig heart, aorta, and kidney. They made a comparison between samples fixed with 10% formaldehyde, samples plastinated and processed in paraffin, samples deplastinated with methylbenzene, and samples deplastinated with dimethylbenzene. However, they did not process the samples with sodium methoxide. The results obtained corresponded to the fact that the samples plastinated and directly subjected to histological processing better preserved the histological characteristics of the tissues, compared to that obtained from the deplastination process with methylbenzene and dimethylbenzene, but because they indicated that the removal of silicone was incomplete, thus generating artifacts in the results obtained. For their part, Baygeldi et al. [98] developed a study similar to that of Ramos et al. [97] but also expanding to the use of 5% sodium methoxide as a deplastination agent in sheep brain samples. In this sense, they indicated that sodium methoxide was a better alternative for the development of deplastination and the subsequent histological analysis of the samples, compared to the use of toluene, with the possibility of carrying out an adequate microscopic examination of the deplastinated tissues.

Rahul et al. [99] developed the deplastination protocol, demonstrating its ability to preserve the histological structures of the samples preserved by the silicone plastination technique, with the aim of performing histopathological analyses. In this way, they used 12 samples obtained in maxillofacial surgeries from patients with oral carcinoma. They then compared samples preserved in 10% formaldehyde and samples preserved by the silicone plastination method (S10, Biodur), which were subsequently subjected to the deplastination process. They applied the deplastination process described by Frierson et al. [93], which consisted of immersing the samples in a 5% sodium methoxide solution and 90% methanol for 48 h. Subsequently, they performed the corresponding histological processing. They indicated that what is fundamental in the deplastination process is the identification of the completion of this process on the sample, and in this case, Francis et al. [95]

identified it when putting a pin through the sample was easy. In addition, they also indicate that it is necessary for the mixture of sodium methoxide and methanol to be fresh, in order to obtain good results in subsequent histological processing. Another important piece of information reported by Rahul et al. [99] consists of the histological staining time, which is longer in samples that have been deplastinated compared to histological staining in traditional samples.

In an article carried out by our research group [100], despite the great progress made in the area of cell and molecular biology in recent years, we report for the first time the analysis of nucleic acids in plastinated tissues. There are very few studies in the literature that investigate the preservation, extraction, and detection of DNA from plastinated samples, and which are presented in this section of the chapter. The objective of our research was to establish a precise protocol for deplastination and extraction of DNA from samples preserved by means of our proposed silicone plastination technique at room temperature [101] and the evaluation of the quantity and quality of the DNA extracted using our proposed protocol. For the development of this research, the samples consisted of a dog (*Canis lupus*) and a Sprague-Dawley rat. The dog was frozen for 1 week at −20 °C and then sagittally cut with a circular saw, while the Sprague-Dawley rat was dissected anatomically. Then, both specimens, without fixing, were subjected to our plastination technique at room temperature [101]. The first step of our plastination technique consisted of dehydration, carried out for 4 weeks, immersing the specimens in 100% acetone at −20 °C, with weekly acetone changes. Dehydration was controlled with an acetonometer calibrated at 15 °C. When an acetone percentage greater than 99.5% was reached, the dehydration phase was terminated. Subsequently, the specimens were degreased, leaving them in 100% acetone for 1 week at room temperature (20 °C). After this period, the forced impregnation phase was continued, for which the samples were immersed for 24 h at 760 mmHg in a silicone mixture (dimethylsiloxane; Dow Corning® 200 Fluid, 50 cst; Merck, Darmstadt, Germany) and catalyst (dibutyltin dimethyl ester; Merck, Darmstadt, Germany), in a ratio of 100:1, inside a vacuum chamber at room temperature (20 °C). The next day, forced vacuum impregnation began, in which acetone (intermediate solvent) was replaced by the impregnation mixture using a vacuum pump, applying active and passive forced impregnation [101] (see Chap. 5). Forced impregnation allows acetone to be extracted from the cell structure and interstitium of the preparation, being replaced by silicone. This process occurs due to the high vapor pressure of the acetone compared to the low vapor pressure of the silicone/catalyst mixture, which causes the acetone to change from a liquid to a gaseous state and be removed, through the generation of bubbles. The pressure was reduced over a period of 5 days from 760 mmHg to 10 mmHg. When the pressure has reached 10 mmHg and without visualization of gaseous extraction (i.e., absence of acetone bubbles), the forced impregnation has finished, having then extracted all the acetone from inside the samples. The samples were then subjected to the curing process, using a curing liquid used (tetraethyl orthosilicate, TEOS; Merck, Darmstadt, Germany) that enhances a three-dimensional cross-reaction by lengthening the silicone chains (joined together by the corresponding catalyst). The specimens were then placed within an airtight chamber and subjected to TEOS, evaporating it within the chamber with the use of a fish tank motor. The rat dried up in 5 days, while the dog dried up in 14 days. These samples were plastinated in 2016, and 3 years later the deplastination process was carried out on samples that were extracted from the two aforementioned specimens: dog liver and rat skeletal muscle. In this way, these portions of liver tissue and muscles were subjected to the deplastination process suggested by Frierson et al. [93], which consisted of immersing the plastinated tissues in 5% sodium methoxide (CH_3NaO) dissolved in absolute methanol. Four rat muscle and dog liver samples were incubated at room temperature, two were incubated for 24 h, and two were incubated for 48 h. In each pair, one was deplastinized with mineral oil and the other without the use of mineral oil. Once the samples were deplastinized, DNA extraction, detection, and evaluation protocols were carried out [100], and then, also for the first time, this extracted genetic material was subjected to processing and analysis by PCR in real time, for the beta-actin gene, to demonstrate its usefulness in subsequent applications. Based on the results obtained from our protocols, it was possible to quantify DNA and demonstrate the obtaining of genetic material in all the proposed protocols. Of these, we can define that the protocol that allowed the extraction of the greatest amount of genetic material was the one in which the samples were incubated for 48 h in CH_3NaO, compared to the 24 h protocols. In addition, those protocols that included mineral oil during incubation (2 min, at 80 °C) allowed obtaining the best yields in DNA extraction, compared to those protocols that did not include mineral oil.

To evaluate the integrity of the extracted genetic material (DNA), 1% agarose gel electrophoresis was used, demonstrating the most cellular tissues, such as the liver, demonstrating a high yield, identifying a marked DNA band, compared to skeletal muscle samples. In addition, excellent results were obtained by real-time PCR analysis of the beta-actin gene, demonstrating successful amplification of the DNA extracted from the samples.

In this way, from our research, we identified for the first time a protocol for extracting DNA from plastinated tissues. For their part, Nagaraj et al. [102] extracted DNA from plastinated tissue, but they do not specify the type of tissue or the plastination technique used; in addition, the DNA quantification method used was spectrophotometry, unlike our quantification method, which was fluorimetry, which is much more specific due to the use of probes that avoid the overestimation of the DNA that can occur with the spectrophotometry technique.

Therefore, from the proposed protocol, we were able to extract fully intact, complete, and non-degraded DNA from plastinated tissues.

In our research we demonstrated the possibility of obtaining DNA from plastinated samples, identifying the conservation capacity of the genetic material that our silicone plastination technique possesses at room temperature [101]. In this way, our work constituted the first report of a complete and non-degraded DNA extraction protocol from plastinated samples that demonstrates that the DNA can be used successfully in downstream applications (such as real-time PCR), thus which opens up many possibilities for plastination in the area of basic and clinical sciences, epidemiology, legal medicine, and forensic sciences.

Education and Training in Anatomy Through Plastination

Nagaraj et al. [103] published a paper aimed at demonstrating the usefulness of applying the silicone plastination technique for the preservation of pathology samples, with the aim of promoting teaching through use of plastinated samples. To do this, they developed the silicone plastination technique, which they defined with four basic steps. The samples that were presented in the publication were mature cystic teratoma of the ovary, ulcerative colitis of the colon, adenocarcinoma of the colon, section of the spleen, and section of the uterine tube. First, they developed the fixation of the samples, with 10% formaldehyde and at room temperature. This process lasted 48 h. Next, they proceeded to dehydrate the samples, using increasing concentrations of ethanol (70%, 80%, 90%, 95%, and three changes of 100%, 48 h at each concentration). The total dehydration time was 2 weeks. Subsequently, the samples were placed in methylene chloride, to favor their defatting, for 2 weeks, with a change of the mixture. After this, the samples are placed in an impregnation mixture made up of silicone (S10) with its corresponding catalyst (S3), at 1% of the total volume. The forced impregnation phase was carried out at $-20\ ^\circ$C and lasted 2 weeks. Once the forced impregnation is finished, the samples are subjected, inside a closed container, to a curing agent (S6) that is vaporized, to develop the curing phase (polymerization), at room temperature, and for 3 weeks. The total time of the plastination process was 16 weeks. The final result was dry, odorless, and non-toxic plastinated samples that maintained their original shape and color.

Bickley et al. [104], from the College of Veterinary Medicine, University of Tennessee, Knoxville, USA, developed research aimed at creating plastinated models of the canine gastrointestinal tract with the aim of allowing teaching anatomy endoscopy by implementing endoscopic techniques in said plastinated models. The importance represented in the years of publication by the advancement of endoscopic treatments, both for diagnosis and treatment in humans and animals, stimulated the authors to develop these novel models to improve knowledge of anatomy from an endoscopic view. Also taking into account the need for training required by those who manipulate endoscopes, in this way, the generation of these plastinated models for training constituted a relevant contribution for the application of these minimally invasive methods, both in human and animal medicine. In this way, the authors developed their research from fresh canine cadaveric samples, from which they extracted the corresponding gastrointestinal tracts, which were washed and dilated according to the corresponding needs, and to avoid distortion of the anatomical tissue structures and characteristics. After proper positioning of the structures, the samples were dilated and consequently fixed with 10% formaldehyde. Once the proper positioning was achieved, the samples were submerged in a 2% formaldehyde solution overnight, to wash the samples with running water the next day and start the plastination process with the cold dehydration stage ($-15\ ^\circ$C) with acetone, for 4–6 weeks. After this stage, the samples were placed in acetone, but at room temperature, to favor the process of defatting the samples, for a total of 2–3 weeks. Next, the samples were immersed in an impregnation mixture made up of silicone

(S10) and its catalyst (S3), to perform the forced impregnation of the samples for 4–6 weeks. Based on what has been indicated, it is understood that the impregnation was carried out cold, since the authors then indicate that the specimens were placed at room temperature to favor the drainage of excess silicone mixture. Air was also introduced through the openings of the gastrointestinal tract in order to extract further excess resin. Subsequently, the samples were subjected to a slow curing process for 3–4 days, with the aim of favoring the final flexibility of the plastinated samples. The S6 agent was used for this curing stage, which was vaporized on the samples inside an airtight container. Then the samples were incorporated into plastic bags for their final curing. During these stages the authors also indicate that the samples were kept connected with air, to maintain the insufflation of the tissues. Likewise, the samples were later placed inside curing chambers subjected to the vaporization of the S6 agent, with a total curing time of 3–4 weeks. Finally, once the plastination process was completed, the samples were visualized and examined with the use of a 9.6 mm endoscope. As general conclusions, the authors managed to generate specimens that correctly maintained the morphological characteristics, especially at the level of their dilation, ensuring an approach to the samples with the use of endoscopic techniques and thus favoring the generation of gastrointestinal anatomy training with an important application in human and animal medicine.

In a research work carried out by our group at the Universidad de La Frontera [105], we developed a modified silicone plastination protocol, at room temperature (also published in [101]), and we applied Wistar rats to generate plastinated training models, in clinical and surgical approaches, thus contributing not only to a more precise training of professionals, using real samples, and recognizing a real anatomy in the specimens, but also to reduce the use of laboratory animals in this type of training, minimizing the number of specimens needed to plastinate. The latter is crucial, since the plastinated samples have a prolonged durability, and allows models to be generated that can be reused a large number of times, and from this derives the real possibility of reducing the number of animals necessary for this type of surgical improvement courses. The samples used in this investigation were five Wistar rats, which were dissected at the level of their thoracic and abdominal cavities to later be fixed with a 5% formaldehyde solution. Next, the silicone plastination process began at room temperature [101], through the development of the dehydration stage, implemented at room temperature (20 °C) for 2 weeks, starting with higher acetone concentrations, 90%, and ending the last two dehydration changes with 100% acetone. When dehydrating at room temperature, the samples were degreased at the same time, which allowed the most important phase of plastination to continue forced impregnation. The samples were placed in an impregnating mixture, consisting of generic silicone and catalyst, in a ratio of 100:1, respectively. The forced impregnation process lasted a total of 5 days, combining phases of active impregnation (8 h) and passive impregnation (16 h), until reaching 5 mmHg and evidencing the absence of bubbles. Once the forced impregnation was finished, the samples were drained, eliminating the excess silicone and positioning them, before beginning their curing. The polymerization (curing) consisted of the application of a curing and hardening agent, by means of vaporization, placing the samples inside an airtight chamber. A fast and superficial curing process was developed, which allowed the external drying of the samples to be completed in 2–3 days. The internal drying process was extended for 3–4 months. In this way, it was possible to plastinate the indicated samples in a total of 32 days, obtaining dry, totally biosafe material for visualizing the thoracic and abdominal anatomy of rats and with the possibility of practicing surgical approaches with a reduced number of samples.

Ottone et al. [106] developed plastinated samples of human placenta, oriented to the implementation of teaching resources for education in gynecology and obstetrics. In this sense, they comment in their publication that the three-dimensional study of the placenta is complex, due to the scarce availability of samples; for them they generated these real plastinated models, to contribute to the provision of new pedagogical tools for undergraduate and postgraduate students aimed at training specialists in obstetrics and gynecology. For the investigation, the placentas were collected 24 h after birth and preserved at 4 °C, to later, after draining the blood, be injected by arterial and venous route, with liquid polymeric latex, in order to be able to highlight, with the venous route is red and the arterial route is blue. Subsequently, the placentas were immersed in 10% formaldehyde to develop their fixation. This fixation process lasted 7 days. Next, they developed the silicone plastination technique (S10) [107], beginning with cold dehydration (−20 °C), using 100% acetone, making five changes every 10–14 days, until achieving a final acetone concentration greater than 99.5%. Once the dehydration was finished, the forced impregnation phase was carried out, immersing the placentas in an impregnation mixture made up of silicone (S10, Biodur) and its corresponding catalyst (S3, Biodur, 1% with respect to the total volume of silicone). The forced impregnation process was carried out cold (−20 °C) and lasted 14 days, until reaching a final pressure of 2–5 mmHg. To finish the plastination technique, the samples were dried by subjecting them to a vaporizing agent (S6, Biodur) inside a hermetic chamber. This curing process lasted 4 weeks, finally obtaining dry and completely plastinated placentas.

Oostrom et al. [108] carried out research aimed at teaching radiological anatomy, through the generation of plastinated hearts, which were scanned by cone beam computed tomography (CBCT). This research constituted the first communication in the literature on the generation of 3D anatomical models from plastinated specimens. Then, from the tomography results, three-dimensional models of the hearts were generated. In this way, the authors developed a silicone plastination technique at room temperature [109], in four human hearts, fixed with formaldehyde. First, the samples were washed with tap water to remove excess formaldehyde. Subsequently, the samples were dehydrated in cold (−25 °C), starting with 98% acetone, reaching a stabilization of the acetone concentration in the fifth week of dehydration (98%), to pass the samples, in the last week, to an acetone with a concentration of 100% and managing to finish the dehydration of the samples when reaching a percentage higher than 99.5% of acetone. Then the samples were submerged in 100% acetone, at room temperature, to favor the defatting of the samples. After this stage was completed, the samples were submerged in an impregnating solution made up of silicone (PR10) and catalyst (Cr20), in a ratio of 100:8, respectively. In this way, the forced impregnation of the samples was carried out in a vacuum chamber at room temperature for a total of 4 days. Finally, the samples were extracted from the vacuum chamber and allowed to drain for 4 days, contributing to the elimination of excess polymer with blotting paper. Then, the samples were covered with a chain crosslinking agent (Ct32), for 3 days, with three applications of the agent per day, placing the samples in plastic bags to reach their final polymerization. The total plastination time was 12 weeks. Once the plastination was finished, the plastinated hearts were scanned by CBCT, the sections reaching a thickness of 0.3 mm. These slices were later incorporated into a 3D software that allowed the generation of virtual models of the plastinated hearts, allowing their manipulation and virtual visualization in 360°. In this way, from 2D CBCT slices of plastinated specimens, three-dimensional volumetric and surface representations of the same plastinated specimens could be generated. This publication demonstrates the possibility of generating libraries of cardiac pathologies, from new plastinated samples, which at the same time could be reconstructed and visualized in three dimensions, allowing the preservation and study of myocardial infarctions, ischemic heart disease, aortic atherosclerosis, pericarditis, and aortic aneurysms, among many other pathologies. A novel application of plastination for the development of anatomy education from radiology is also demonstrated, allowing the student to examine radiological studies and immediately compare them with the plastinated, normal, or pathological sample and, in turn, integrate all this information with 2D reconstructions of CBCT and 3D.

References

1. Sittel C, Eckel HE, Sprinzl GM, Stennert E. Plastination of the larynx for whole-organ sectioning. Eur Arch Otorhinolaryngol. 1997;254(Suppl 1):S93–6. https://doi.org/10.1007/BF02439734.
2. Johnson GM, Zhang M, Jones DG. The fine connective tissue architecture of the human ligamentum nuchae. Spine. 2000;25(1):5–9. https://doi.org/10.1097/00007632-200001010-00003.
3. Zhang M, An PC. Liliequist's membrane is a fold of the arachnoid mater: study using sheet plastination and scanning electron microscopy. Neurosurgery. 2000;47:902–8; discussion 8-9. https://doi.org/10.1097/00006123-200010000-00021.
4. An P-C, Zhang M. A technique for preserving the subarachnoid space and its contents in a natural state with different colours. J Int Soc Plast. 1999;14:12–7.
5. Leaper M, Zhang M, Dawes PJ. An anatomical protrusion exists on the posterior hypopharyngeal wall in some elderly cadavers. Dysphagia. 2005;20(1):8–14. https://doi.org/10.1007/s00455-004-0018-1.
6. Sora MC, Strobl B, Staykov D, Traxler H. Optic nerve compression analyzed by using plastination. Surg Radiol Anat. 2002;24(3-4):205–8. https://doi.org/10.1007/s00276-002-0037-2.
7. Zhang M, Lee AS. The investing layer of the deep cervical fascia does not exist between the sternocleidomastoid and trapezius muscles. Otolaryngol Head Neck Surg. 2002;127(5):452–4. https://doi.org/10.1067/mhn.2002.129823.
8. Zhang WG, Zhang SX, Wu BH. A study on the sectional anatomy of the oculomotor nerve and its related blood vessels with plastination and MRI. Surg Radiol Anat. 2002;24(5):277–84. https://doi.org/10.1007/s00276-002-0052-3.

9. Qiu MG, Zhang SX, Liu ZJ, Tan LW, Wang YS, Deng JH, Tang ZS. Three-dimensional computational reconstruction of lateral skull base with plastinated slices. Anat Rec A Discov Mol Cell Evol Biol. 2004;278(1):437–42. https://doi.org/10.1002/ar.a.20023.

10. Nash L, Nicholson H, Lee AS, Johnson GM, Zhang M. Configuration of the connective tissue in the posterior atlanto-occipital interspace: a sheet plastination and confocal microscopy study. Spine. 2005;30(12):1359–66. https://doi.org/10.1097/01.brs.0000166159.31329.92.

11. Nash L, Nicholson HD, Zhang M. Does the investing layer of the deep cervical fascia exist? Anesthesiology. 2005;103(5):962–8. https://doi.org/10.1097/00000542-200511000-00010.

12. Nash LG, Phillips MN, Nicholson H, Barnett R, Zhang M. Skin ligaments: regional distribution and variation in morphology. Clin Anat. 2004;17(4):287–93. https://doi.org/10.1002/ca.10203.

13. Chen S, Wang H, Fong AH, Zhang M. Micro-CT visualization of the cricothyroid joint cavity in cadavers. Laryngoscope. 2012;122(3):614–21. https://doi.org/10.1002/lary.22504.

14. Liu M, Chen S, Liang L, Xu W, Zhang M. Microcomputed tomography visualization of the cricoarytenoid joint cavity in cadavers. J Voice. 2013;27(6):778–85. https://doi.org/10.1016/j.jvoice.2013.05.010.

15. Diao Y, Liang L, Yu C, Zhang M. Is there an identifiable intact medial wall of the cavernous sinus? Macro- and microscopic anatomical study using sheet plastination. Neurosurgery. 2013;73(1):106–9; discussion ons110. https://doi.org/10.1227/NEU.0b013e3182889f2b.

16. Arnts H, Kleinnijenhuis M, Kooloos JG, Schepens-Franke AN, van Cappellen, van Walsum AM. Combining fiber dissection, plastination, and tractography for neuroanatomical education: revealing the cerebellar nuclei and their white matter connection. Anat Sci Educ. 2014;7(1):47–55. https://doi.org/10.1002/ase.1385.

17. De Castro I, de Christoph DH, dos Santos DP, Landeiro JA. Internal structure of the cerebral hemispheres: an introduction of fiber dissection technique. Arq Neuropsiquiatr. 2005;63(2A):252–8. https://doi.org/10.1590/s0004-282x2005000200011.

18. DeJong K, Henry RW. Silicone plastination of biological tissue: cold-temperature technique BiodurTM S10/S15 technique and products. J Int Soc Plast. 2007;22:2–14.

19. Liang L, Diao Y, Xu Q, Zhang M. Transcranial segment of the trigeminal nerve: macro-/microscopic anatomical study using sheet plastination. Acta Neurochir. 2014;156(3):605–12. https://doi.org/10.1007/s00701-013-1920-9.

20. Scali F, Nash LG, Pontell ME. Defining the morphology and distribution of the alar fascia: a sheet plastination investigation. Ann Otol Rhinol Laryngol. 2015;124(10):814–9. https://doi.org/10.1177/0003489415588129.

21. Scali F, Pontell ME, Nash LG, Enix DE. Investigation of meningomyovertebral structures within the upper cervical epidural space: a sheet plastination study with clinical implications. Spine J. 2015;15(11):2417–24. https://doi.org/10.1016/j.spinee.2015.07.438.

22. Johnson GM, Zhang M. Regional differences within the human supraspinous and interspinous ligaments: a sheet plastination study. Eur Spine J. 2002;11(4):382–8. https://doi.org/10.1007/s00586-001-0378-2.

23. Bernal-Mañas CM, González-Sequeros O, Moreno-Cascales M, Sarria-Cabrera R, Latorre-Reviriego RM. New anatomo-radiological findings of the lateral pterygoid muscle. Surg Radiol Anat. 2016;38(9):1033–43. https://doi.org/10.1007/s00276-016-1665-2.

24. Thorpe Lowis CG, Zhang M, Amin NF. Fine configuration of thoracic type II meningeal cysts: macro- and microscopic cadaveric study using epoxy sheet plastination. Spine. 2016;41(20):E1195–200. https://doi.org/10.1097/BRS.0000000000001587.

25. Liang L, Qu L, Chu X, Liu Q, Lin G, Wang F, Xu S. Meningeal architecture of the jugular foramen: an anatomic study using plastinated histologic sections. World Neurosurg. 2019;127:e809–17. https://doi.org/10.1016/j.wneu.2019.03.272.

26. Wu X, Ding H, Yang L, Chu X, Xie S, Bao Y, Wu J, Yang Y, Zhou L, Li M, Li SY, Tang B, Xiao L, Zhong C, Liang L, Hong T. Invasive corridor of clivus extension in pituitary adenoma: bony anatomic consideration, surgical outcome and technical nuances. Front Oncol. 2021;11:689943. https://doi.org/10.3389/fonc.2021.689943.

27. Didenko M, Starchik D, Natarova S, Pasenov G, Khubulava G. Morphometric characteristics of the right atrium and Bachmann's bundle region: implications for atrial pacing. P1525, ii263, Poster Session, 2013.
28. Starchik D, Shishkevich A, Sora MC. Visualization of metal stents in coronary arteries with epoxy plastination. J Plast. 2020;32(2):20. https://doi.org/10.56507/EAOL4915.
29. Skalkos E, Williams G, Baptista CAC. The E12 technique as an accessory tool for the study of myocardial fiber structure analysis in MRI. J Int Soc Plast. 1999;14:18–21.
30. Weber W, Henry RW. Sheet plastination of body slices - E12 technique, filling method. J Int Soc Plast. 1993;7:16–2.
31. Porzionato A, Macchi V, Gardi M, Parenti A, De Caro R. Histotopographic study of the recto-urethralis muscle. Clin Anat. 2005;18(7):510–7. https://doi.org/10.1002/ca.20184.
32. Sebe P, Oswald J, Fritsch H, Aigner F, Bartsch G, Radmayr C. An embryological study of fetal development of the rectourethralis muscle–does it really exist? J Urol. 2005;173(2):583–6. https://doi.org/10.1097/01.ju.0000151248.37875.24.
33. Fritsch H. Developmental changes in the retrorectal region of the human fetus. Anat Embryol. 1988;177(6):513–22. https://doi.org/10.1007/BF00305138.
34. Sebe P, Fritsch H, Oswald J, Schwentner C, Lunacek A, Bartsch G, Radmayr C. Fetal development of the female external urinary sphincter complex: an anatomical and histological study. J Urol. 2005;173(5):1738–42; discussion 1742. https://doi.org/10.1097/01.ju.0000154616.51979.da.
35. Fritsch H. Staining of different tissues in thick epoxy resin-impregnated sections of human fetuses. Stain Technol. 1989;64(2):75–9. https://doi.org/10.3109/10520298909108049.
36. Lunacek A, Schwentner C, Fritsch H, Bartsch G, Strasser H. Anatomical radical retropubic prostatectomy: 'curtain dissection' of the neurovascular bundle. BJU Int. 2005;95(9):1226–31. https://doi.org/10.1111/j.1464-410X.2005.05510.x.
37. Fritsch H, Pinggera GM, Lienemann A, Mitterberger M, Bartsch G, Strasser H. What are the supportive structures of the female urethra? Neurourol Urodyn. 2006;25(2):128–34. https://doi.org/10.1002/nau.20133.
38. von Hagens G, Tiedemann K, Kriz W. The current potential of plastination. Anat Embryol. 1987;175(4):411–21. https://doi.org/10.1007/BF00309677.
39. Macchi V, Porzionato A, Stecco C, Vigato E, Parenti A, De Caro R. Histo-topographic study of the longitudinal anal muscle. Clin Anat. 2008;21(5):447–52. https://doi.org/10.1002/ca.20633.
40. Al-Ali S, Blyth P, Beatty S, Duang A, Parry B, Bissett IP. Correlation between gross anatomical topography, sectional sheet plastination, microscopic anatomy and endoanal sonography of the anal sphincter complex in human males. J Anat. 2009;215(2):212–20. https://doi.org/10.1111/j.1469-7580.2009.01091.x.
41. Cook P, Al-Ali S. Submacroscopic interpretation of human sectional anatomy using plastinated E12 sections. J Int Soc Plast. 1997;12:17–27.
42. Kaulhausen T, Siewe J, Eysel P, Knifka J, Notermans HP, Koebke J, Sobottke R. The role of the inter-/supraspinous ligament complex in stand-alone interspinous process devices: a biomechanical and anatomic study. J Neurol Surg A Cent Eur Neurosurg. 2012;73(2):65–72. https://doi.org/10.1055/s-0031-1297250.
43. Xu Z, Chapuis PH, Bokey L, Zhang M. Nature and architecture of the puboprostatic ligament: a macro- and microscopic cadaveric study using epoxy sheet plastination. Urology. 2017;110:263.e1–8. https://doi.org/10.1016/j.urology.2017.08.018.
44. Xu Z, Tu L, Zheng Y, Ma X, Zhang H, Zhang M. Fine architecture of the fascial planes around the lateral femoral cutaneous nerve at its pelvic exit: an epoxy sheet plastination and confocal microscopy study. J Neurosurg. 2018;131(6):1860–8. https://doi.org/10.3171/2018.7.JNS181596.
45. Sora MC, Genser-Strobl B. The sectional anatomy of the carpal tunnel and its related neurovascular structures studied by using plastination. Eur J Neurol. 2005;12(5):380–4. https://doi.org/10.1111/j.1468-1331.2004.01034.x.

46. Koslowsky TC, Berger V, Hopf JC, Müller LP. Presentation of the vascular supply of the proximal ulna using a sequential plastination technique. Surg Radiol Anat. 2015;37(7):749–55. https://doi.org/10.1007/s00276-015-1476-x.

47. Xu Z, Chapuis PH, Bokey L, Zhang M. Denonvilliers' fascia in men: a sheet plastination and confocal microscopy study of the prerectal space and the presence of an optimal anterior plane when mobilizing the rectum for cancer. Color Dis. 2017. https://doi.org/10.1111/codi.13906.

48. Fritsch H. Sectional anatomy of connective tissue structures in the hindfoot of the newborn child and the adult. Anat Rec. 1996;246(1):147–54. https://doi.org/10.1002/(SICI)1097-0185(199609)246:1<147::AID-AR16>3.0.CO;2-P.

49. Sora MC, Genser-Strobl B, Radu J, Lozanoff S. Three-dimensional reconstruction of the ankle by means of ultrathin slice plastination. Clin Anat. 2007;20(2):196–200. https://doi.org/10.1002/ca.20335.

50. von Hagens G, editor. Heidelberg plastination folder. Collection of technical leaflets of plastination. Heidelberg: Biodur Products GmbH; 1986.

51. Sora MC, Jilavu R, Grübl A, Genser-Strobl B, Staykov D, Seicean A. The posteromedial neurovascular bundle of the ankle: an anatomic study using plastinated cross sections. Arthroscopy. 2008;24(3):258–263.e1. https://doi.org/10.1016/j.arthro.2007.08.030.

52. Rath B, Notermans HP, Frank D, Walpert J, Deschner J, Luering CM, Koeck FX, Koebke J. Arterial anatomy of the hallucal sesamoids. Clin Anat. 2009;22(6):755–60. https://doi.org/10.1002/ca.20843.

53. Rath B, Notermans HP, Franzen J, Knifka J, Walpert J, Frank D, Koebke J. The microvascular anatomy of the metatarsal bones: a plastination study. Surg Radiol Anat. 2009;31(4):271–7. https://doi.org/10.1007/s00276-008-0441-3.

54. Faymonville C, Andermahr J, Seidel U, Müller LP, Skouras E, Eysel P, Stein G. Compartments of the foot: topographic anatomy. Surg Radiol Anat. 2012;34(10):929–33. https://doi.org/10.1007/s00276-012-0982-3.

55. Sora MC, Strobl B, Staykov D, Förster-Streffleur S. Evaluation of the ankle syndesmosis: a plastination slices study. Clin Anat. 2004;17(6):513–7. https://doi.org/10.1002/ca.20019.

56. Oppermann J, Franzen J, Spies C, Faymonville C, Knifka J, Stein G, Bredow J. The microvascular anatomy of the talus: a plastination study on the influence of total ankle replacement. Surg Radiol Anat. 2014;36(5):487–94. https://doi.org/10.1007/s00276-013-1219-9.

57. Villamonte-Chevalier AA, Soler M, Sarria R, Agut A, Gielen I, Latorre R. Ultrasonographic and anatomic study of the canine elbow joint. Vet Surg. 2015;44(4):485–93. https://doi.org/10.1111/j.1532-950X.2014.12249.x.

58. Latorre RM, Reed RB, Gil F, Lopez-Albors O, Ayala MD, Martinez-Gomariz F, Henry RW. Epoxy impregnation without hardener: to decrease yellowing, to delay casting, and to aid bubble removal. J Int Soc Plast. 2002;17:17–22.

59. Xu Z, Mei B, Liu M, Tu L, Zhang H, Zhang M. Fibrous configuration of the fascia iliaca compartment: an epoxy sheet plastination and confocal microscopy study. Sci Rep. 2020;10(1):1548. https://doi.org/10.1038/s41598-020-58519-0.

60. Ottone NE, Baptista CAC, Latorre R, Bianchi HF, Del Sol M, Fuentes R. E12 sheet plastination: techniques and applications. Clin Anat. 2018;31(5):742–56. https://doi.org/10.1002/ca.23008.

61. Lane A. Sectional anatomy: standardized methodology. J Int Soc Plast. 1990;4:16–22.

62. Beyersdorff D, Schiemann T, Taupitz M, Kooijman H, Hamm B, Nicolas V. Sectional depiction of the pelvic floor by CT, MR imaging and sheet plastination: computer-aided correlation and 3D model. Eur Radiol. 2001;11(4):659–64. https://doi.org/10.1007/s003300000561.

63. Steinke H. Plastinated body slices for verification of magnetic resonance tomography images. Ann Anat. 2001;183(3):275–81. https://doi.org/10.1016/S0940-9602(01)80234-X.

64. Sora MC, Brugger PC, Strobl B. Shrinkage during E12 plastination. J Int Soc Plast. 2002;17:23–7.

65. Thomas M, Steinke H, Schulz T. A direct comparison of MR images and thin-layer plastination of the shoulder in the apprehension-test position. Surg Radiol Anat. 2004;26(2):110–7. https://doi.org/10.1007/s00276-003-0193-z.

66. Sora MC, Cook P. Epoxy plastination of biological tissue: E12 technique. J Int Soc Plast. 2007;22:31–9.

67. Soal S, Pollard M, Burland G, Lissaman R, Wafer M, Stringer MD. Rapid ultrathin slice plastination of embalmed specimens with minimal tissue loss. Clin Anat. 2010;23(5):539–44. https://doi.org/10.1002/ca.20972.

68. Steinke H, Hammer N, Slowik V, Stadler J, Josten C, Böhme J, Spanel-Borowski K. Novel insights into the sacroiliac joint ligaments. Spine. 2010;35(3):257–63. https://doi.org/10.1097/BRS.0b013e3181b7c675.

69. Sora MC, Jilavu R, Matusz P. Computer aided three-dimensional reconstruction and modeling of the pelvis, by using plastinated cross sections, as a powerful tool for morphological investigations. Surg Radiol Anat. 2012;34(8):731–6. https://doi.org/10.1007/s00276-011-0862-2.

70. Hammer N, Hirschfeld U, Strunz H, Werner M, Wolfskämpf T, Löffler S. Can the diagnostics of triangular fibrocartilage complex lesions be improved by MRI-based soft-tissue reconstruction? An imaging-based workup and case presentation. Biomed Res Int. 2017;2017:5870875. https://doi.org/10.1155/2017/5870875.

71. Hedderwick M, Stringer MD, McRedmond L, Meikle GR, Woodley SJ. The oblique popliteal ligament: an anatomic and MRI investigation. Surg Radiol Anat. 2017;39(9):1017–27. https://doi.org/10.1007/s00276-017-1838-7.

72. Basa RM, Podadera JM, Burland G, Johnson KA. High field magnetic resonance imaging anatomy of feline carpal ligaments is comparable to plastinated specimen anatomy. Vet Radiol Ultrasound. 2018;59(5):597–606. https://doi.org/10.1111/vru.12667.

73. Elliott JM, Cornwall J, Kennedy E, Abbott R, Crawford RJ. Towards defining muscular regions of interest from axial magnetic resonance imaging with anatomical cross-reference: part II - cervical spine musculature. BMC Musculoskelet Disord. 2018;19(1):171. https://doi.org/10.1186/s12891-018-2074-y.

74. Wiersbicki D, Völker A, Heyde CE, Steinke H. Ligamental compartments and their relation to the passing spinal nerves are detectable with MRI inside the lumbar neural foramina. Eur Spine J. 2019;28(8):1811–20. https://doi.org/10.1007/s00586-019-06024-y.

75. Singh A, Zwirner J, Templer F, Kieser D, Klima S, Hammer N. On the morphological relations of the Achilles tendon and plantar fascia via the calcaneus: a cadaveric study. Sci Rep. 2021;11(1):5986. https://doi.org/10.1038/s41598-021-85251-0.

76. Sha Y, Zhang SX, Liu ZJ, Tan LW, Wu XY, Wan YS, Deng JH, Tang ZS. Computerized 3D-reconstructions of the ligaments of the lateral aspect of ankle and subtalar joints. Surg Radiol Anat. 2001;23(2):111–4. https://doi.org/10.1007/s00276-001-0111-1.

77. Qiu MG, Zhang SX, Liu ZJ, Tan LW, Wang YS, Deng JH, Tang ZS. Plastination and computerized 3D reconstruction of the temporal bone. Clin Anat. 2003;16(4):300–3. https://doi.org/10.1002/ca.10076.

78. Xu Z, Lin G, Zhang H, Xu S, Zhang M. Three-dimensional architecture of the neurovascular and adipose zones of the upper and lower lumbar intervertebral foramina: an epoxy sheet plastination study. J Neurosurg Spine. 2020;10:1–11. https://doi.org/10.3171/2019.10.SPINE191164.

79. Ma C, Zhu X, Chu X, Xu L, Zhang W, Xu S, Liang L. Formation and fixation of the annulus of zinn and relation with extraocular muscles: a plastinated histologic study and its clinical significance. Invest Ophthalmol Vis Sci. 2022;63(12):16. https://doi.org/10.1167/iovs.63.12.16.

80. Shi K, Li Z, Wu X, Ma C, Zhu X, Xu L, Sun Z, Xu S, Liang L. The medial wall and medial compartment of the cavernous sinus: an anatomic study using plastinated histological sections. Neurosurg Rev. 2022;45(5):3381–91. https://doi.org/10.1007/s10143-022-01846-9.

81. Texeira D, Cheung A, Naveed H, Adds J, P. Three-dimensional reconstruction of the orbital retrobulbar vasculature. Orbit. 2022;41(4):469–75. https://doi.org/10.1080/01676830.2021.1955395.

82. Raoof A, Henry RW, Reed RB. Silicone plastination of biological tissue: room temperature technique—Dow/Corcoran technique and products. J Int Soc Plast. 2007;22:21–5.

83. Hermans JJ, Wentink N, Kleinrensink GJ, Beumer A. MR-plastination-arthrography: a new technique used to study the distal tibiofibular syndesmosis. Skelet Radiol. 2009;38:697–701. https://doi.org/10.1007/s00256-008-0631-4.

84. Sora MC, Matusz P. General considerations regarding the thin slice plastination technique. Clin Anat. 2010;23(6):734–6; author reply 737-8. https://doi.org/10.1002/ca.21013.

85. Adds PJ, Al-Rekabi A. 3-D reconstruction of the ethmoidal arteries of the medial orbital wall using BiodurVR E12. J Plast. 2014;26:5–10.

86. Richardson KC, Jarrett L, Finke EH. Embedding in epoxy resins for ultrathin sectioning in electron microscopy. Stain Technol. 1960;35:313–25. https://doi.org/10.3109/10520296009114754.

87. Laczko J, Levai G. A simple differential staining method for semi-thin sections of ossifying cartilage and bone tissues embedded in epoxy resin. Mikroskopie. 1975;31:1–4.

88. von Hagens G. Impregnation of soft biological specimens with thermosetting resins and elastomers. Anat Rec. 1979;194(2):247–55. https://doi.org/10.1002/ar.1091940206.

89. Brökelmann J, Prondzinsky R. Herstellung von Großschnitten plastinierter Gewebe mittels einer Diamantdrahtsäge. Verh Anat Ges. 1985;79:199–200. https://doi.org/10.1007/978-3-642-74784-7_193.

90. Fritsch H, Hötzinger H. Tomographical anatomy of the pelvis, visceral pelvic connective tissue, and its compartments. Clin Anat. 1995;8(1):17–24. https://doi.org/10.1002/ca.980080103.

91. Konschake M, Fritsch H. Anatomical mapping of the nasal muscles and application to cosmetic surgery. Clin Anat. 2014;27(8):1178–84. https://doi.org/10.1002/ca.22418.

92. Eckel HE, Sittel C, Walger M, Sprinzl G, Koebke J. Plastination: a new approach to morphological research and instruction with excised larynges. Ann Otol Rhinol Laryngol. 1993;102(9):660–5. https://doi.org/10.1177/000348949310200902.

93. Johnson G, Zhang M, Barnett R. A comparison between epoxy resin slices and histology sections in the study of spinal connective tissue structure. J Int Soc Plast. 2000;15(1):10–3.

94. Frierson H, Walker AN, Jackson RL, Powell. Technical communication: DNA ploidy analysis of plastinated tissue. J Int Soc Plast. 1988;2(2):13–6.

95. Grondin G, Grondin GG, Talbot BG. A study of criteria permitting the use of plastinated specimens for light and electron microscopy. Biotech Histochem. 1994;69(4):219–34. https://doi.org/10.3109/10520299409106291.

96. Francis DV, Rabi S. Deplastination: making plastinates histo-pathologically relevant. J Anat Soc India. 2018;67(1):77–9.

97. Ripani M, Boccia L, Cervone R, Macciucca DV. Light microscopy of plastinated tissue. Can plastinated organs be considered viable for structural observation. J Int Soc Plast. 1996;11:28–30.

98. Ramos M, Paula T, Zerlotini M, Silva V, Carazo L, Paula M, Silva FFR, Santana ML, Silva L, Ferreira LBC, Ramos M. A comparison of different deplastination methodologies for preparing histological sections of material plastinated with Biodur® S10/S3. J Plast. 2018;30:10–5. https://doi.org/10.56507/OHLF5315.

99. Baygeldi SB, Güzel BC, Şeker U. Colorimetric evaluation of cross-sectional silicone plastination of the Total head region of sheep and deplastination of the histological sections of brain tissue. Anat Histol Embryol. 2022;51(4):542–8. https://doi.org/10.1111/ahe.12827.

100. Rahul TG, Francis DV, Pandit S, Suganthy J. Deplastination: preservation of histological structures and its anticipated role in the field of histopathology. Clin Anat. 2020;33(1):108–12. https://doi.org/10.1002/ca.23477.

101. Ottone NE, Baptista CAC, Del Sol M, Muñoz OM. Extraction of DNA from plastinated tissues. Forensic Sci Int. 2020;309:110199. https://doi.org/10.1016/j.forsciint.2020.110199.

102. Ottone NE, Cirigliano V, Bianchi HF, Medan CD, Algieri RD, Borges Brum G, Fuentes R. New contributions to the development of a plastination technique at room temperature with silicone. Anat Sci Int. 2015;90(2):126–35. https://doi.org/10.1007/s12565-014-0258-6.
103. Nagaraj S, Nivargi S, Nanjappa L, Subbarayappa PKKK. A simple, cost-effective, fast and easy to isolate DNA protocol from human simple. Int J Biol Med Res. 2018;96:508–6510.
104. Bickley HC, Walker AN, Jackson RL, Donner RS. Preservation of pathology specimens by silicone plastination. An innovative adjunct to pathology education. Am J Clin Pathol. 1987;88(2):220–3. https://doi.org/10.1093/ajcp/88.2.220.
105. Janick L, DeNovo RC, Henry RW. Plastinated canine gastrointestinal tracts used to facilitate teaching of endoscopic technique and anatomy. Acta Anat. 1997;158(1):48–53. https://doi.org/10.1159/000147910.
106. Ottone NE, Cirigliano V, Lewicki M, Bianchi H, Aja-Guardiola S, Algieri RD, Cantin M, Fuentes R. Plastination technique in laboratory rats: an alternative resource for teaching, surgical training and research development. Int J Morphol. 2014;32(4):1430–5. https://doi.org/10.4067/S0717-95022014000400048.
107. McRae KE, Davies GA, Easteal RA, Smith GN. Creation of plastinated placentas as a novel teaching resource for medical education in obstetrics and gynaecology. Placenta. 2015;36(9):1045–51. https://doi.org/10.1016/j.placenta.2015.06.018.
108. Oostrom K, von Hagens G. Plastination of the human placenta. J Int Soc Plast. 1988;2(1):18–23.
109. Chang CW, Atkinson G, Gandhi N, Farrell ML, Labrash S, Smith AB, Norton NS, Matsui T, Lozanoff S. Cone beam computed tomography of plastinated hearts for instruction of radiological anatomy. Surg Radiol Anat. 2016;38(7):843–53. https://doi.org/10.1007/s00276-016-1645-6.

Chapter 9
Biosafety Issues Associated with Plastination

Plastination is a technique used to preserve biological specimens by replacing water and lipids with plastic, generating totally biosecure samples [1–5]. The resulting specimens can be used for educational and research purposes in various fields, including anatomy, pathology, and biology. However, the chemical products used during the plastination technique, and in some cases its residues, determine the need for safety and health protection approaches [1–16]. Safety in Plastination refers to the measures taken to ensure the health and safety of individuals involved in the process, as well as those who may come into contact with the preserved specimens [2, 5, 17–24].

Setting up a plastination laboratory requires careful planning and attention to safety protocols. Safety protocols and regulations, oriented toward the handling, transfer, and disposal of chemicals, as well as the start-up of laboratories and the establishment of safety measures, will be different according to current legislation, whether at a local, national, and international level, for which will be very important to be aware of and comply with these regulations. Here are some general steps that might be involved in the installation of a laboratory of plastination [17–22].

Before starting the installation, you need to decide what the laboratory will be used for. Will it be for research or educational purposes or both? What type of specimens will be prepared? These factors will affect the equipment and materials you'll need on the size of the laboratory (small, medium, or large).

The laboratory should be located in a suitable area with sufficient space and proper ventilation. It should also be located near educational or scientific institutions that can benefit from the use of the laboratory. The laboratory should be designed to maximize space efficiency and provide a safe and functional environment for researchers and students. Consider factors such as lighting, ventilation, plumbing, and electrical systems.

The first step is to determine the requirements for the laboratory, including the space needed for the equipment; the number of workstations, ventilation, and electrical requirements; and safety protocols.

© The Author(s), under exclusive license to Springer Nature Switzerland AG 2023

N. E. Ottone, *Advances in Plastination Techniques*, https://doi.org/10.1007/978-3-031-45701-2_9

The next step is to select the appropriate equipment for the laboratory, including a vacuum chamber, a freezer, a fume hood, a ventilation system, and other necessary tools and supplies. It is important to select equipment that meets safety standards and is compatible with the type of plastination being performed.

Before installing the equipment, the laboratory needs to be prepared to meet safety standards. This may include installing safety features such as emergency eye wash stations, fire extinguishers, and safety showers. The laboratory may also need to be equipped with protective clothing, such as gloves, aprons, and safety glasses.

Once the laboratory is prepared, the equipment can be installed. This includes setting up the vacuum chamber, the freezer, and any other necessary equipment. The ventilation system should be installed to ensure proper air flow and to minimize exposure to fumes and gases and equipped with filters that capture organic vapors to avoid environmental contamination specimens [17, 22–25].

It is essential to establish safety protocols for the laboratory, including procedures for handling and disposing of chemicals, proper use of equipment, and protocols for emergency situations [26]. All personnel working in the laboratory should receive appropriate safety training. The laboratory should establish and enforce strict safety procedures to ensure the health and well-being of laboratory staff and visitors. Plastination involves the use of chemicals and equipment that can be hazardous if not handled properly. It's important to implement safety procedures and provide appropriate personal protective equipment to ensure the safety of staff and students [17–22].

Before beginning work with biological specimens, it is important to test the equipment to ensure that it is functioning properly and meets safety standards. This may include testing the vacuum chamber, the freezer, and the ventilation system [17–22].

Once the laboratory is fully equipped and safety protocols are in place, the process of plastination can begin. Biological specimens are prepared and processed according to established protocols, and the resulting specimens are then cured in plastic to create a permanent, preserved specimen. Once the plastination process is complete, the specimens need to be stored in a safe and secure area. Depending on the laboratory's purpose, you may also need to set up display areas for educational purposes [17–22].

Overall, the installation of a laboratory of plastination requires careful planning and attention to safety protocols to ensure that the laboratory meets safety standards and produces high-quality specimens. It is essential to work with experienced professionals to establish and maintain a safe and effective laboratory [17–22].

Types of Hazards

One of the primary safety concerns in plastination protocols is exposure to the chemicals used in the process [17–22, 26]. Many of these chemicals, such as formaldehyde and acetone, can be hazardous to human health if not handled properly.

Since this process involves working with potentially hazardous chemicals and biological materials, safety is of utmost importance to ensure the health and well-being of those involved in the process. According to publications made on safe work in plastination, carried out by Prof. Volker Schill in 2018 [23] and 2019 [24], it is essential to recognize at least four types of dangers when implementing the plastination technique:

– Health hazards
– Physical hazards
– Biological hazards
– Environmental hazards

In relation to **health** and **biological hazards**, this refers to the health of the personnel in charge of the technique, as well as those who could collaborate in its development, from anatomical dissection, which would consist of embalming the samples and then proceed to their dissection. In all these processes, technical personnel, students, and academics may be affected, for example, by coming into contact with embalming fluids, which may cause situations of acute toxicity, sensitization, and long-term potential carcinogenic effects, as occurs with formaldehyde [27, 28] (see Chap. 3). Likewise, at the beginning of the plastination process, the dehydration stage can include a **health hazard**, due to contact with this chemical component without the appropriate safety measures. Therefore, appropriate personal protective equipment (PPE), such as gloves and respirators, should be worn during the process. All individuals involved in the plastination process must wear appropriate PPE, including gloves, eye protection, lab coats or gowns, and respirators, as needed [17–22]. PPE helps to protect against exposure to chemicals and biological materials. But it is also essential to have a proper ventilation system [17–22]. Plastination involves working with chemicals that can release vapors and fumes, which can be harmful if inhaled [26]. Thus, adequate ventilation is necessary to ensure that the air quality in the work area remains safe. Since acetone, ethanol, and other solvent vapors are heavier than air, it is recommended that the exhaust system be located close to ground level, no more than 20 cm from the ground. In addition, as Schill [23] recommends, when working with resins such as polyester or epoxy (especially the latter when used to perform vascular injections), a ventilation system with a movable arm is advisable to ensure localized vapor extraction.

In turn, this adequate ventilation system would be necessary to avoid other types of **hazards**, such as explosions. In this sense, effective and adequate removal of acetone vapor from technical environments will be crucial to avoid the formation of a dangerous mixture of air and vapor, which could lead to an explosion [17–22]. This is essential, especially when installing the refrigeration system necessary to develop the cold plastination technique, in which the vacuum chambers are installed inside freezers, and thus it becomes imperative to remove the compressor from the freezer and place it in a room next to the laboratory, extending its connections, in order to avoid a possible interaction between the acetone vapor and the on and off compressor, which could cause an explosion. This is not only avoided with an adequate ventilation system, but also with hermetic and totally safe connections that prevent the escape of acetone vapor, as well as vacuum chambers correctly installed and without leaks at the level of the safety glass that is used as a lid in these vacuum chambers [17–22]. In this way, and following what was indicated by Schill [23] in relation to **physical hazards**, the formation of the air-acetone vapor mixture cannot be completely avoided, especially during dehydration and degreasing stages that practically require to use open containers, and therefore at this stage the personal protective equipment (PPE) of the personnel involved will be crucial, in addition to the safety measures mentioned and associated with ventilation of the environment [17–22]. For this reason, in these cases, flammable liquids (acetone) will be handled in various circumstances: storage/disposal, transport or transfer between laboratories of the institution, its handling when filling/emptying containers, the moment of the introduction of the samples to start dehydration or defatting, as well as extracting them at the end of these processes or simply by taking acetone samples to control the concentration with an acetonometer [23]. In addition, it is very important to avoid the generation of sparks inside the plastination laboratory, for which the installation of a system that ensures the ground connection of all equipment is recommended, thus preventing electrostatic discharges, with their explosive potential [23, 24]. In this way, it will be crucial to comply with the regulations indicated above, regarding the handling of this flammable liquid, due to the probability of evaporation and the risk of explosion when coming into contact, in particular with heat sources, such as, in this case, the refrigerator compressors, being their transfer to an adjoining room the ideal solution to avoid this **physical hazard** [17–22].

Other situations associated with **physical hazards** during the development of the plastination process are associated with the use of tempered glass plates, which are applied as covers for the vacuum chambers, in this case there being the danger of breaking them with generation of injuries to users, especially when they are manipulated when opening and closing the vacuum chambers, as well as in situations of "homemade" construction of the vacuum chambers, with failures in the weather stripping on which the glass rests, and in this way, alterations can occur in the sealing system when generating vacuum, with a massive and sudden loss of vacuum due to leaks, which can cause the glass to be sucked into the vacuum chamber and explode [17–22]. There is also the possibility of suffering injuries when handling materials that are placed at extremely low temperatures, being able to suffer cryogenic burns [23], such as during the cold dehydration process or low-temperature plastination. In all these cases, personal protection is essential, using gloves, overalls, and facial protection (PPE).

Another safety consideration is the handling, storage, and disposal of the chemicals used in the plastination process. These chemicals can be harmful to the environment and must be handled in accordance with local regulations and guidelines [17–22].

All chemicals used in the plastination process should be stored and handled in accordance with their respective Safety Data Sheets (SDS). The SDS provides information on the potential hazards of the chemical, proper handling procedures, and emergency response protocols [26].

Proper disposal of waste generated during the plastination process is crucial to prevent environmental contamination and protect public health. Waste should be segregated and disposed of in accordance with applicable regulations and guidelines [17–22].

In addition, there are safety concerns associated with the use of power tools and other equipment used to prepare specimens for plastination. Proper training and supervision are necessary to ensure that these tools are used safely [17–22].

Plastination involves the use of **potentially hazardous chemicals**, such as formaldehyde, acetone, and other solvents. Exposure to these chemicals can cause respiratory problems, skin irritation, and other health problems for workers who handle them. In the case of formaldehyde [27, 28], for example, it is a known carcinogen and can cause respiratory problems if inhaled in large quantities. Similarly, acetone and other solvents used in the process can be harmful if ingested or absorbed through the skin. Therefore, it is important to follow strict safety protocols and use appropriate protective equipment when working with these chemicals. In case of an accident or exposure to hazardous materials, an emergency response plan should be in place. All individuals involved in the plastination process should be trained on the proper response procedures and emergency contact information [17–22].

Also there is the risk of explosion or fire during the process due to the use of flammable chemicals such as acetone. These chemicals are highly volatile and can ignite if exposed to an ignition source such as a spark or flame. In this way, the electrical installation (plugs, outlets, switches, and motors) must be explosion-proof or located outside of the laboratory [25].

There is the risk of contamination of the specimens if proper hygiene practices are not followed during the preparation process. Bacteria and fungi can grow on the specimens if they are not properly cleaned and disinfected, which can compromise the integrity of the plastination process and make the specimens unsuitable for use [17–22].

Overall, safety in plastination protocols requires a combination of appropriate PPE, safe handling and disposal of chemicals, and proper training and supervision of personnel. By following these protocols, the risks associated with the process can be minimized, and both workers and the environment can be protected. Safety in plastination protocols is essential to protect the health and well-being of those involved in the process and prevent environmental contamination [17–22] (Table 9.1).

Lab Set-up Brief Conclusions

When installing a plastination laboratory, it is essential to take into account all safety regulations, both for the transfer, handling, and storage of the chemicals used in the technique, as well as all safety and health protection measures of who implement the plastination technique [2, 5, 17–25, 27–31]. In this sense, it will be of special relevance to have specially adapted spaces, from the point of view of

Table 9.1 Dangers during the different stages of plastination [23, 24]

Dehydration/defatting
Storage of large quantities of acetone
Risk of explosion, in case of keeping the freezer compressor inside
Mixture of acetone vapor with air and explosive atmosphere (acetone flash point is -20 °C)
Do not place sources of ignition inside the room where the dehydration stage is carried out (cell phones or computers are prohibited inside these rooms, as they are common sources of dangerous ignition)
Connections, lights, all kinds of devices inside the plastination rooms, especially those that contain acetone, must be explosion-proof
Special care with security cabinets, stoves, Bunsen burners, static electricity
Acetone at -20 °C (dehydration): 15 mmHg vapor pressure. Acetone at $+20$ °C (defatting): 180 mmHg vapor pressure. The higher vapor pressure leads to higher evaporation. Methylene chloride at $+20$ °C (defatting): 352 mmHg vapor pressure. Very high level of evaporation. It is essential with this solvent to use protection against inhalation, due to its carcinogenic potential.

Forced impregnation
Silicones have no adverse effects, except for their impurities at certain high levels, which could cause reproductive toxicity.
The catalysts, which are used at 1% of the total volume of impregnation, can have acute and chronic effects on health, for which it is suggested to avoid contact with the skin and eyes
These components, having a low vapor pressure (less than 0.0750 mmHg), do not cause inhalation complications when working regularly.

Curing
For this process, a TEOS (tetraethyl orthosilicate) is used, which is vaporized inside a curing chamber
Despite its high flash point, it is a flammable liquid; therefore protection against explosions is required, suggesting its use under an extraction hood, which would also be useful to avoid respiratory irritation.
Likewise, because it causes an ocular increase, the use of facial protection is recommended.

electricity, for the development of each of the plastination steps, preferring to allocate separate rooms for each of the stages (dehydration/defatting, forced impregnation, and curing). Likewise, it will be essential to move and locate the refrigerated compressors as well as the vacuum pumps to separate rooms from those in which the vacuum chambers are located (both those intended for the room temperature process and the cold temperature) [17–22]. Regarding the electrical system, in case of requiring the presence of equipment inside these rooms, it will be necessary that they be explosion-proof equipment, as well as the need to have led lighting and an air extraction/ventilation system located at ground level (no more than 20 cm from it) due to the greater weight of solvent vapors compared to air [17–22]. Likewise, in the case of the curing stage, the use of air extraction hoods will be important, also being necessary, in case of developing the degreasing with methylene chloride. In all these cases, it will be imperative that the personnel use face protection masks, respirators (essential in the dehydration/degreasing and curing stages), aprons, and suitable gloves (PPE). From this, the possibility of suffering accidents or illnesses on the part of health personnel can be significantly reduced, which should be the number one priority when developing the plastination technique [17–22].

Some Characteristics of Acetone, Methylene Chloride, and Formaldehyde

Acetone [32]

Acetone is a colorless, volatile, and flammable organic solvent with a distinctive odor. Here are some of its key characteristics:

Solvent properties: Acetone has strong solvent properties and can dissolve many organic substances, including fats, oils, and resins.

Volatility: Acetone is highly volatile, meaning it evaporates quickly at room temperature.

Low boiling point: Acetone has a boiling point of 56°C, which makes it easy to evaporate and remove from the specimen.

Flammability: Acetone is flammable and must be handled with care. It has a flash point of -20°C, meaning it can easily ignite.

Acetone is used extensively in the process of plastination, which is a technique used to preserve biological specimens for study and teaching purposes. In this process, acetone is used to remove water and lipids from the specimen, which are replaced with a polymer material, such as silicone or epoxy. Acetone is used as a dehydrating agent because of its strong solvent properties, low viscosity, and high volatility, which allows it to quickly remove water from the tissue, which is a crucial step in the preservation process.

First Aid Measures

General notes: Take off contaminated clothing.

Following inhalation: Provide fresh air. In all cases of doubt, or when symptoms persist, seek medical advice.

Following skin contact: Rinse skin with water/shower.

Following eye contact: Irrigate copiously with clean, fresh water for at least 10 minutes, holding the eyelids apart. In case of eye irritation, consult an ophthalmologist.

Following ingestion: Rinse mouth. Call a doctor if you feel unwell.

Most important symptoms and effects, both acute and delayed: Irritation, nausea, vomiting, gastrointestinal complaints, headache, vertigo, dizziness, drowsiness, narcosis.

Firefighting Measures

Suitable extinguishing media: Coordinate firefighting measures to the fire surroundings. Water spray, alcohol resistant foam, dry extinguishing powder, BC-powder, carbon dioxide (CO_2).

Unsuitable extinguishing media: Water jet.

Special hazards arising from the substance or mixture: Combustible. In the case of insufficient ventilation and/or in use, may form flammable/explosive vapor air mixture. Solvent vapors are heavier than air and may spread along floors. Places which are not ventilated, e.g., unventilated below ground level areas such as trenches, conduits, and shafts, are particularly prone to the presence of flammable substances or mixtures. Vapors are heavier than air, spread along floors, and form explosive mixtures with air. Vapors may form explosive mixtures with air.

Hazardous combustion products: In the case of fire may be liberated: carbon monoxide (CO), carbon dioxide (CO_2).

Advice for firefighters: In the case of fire and/or explosion, do not breathe fumes. Fight fire with normal precautions from a reasonable distance. Wear self-contained breathing apparatus.

Accidental Release Measures

Personal precautions, protective equipment, and emergency procedures

For non-emergency personnel: Avoid contact with the skin, eyes, and clothes. Do not breathe vapor/spray. Avoidance of ignition.

Sources

Environmental precautions: Keep away from drains, surface, and groundwater. Danger of explosion.

Methods and material for containment and cleaning up. Advice on how to contain a spill: Covering of drains.

Advice on how to clean up a spill: Absorb with liquid-binding material (sand, diatomaceous earth, acid- or universal binding agents).

Other information relating to spills and releases: Place in appropriate containers for disposal. Ventilate affected area.

Handling and storage.

Precautions for safe handling: Provision of sufficient ventilation. When not in use, keep containers tightly closed.

Measures to prevent fire as well as aerosol and dust generation: Take precautionary measures against static discharge. Due to danger of explosion, prevent leakage of vapors into cellars, flues, and ditches.

Advice on general occupational hygiene: Wash hands before breaks and after work. Keep away from food, drink, and animal feeding stuffs. When using do not smoke.

Conditions for safe storage, including any incompatibilities: Store in a well-ventilated place. Keep container tightly closed.

Incompatible substances or mixtures: Observe hints for combined storage.

Consideration of other advice: Ground/bond container and receiving equipment.

Ventilation requirements: Use local and general ventilation.

Specific designs for storage rooms or vessels: Recommended storage temperature, 15–25 °C.

Other Physical and Chemical Properties

Physical state: liquid

Color: colorless

Odor: mild sweet—fruity

Melting point/freezing point: −94.8 °C (ECHA)

Boiling point or initial boiling point and boiling range: 56.05 °C (ECHA)

Flammability: flammable liquid in accordance with GHS criteria
Lower and upper explosion limit: 2.6 vol% (LEL)–12.8 vol% (UEL)
Flash point: -17 °C (ECHA)
Auto-ignition temperature: 465 °C (ECHA)
Decomposition temperature: not relevant
pH (value): 5–6 (in aqueous solution: 395 g/l, 20 °C)
Kinematic viscosity: 0.4051 mm^2/s at 20 °C
Dynamic viscosity: 0.32 mPa s at 20 °C
Solubility(ies): Water solubility miscible in any proportion

Methylene Chloride [33]

Methylene chloride (CH2Cl2) is a colorless liquid that can damage the eyes, skin, liver, and heart. Exposure may cause drowsiness, dizziness, numbness and tingling in the extremities, and nausea. It can cause cancer. Severe exposure can cause unconsciousness and death.

Potential acute health effects: Very hazardous in case of eye contact (irritant), of ingestion, of inhalation. Hazardous in case of skin contact (irritant, permeator). Inflammation of the eye is characterized by redness, watering, and itching.

Related to potential chronic health effects, specifically carcinogenic effects, methylene chloride is classified + (Proven) by OSHA and Classified 2B (Possible for human) by IARC. There is no information available associated with mutagenic and teratogenic effects.

The substance is toxic to the lungs, nervous system, liver, mucous membranes, and central nervous system (CNS). Repeated or prolonged exposure to the substance can produce target organ damage.

Flammability of the product: May be combustible at high temperature.

Auto-ignition temperature: 556 °C (1032.8 °F)

Flammable limits: Lower, 12 %; upper, 19%

Products of combustion: These products are carbon oxides (CO, CO_2), halogenated compounds.

In the case of small fire, use dry chemical powder. In the case of large fire, use water spray, fog, or foam. Do not use water jet.

Related to accidental release measures, in cases of small spill: Dilute with water and mop up, or absorb with an inert dry material, and place in an appropriate waste disposal container. In cases of large spill: Absorb with an inert material, and put the spilled material in an appropriate waste disposal. Be careful that the product is not present at a concentration level above TLV. Check TLV on the MSDS and with local authorities.

Personal Protection

Engineering controls: Provide exhaust ventilation or other engineering controls to keep the airborne concentrations of vapors below their respective threshold limit

value. Ensure that eyewash stations and safety showers are proximal to the worksta-
tion location.

Personal protection: Splash goggles. Lab coat. Vapor respirator. Be sure to use an
approved/certified respirator or equivalent. Gloves.

Personal protection in case of a large spill: Splash goggles. Full suit. Vapor respirator. Boots. Gloves. A self-contained breathing apparatus should be used to avoid
inhalation of the product. Suggested protective clothing might not be sufficient;
consult a specialist before handling this product.

Exposure limits: TWA, 50 from ACGIH (TLV) [United States]; TWA, 174 from
ACGIH (TLV) [United States]. Consult local authorities for acceptable exposure limits.

Some Physical and Chemical Properties
Physical state and appearance: Liquid
  Molecular weight: 84.93g/mole
  pH (1% soln/water): Not available.
  Boiling point: 39.75 °C (103.5°F)
  Melting point: −96.7 °C (−142.1 °F)
  Critical temperature: Not available
  Specific gravity: 1.3266 (water = 1)
  Vapor pressure: 46.5 kPa (at 20 °C)
  Vapor density: 2.93 (air = 1)

Formaldehyde [34]

Formaldehyde, also known as formalin, is a colorless gas with a pungent odor. Its
chemical formula is HCHO, and it is the simplest aldehyde. In plastination, formaldehyde is used as a fixative to preserve biological tissues by cross-linking their
proteins. The process involves removing water and lipids from the tissues and
replacing them with a polymer, such as silicone or epoxy resin. Formaldehyde is
used to fix the tissues in their natural state, preventing decay and degradation
[27, 28].

Some of the characteristics of formaldehyde include the following:

Highly reactive: Formaldehyde is a highly reactive molecule due to the presence
of the carbonyl group. It readily reacts with various biological molecules such as
proteins, nucleic acids, and lipids.

Water-soluble: Formaldehyde is soluble in water and forms a colorless solution
when dissolved in it.

Toxic: Formaldehyde is toxic and can cause irritation of the eyes, nose, and
throat. It can also cause respiratory problems and has been classified as a carcinogen
by the International Agency for Research on Cancer (IARC).

In cadaveric fixation, formaldehyde is commonly used as a fixative to preserve tissues and prevent decomposition [27, 28]. The application of formaldehyde in cadaveric fixation involves the following steps:

Immersion: The cadaver is immersed in a solution of formaldehyde to ensure that all the tissues are fixed.

Perfusion: In some cases, formaldehyde is also infused into the blood vessels of the cadaver to ensure that all the tissues are thoroughly fixed.

Cross-linking: Formaldehyde reacts with the proteins in the tissues, forming stable cross-links that prevent further decomposition.

Dehydration: After fixation, the tissues are dehydrated to remove the water and make them suitable for further processing.

The most common use of formaldehyde is in the preservation of biological tissues, including cadavers, which is a highly effective fixative for cadaveric preservation, although there are concerns about its toxicity and potential health risks for those who handle it. The process of cadaveric fixation involves replacing the water in the body tissues with a solution containing formaldehyde, preventing bacterial growth and tissue decay. Formaldehyde cross-links with the amino acids of proteins in the tissue, forming a complex network that stabilizes the tissue and prevents its breakdown. The use of formaldehyde in cadaveric fixation has been widely employed in anatomical and medical research and education for centuries.

References

1. von Hagens G. Impregnation of soft biological specimens with thermosetting resins and elastomers. Anat Rec. 1979;194(2):247–55. https://doi.org/10.1002/ar.1091940206.
2. von Hagens G. Heidelberg plastination folder. Collection of technical leaflets of plastination. Heidelberg: Biodur Products GmbH; 1986.
3. von Hagens G, Tiedemann K, Kriz W. The current potential of plastination. Anat Embryol. 1987;175(4):411–21. https://doi.org/10.1007/BF00309677.
4. Ottone NE, von Hagens G. Historical review and technical development. Rev Argent Anat Online. 2013;4(2):70–6. https://www.revista-anatomia.com.ar/archivos-parciales/2013-2-revista-argentina-de-anatomia-online-f.pdf.
5. Ottone NE, Cirigliano V, Bianchi HF, Medan CD, Algieri RD, Borges Brum G, Fuentes R. New contributions to the development of a plastination technique at room temperature with silicone. Anat Sci Int. 2015;90(2):126–35. https://doi.org/10.1007/s12565-014-0258-6.
6. Ottone NE. Micro-plastination. Technique for obtaining slices below 250 µm for the visualization of microanatomy in morphological and pathological experimental protocols. Int J Morphol. 2020;38(2):389–91. https://doi.org/10.4067/S0717-95022020000200389.
7. Ottone NE, Baptista CAC, Del Sol M, Muñoz OM. Extraction of DNA from plastinated tissues. Forensic Sci Int. 2020;309:110199. https://doi.org/10.1016/j.forsciint.2020.110199.
8. Ottone NE, Cirigliano V, Lewicki M, Bianchi HF, Aja Guardiola S, Algieri RD, Cantin M, Fuentes R. Plastination technique in laboratory rats: an alternative resource for teaching surgical training and research development. Int J Morphol. 2014;32(4):1430–5. https://doi.org/10.4067/S0717-95022014000400048.
9. Ottone NE, del Sol M, Fuentes R. Report on a sheet plastination technique using commercial epoxy resin. Int J Morphol. 2016;34(3):1039–43. https://doi.org/10.4067/S0717-95022016000300036.

10. Ottone NE, Guerrero M, Alarcón E, Navarro P. Statistical analysis of shrinkage levels of human brain slices preserved by sheet plastination technique with polyester resin. Int J Morphol. 2020;38(1):13–6. https://doi.org/10.4067/S0717-95022020000100013.

11. Ottone NE, Vargas CA, Veuthey C, del Sol M, Fuentes F. Epoxy sheet plastination on a rabbit head–new faster protocol with Biodur® E12/E1. Int J Morphol. 2018;36(2):441–6. https://doi.org/10.4067/S0717-95022018000200441.

12. Ottone NE, Baptista CAC, Latorre R, Bianchi HF, Del Sol M, Fuentes R. E12 sheet plastination: techniques and applications. Clin Anat. 2018;31(5):742–56. https://doi.org/10.1002/ca.23008.

13. Guerrero M, Vargas C, Alarcón E, del Sol M, Ottone NE. Development of a sheet plastination protocol with polyester resin applied to human brain slices. Int J Morphol. 2019;37(4):1557–63. https://doi.org/10.4067/S0717-95022019000401557.

14. Ottone NE, Guzmán D, Bianchi HF, del Sol M. Anatomical variations of radial and ulnar arteries in plastinated upper limbs. Int J Morphol. 2023;41(2):548–54. https://doi.org/10.4067/S0717-95022023000200548.

15. Toaquiza AB, Gómez C, Ottone NE, Revelo-Cueva M. Conservation of organs (heart, brain and kidney) of canine by cold-temperature silicone plastination done at an animal anatomy laboratory in Ecuador. Int J Morphol. 2023;41:1004–8.

16. Rodriguez-Torrez VH, Ottone NE. First plastination experience at 4,150 meters above sea level, in the height of La Paz, Bolivia. Int J Morphol. 2023;34:2.

17. Ottone NE. Lecture: "Biosecurity in the Plastination Laboratory". Plastination Day. Argentine Association of Anatomy. Pan American Association of Anatomical Techniques. Faculty of Veterinary Sciences, University of Buenos Aires, Argentina. September 23 and 24, 2015.

18. Ottone NE. Lecture: "Fundamentals and applications of the plastination technique". In: "Round table on anatomical techniques. Present and future in cadaveric fixation and conservation" – VII clinical anatomy teaching and research conference, I congress of the international forum of biomedical sciences – Organized by the Argentine Association of Clinical Anatomy. Online. November 10, 2022.

19. Ottone NE. Lecture: "Plastination techniques" – In: "Anatomical Techniques Seminar", organized by the Post Graduation Program in Domestic and Wild Animals Anatomy, Faculty of Veterinary Medicine and Zootechnics, Department of Surgery, University of São Paulo. Sao Paulo, Brazil. Online. March 7, 2022.

20. Ottone NE. Lecture: "Anatomical Techniques for the Laboratory" – II Ecuadorian Congress of Morphological Sciences and I Student Scientific Conference on Morphology, of the Ecuadorian Society of Morfunctional Sciences, sponsored by the Association of Ecuadorian Faculties of Medical and Health Sciences AFEME, the Scientific Association of Medical Students ASOCEM and the Pan-American Association of Anatomy. Quito, Ecuador. Online. March 24 to 28, 2021.

21. Ottone NE. Lecture: "Fundamentals and Applications of Plastination". XIX Pan American Congress of Anatomy - 21st Congress of Anatomy of the Southern Cone - 7th International Congress of Anatomy - 56th Argentine Congress of Anatomy - 40th Chilean Congress of Anatomy - 1st Ecuadorian Congress of Morphological Sciences - 17th Ibero-Latin American Symposium of Anatomical, Histological, Embryological Terminology - 7th Argentine Congress of Anatomical Techniques - 11th Argentine Conference on Anatomy for Health Sciences Students - I South American Meeting of the International Society for Plastination (ISP). Faculty of Medicine, University of Buenos Aires (UBA) - Buenos Aires, Argentina. May 27 to 31, 2019.

22. Ottone NE. Lecture: "Fundamentals of Plastination". Course of Advanced Anatomical Techniques. Doctorate in Morphological Sciences. Faculty of Medicine, University of La Frontera. Temuco, Chile. May 2, 2018.

23. Schill VK. General issues of safety in plastination. J Plastination. 2018;30(1):17. https://doi.org/10.56507/NJCY9228.

24. Schill VK. Work safety in plastination. Anat Histol Embryol. 2019;48(6):584–90. https://doi.org/10.1111/ahe.12473.
25. Popp AI, Lodovichi MV, Castillo DF, Casanave EB, Sidorkewicj NS. Adaptation and transformation of existing space into a plastination laboratory: experience at the Universidad Nacional del Sur, Argentina. J Plastination. 2022;34(2):22. https://doi.org/10.56507/RRFM4062.
26. United Nations. Globally harmonized system of classification and labelling of chemicals (GHS). 7th ed. New York: United Nations; 2017. https://www.unece.org/fileadmin/DAM/trans/danger/publi/ghs/ghs_rev07/English/ST_SG_AC10_30_Rev7e.pdf. Accessed 13 October 2017.
27. Adamović D, Čepić Z, Adamović S, Stošić M, Obrovski B, Morača S, Vojinović MM. Occupational exposure to formaldehyde and cancer risk assessment in an anatomy laboratory. Int J Environ Res Public Health. 2021;18(21):11198. https://doi.org/10.3390/ijerph182111198.
28. Oostrom K. Fixation of tissue for plastination: general principles. J Int Soc Plastination. 1987;1:3.
29. Sawad AA, Al-Asadi FS. Establishing a plastination laboratory at the College of Veterinary Medicine, University of Basra, Iraq. J Plastination. 2014;26(2):14. https://doi.org/10.56507/YMVO2353.
30. Zerlotini MF, Paula TAR, Ramos ML, Silva FFR, Santana ML, Figueira MP, Silva LC, Silva VHD, Bustamante LRC. Establishment and operationalization of a low-budget plastination laboratory in the Veterinary Morphology Section of the Federal University of Viçosa, Minas Gerais – Brazil. J Plastination. 2020;32(1):20. https://doi.org/10.56507/WPBG4492.
31. Gubbins RBG. Design of a plastination laboratory. J Int Soc Plastination. 1990;4:24–7.
32. Acetone ≥99.5%, for synthesis. Safety data sheet Safety data sheet. Web Site, Carl Roth, 2022. https://www.carlroth.com/medias/SDB-5025-AU-EN.pdf.
33. Methylene Chloride MSDS. Material safety data sheet. Web Site, ScienceLab.com. 2023. https://www.jscc.edu/about-jackson-state/administration/safety-and-security/sds-files/methylene-chloride.pdf.
34. Formaldehyde, 37% w/w. Safety Data Sheet. Web Site, Val Tech Diagnostics Inc., 2023. http://www.labchem.com/tools/msds/msds/VT310.pdf.

Chapter 10
Ethical Considerations in Plastination

Introduction

Through plastination, organic substances are preserved by means of plastic materials, which keep the cells and the relief of the surfaces unaltered up to the microscopic level, that is, identical to the natural state of the body before the process. The end result is dry, odorless human parts that can be positioned in such a way as to allow three-dimensional visualization of the body and its individual variations, previously restricted to the skeleton. During the elaboration process, while the silicone is incorporated, the pieces can be manipulated in various positions, so that the final result moves them away from the corpses and brings them closer to the spatial plasticity of the movements of living human and animal bodies [1–3].

Plastination is a process of preserving biological specimens by replacing water and fats with plastics, resulting in dry, odorless, and durable specimens that can be used for educational and scientific purposes. As with any practice involving biological materials, there are ethical considerations that must be taken into account when it comes to plastination.

One of the primary ethical concerns with plastination is the source of the specimens. Plastination requires the use of human or animal bodies, and there are concerns about the ethical implications of using cadavers that have not been voluntarily donated for scientific purposes. In anatomy in general, not only in plastination, the bodies have been obtained from unclaimed or unidentified individuals, which raises questions about the ethical implications of using such specimens. It is important that the bodies used for plastination have been obtained ethically, with informed consent from donors or their next of kin. This means that plastination facilities must have clear policies and procedures in place for obtaining bodies or body parts and must follow all relevant laws and regulations [4–6].

Another ethical issue surrounding plastination is the potential for the dehumanization of the bodies. When bodies are transformed into plastinated specimens, they

© The Author(s), under exclusive license to Springer Nature
Switzerland AG 2023
N. E. Ottone, *Advances in Plastination Techniques*,
https://doi.org/10.1007/978-3-031-45701-2_10

can become objects of study or curiosity rather than being seen as the remains of human beings who had families, friends, and lives. It is important to treat these specimens with respect and to remember that they were once living beings [4–6].

Ethics in plastination also involves the cultural and religious sensitivities of different communities and the educational and scientific value of the specimens.

Cultural and religious sensitivities must also be taken into account when using plastinated specimens for educational or public exhibition purposes. While such exhibits can be educational and informative, it is essential to ensure that they are presented in a respectful and dignified manner. Some communities may object to the display of human bodies or body parts or to the use of plastination techniques that alter the appearance of the body. It is important to respect these sensitivities and to ensure that any display or use of plastinated specimens is done in a culturally sensitive and appropriate way.

The educational and scientific value of plastination must also be considered. While plastination can be a valuable tool for teaching anatomy and physiology, it is important to ensure that its use is justified by the educational or scientific benefits it provides. Plastination facilities should also ensure that their specimens are used in ways that are consistent with ethical principles and best practices in anatomy education.

There are also concerns about the use of plastinated specimens in commercial settings, such as in exhibitions or for sale to collectors. In some cases it is believed that the use of plastinated specimens for commercial purposes is unethical and that it trivializes the human body.

It is also crucial to consider the long-term effects of plastination on the environment. The process involves the use of chemicals, which can have an adverse impact on the environment if not managed properly. It is important to ensure that proper disposal procedures are followed and that the chemicals used in the process do not harm the environment.

Overall, the ethical considerations surrounding plastination are complex and require careful consideration; plastination is a complex and multifaceted issue that requires careful consideration of a range of factors. It is important to ensure that the bodies used for plastination have been ethically sourced and that the specimens are treated with respect and used for legitimate scientific or educational purposes. By following ethical guidelines and best practices, plastination can be a valuable tool for educating and advancing our understanding of the human body while respecting the rights and dignity of donors and their families.

Body Donation: Procurement of Tissue—Consent and Privacy

One of the primary ethical concerns regarding plastination is the procurement of human and animal tissue. While most tissue used in plastination comes from voluntary donations, there have been cases of unethical procurement practices, such as unclaimed or unidentified individuals, that many laws worldwide authorize, but that do not follow the recommendations, for example, issued by the International Federation of Associations of Anatomists (IFAA) [7].These practices raise serious questions about the dignity and autonomy of the deceased and the responsibility of those who could profit from their bodies.

In addition to the ethical implications of these practices, there are also concerns about the quality and safety of the tissue used in plastination. The use of tissue from unregulated sources may pose health risks to those who handle the specimens, and there is a risk of contamination or mislabeling of specimens that can compromise their scientific value.

Another ethical issue surrounding plastination is the issue of consent and privacy. Donors who choose to donate their bodies for scientific and medical purposes may not have explicitly consented to their bodies being used for plastination or displayed publicly. Even if donors have given their consent, there are questions about whether they fully understood the implications of their decision, particularly with regard to the potential for their bodies to be used for commercial purposes or sensationalized displays [4–6].

Related with that, it is very important to record our support for the recommendations made by IFAA in relation to good practices for the donation and study of human bodies and tissues for anatomical examination [7].

IFAA established that institutions (university, medical school, laboratory, or anatomy department) depend on organ donations from the general public and are grateful to donors [7]. Cadaver studies are important in teaching, advanced training, and research in the medical and anatomical sciences. However, it is imperative that institutions follow procedures of the highest ethical standards so that donors have full confidence in their donation decisions. About that transparency is crucial, about the use of human resources and procedures by publicly trusted institutions increases public trust and, in turn, increases public support for organ donation.

Below we share the 11 recommendations that the IFAA makes related to donation and study of human bodies and tissues for anatomical examination [7]:

1. Informed consent from donors must be obtained in writing before any bequest can be accepted (this excludes the use of unclaimed bodies). Consent forms should take into account the following:

   (a) Donors must be entirely free in their decision to donate; this excludes donation by minors and prisoners condemned to death.
   (b) Although not essential, good practice is encouraged by having the next of kin also sign the form.
   (c) Whether the donor consents to their medical records being accessed.

2. There should be no commercialization in relation to bequests of human remains for anatomical education and research. This applies to the bequest process itself, where the decision to donate should be free from financial considerations, and also to the uses to which the remains are put following bequest. If bodies, body parts, or plastinated specimens are to be supplied to other institutions for educational or research purposes, this may not yield commercial gain. However,

charging for real costs incurred, including the cost of maintaining a body donation program and preparation and transport costs, is considered appropriate.
Payment for human material per se is not acceptable.

3. There needs to be an urgent move toward the establishment of guidelines regu-
   lating the transport of human bodies, or body parts, within and between
   countries.
4. Specimens must be treated with respect at all times. This includes, but is not
   limited to, storing and displaying human and non-human animal parts separately.
5. The normal practice is to retain donor anonymity. Any exceptions to this should
   be formally agreed to beforehand by the bequestee and, if appropriate,
   the family.
6. Limits need to be placed on the extent to which images or other artifacts produced from donations are placed in the public domain, including in social
   media, both to respect the privacy of the donor (and their surviving relatives)
   and to prevent arousing morbid curiosity. No individual should be identifiable
   in images.
7. A clear and rigorous legal framework should be established on a national and/
   or state level. This legal framework should detail:

   (a) The procedures to be followed in accepting bequests of human remains for
       anatomical examination, including who is responsible for human remains
       after death
   (b) The formal recognition of institutions which may accept bequests, which in
       some jurisdictions may involve licensing
   (c) The safe and secure storage of human remains within institutions
   (d) The length of time such remains will be retained by the institution
   (e) The procedures to be followed in disposing of remains once the anatomical
       examination is complete and they are no longer required for anatomical
       education and research

8. Institutional procedures should be formally established by an oversight committee, which shall review the body donation program at regular intervals. Such
   procedures should include the following:

   (a) Copies of the bequest should be retained both by the donor and by the insti-
       tution for whom the bequest is intended.
   (b) Records should be kept for a minimum of 20 years from the date of dis-
       posal to ensure that human material can be identified as originating from a
       specific donor.
   (c) Good conservation procedures should be employed throughout the entire
       period during which the human remains are retained to ensure that the most
       effective use is made of any bequest received.
   (d) Efficient tracking procedures should ensure that the identity and location of
       all body parts from an individual donor are known at all times.
   (e) Facilities where cadavers are used must be appropriate for the storage of
       human remains and secured from entry by unauthorized personnel.

9. There needs to be transparency between the institution and potential donors and their relatives at every stage, from the receipt of an initial enquiry to the final disposal of the remains. The clear communication of information should include but not necessarily be limited to the production of an information leaflet (hard copy and/or digital), which could also help publicize anatomical bequests and increase the supply of donors. This should set out the following:

   (a) The procedures relating to registering bequests, acceptance criteria, the procedures to be followed after death (including under what circumstances a bequest might be declined), and the procedures relating to disposal of the human remains. Sufficient grounds for rejection could include, but need not be limited to:

   > The physical condition of the body
   > The virological or microbiological status of the donor in life
   > The existence of other diseases (e.g., neurological pathology) that might expose staff or students handling the body to unacceptable risks
   > Body weight over a specified limit
   > The possible over-supply of donations at that institution at that time
   > Place of death outside the designated area from which bodies are obtained

   (b) The range of uses of donated bodies at that institution.
   (c) Possible costs, if any, that might be incurred by the bequestee's family in making a bequest and the costs to be met by the institution accepting the bequest.
   (d) Whether the donor's anonymity will be preserved and whether their medical history accessed.
   (e) Whether the body or body parts might be supplied to another institution.
   (f) The maximum length of time the body will be retained, including any legally sanctioned possibility of indefinite retention of body parts. The relatives of the donor should be given the option of being informed in due course of the date when the remains will be disposed of.
   (g) Donors should be strongly encouraged to discuss their intentions with their relatives to ensure that their relatives are familiar with their wishes and that as far as possible those wishes will be carried out after death.

10. Special lectures/tutorials in ethics relating to the bequest of human remains should be made available to all students studying anatomy. This is to encourage the development of appropriate sensitivities in relation to the conduct and respect that is expected of those handling human remains used for purposes of anatomical education and research.

11. Institutions should be encouraged to hold services of thanksgiving or commemoration for those who have donated their bodies for medical education and research, to which can be invited relatives of the deceased, along with staff and students.

In this sense, according to Bin et al. [8] at the center of the ethical debate on the displays of plastinated corpses is whether the donors were able to express free and fully informed consent for the use of their bodies for display purposes. If such consent has not been made absolutely and unequivocally clear, any further discussion or extrapolations regarding any other use of the human body is necessarily invalidated [9]. However, even in the face of the clear consent granted by the donors to be able to use their bodies in the Body Worlds exhibitions and recognizing the respect for individual freedom exercised in this act of altruistic and voluntary donation by the donor, debates associated with morality in the display of the human body could be raised. In this sense, as reported by Bin et al. [8], anatomists from different countries have reacted in various ways to this global phenomenon. The German Anatomical Society initially tried to prevent the show in 1997 [8]. In 2008, the American Association of Anatomists officially endorsed intense exposures on the fact that new ones had informed consent [10]. It has been bequeathed that the mission of the cadaver exhibits was based on health education with the aim of giving laymen the opportunity to better understand the human body and its functions. However, being configured to show the naturalness, individuality, and anatomical beauty of the human body, a purely artistic aspiration seems to be at stake, possibly only combined with scientific purposes [8]. For their part, the British Anatomy Association also expressed their concern that an exhibition of human bodies could emphasize a kind of sensationalism and trivialize cadaveric dissection as a mere spectacle, to the detriment of medical education [8]. That was related on the fact that in some countries, the display of bodies or body parts is generally and conveniently limited by law to certain educational areas and especially only those following certain educational programs [11]. And for this reason, in these countries, the exhibitions were organized in collaboration with educational institutions, which housed the exhibitions in their facilities, as well as providing academic and scientific support for the exhibition of plastinated human bodies. This gives ethical support to the possibility of making real plastinated human bodies available to the general population. An example consisted of the exhibition "Body Worlds Vital" organized at Universum, a Sciences Museum of National Autonomous University of Mexico [12] and which was a clear example of the conjunction between the exhibition and the academy in support of the need to provide society with knowledge of the anatomy of the human and animal body. This exhibition closed in 2014. On the other hand, in 2019 the first Body Worlds exhibition in the Middle East was held in Abu Dhabi, United Arab Emirates, with the support of Khalifa University [13], currently having a stable museum with an exhibition of plastinated specimens [14].

It is very important to consider the importance of thanking donors and their families for the generous and altruistic act of donation. In this sense, in 2022, 22 editors-in-chief (from 17 different countries) of the most important anatomy journals in the world published a joint scientific article, in which they established the need for all scientific articles that use donated human samples to incorporate a statement of appreciation and gratitude for the donation of bodies for science [15].

According to Berriz [16], in 2017, von Hagens considers himself a continuation of the tradition of anatomical artists from whose works a better understanding of human body structure was achieved. In this way, Leonardo da Vinci was one of such artists, who prepared parts of the human body to serve as models for his anatomical drawings, with great realism. Another example was André Vesalio, physician of Carlos V and Felipe II, who, in his work De fábrica humani corporis, describes the anatomy of the human body in a systematic way from public dissections carried out in his Anatomy Theatre, concept reworked by von Hagens in the form of a traveling exhibition [16]. Other important influences on the work of von Hagens have been the drawings of the eighteenth-century German anatomist Bernhard Siegfried Albinus, in which bodies are presented in an aesthetic and dynamic way, facilitating their educational potential [16].

According to Bianucci et al. [17], plastination, as a method of preserving human remains, has enabled public display of skinless stuffed bodies in a series of popular international exhibitions titled Body Worlds. Also according to Bianucci et al. [17], these spectacular exhibits are intended to be educational, democratizing the study of anatomy and freeing it from the traditional confines of professional medical study. However, Body Worlds has raised several ethical objections to its commercial purpose, the sourcing of some bodies, and the arrangement of bodies in poses or dissections that some viewers find offensive.

According to Kruse [18], and related to the first exhibition on Manheim, Germany (see Chap. 2), the exhibition exceeded the most optimistic expectations, having been visited by approximately 780,000 people during a period of 4 months, opening also the Museum at night. For those who visit the exhibition and do not have previous knowledge of Anatomy, these organs are perfectly understandable, without the need for extensive explanations, allowing them to obtain knowledge about organs, muscles, and bones. After the initial shock, revealed by the expressions and astonishment stamped on the faces of the men and women photographed visiting the exhibition, an artistic effect is produced, produced by the force of the imaginary of authenticity and postures.

Kruse [18] also considers that plastinated bodies have become paradoxical living dead, condemned to inhabit eternally on the other side of monitor screens and museum rooms, like "postmodern mummies."

Prof. Gunther von Hagens created Plastination in 1977, patented his technique in 1983, and began using it to create specimens for medical education and research. He also began to exhibit his plastinated specimens publicly, beginning with an exhibition in Japan in 1995. The exhibits, which feature whole bodies or body parts, such as organs or muscles, have generated controversy and criticism.

In this way, it is very important to consider the information about ethics that is made from "von Hagens Plastination" [19]:

– The donation of a body for Plastination to the Institute for Plastination directed by Prof. von Hagens is not a contract, but a declaration of intent. All anatomical specimens corresponded to individuals who donated their bodies with voluntary legal consent while alive, for the education of future generations. The donor declares while he is alive that his body, after death, will be donated and transferred to the Plastination Institute. From the same Institute of Plastination of

Prof. von Hagens, they inform that the consent form consists of a 30-page document in which all the information for the donor is provided in detail, where he is explicitly informed about the future use of his body, including the production of teaching specimens for sale. Likewise, from the Plastination Institute of Prof. von Hagens, the regular holding of meetings with body donors is reported.

In this sense, what is explained by Prof. von Hagens' Plastination Institute is what is done in all the University Institutions worldwide that have body donation programs for science and research and which basically consists of the declaration free and without any type of pressure (financial or social) on the part of the donor to provide his body in freedom to the determined institution for its use in practice and the teaching of anatomy. It is also very important to provide fully complete and detailed information on the future application that the donated person's body will have, so that the final decision to donate is made with all the information required to understand the importance of this practice in the future training of professionals in the health area, as well as the possibility of contributing to the knowledge of the general population regarding the importance of caring for one's own body to enhance the health of the community [19].

It is also important to take into account the restrictions established in relation to the commercialization of plastinated specimens, which von Hagens Plastination identifies [19]. Just as clarification, it is interesting to comment that this information is provided in this way since this company provides clear and precise information in relation to these topics associated with ethics and marketing, which we consider crucial to expand its information and also to consider its respect to the regulations and recommendations made by IFAA and that we previously shared in this chapter [7]. Therefore, in relation to commercialization, the "Qualifed Users" are defined as institutions or individuals using samples exclusively for educational and research purposes or for medical, diagnostic, and therapeutic education. Eligible users include universities, hospitals, schools, and museums or medical researchers, professors, lecturers, and others involved in medical and educational research projects [19]. This means that only plastinated specimens are sold to these eligible qualified users.

Despite the controversy, plastination has become an important tool in medical education and research. In the words of Prof. Gunther von Hagens [12], "The human body is the last remaining nature in a man-made environment," he says. "I hope the exhibitions will be places of enlightenment and contemplation, even philosophical and religious self-recognition, and open to interpretation regardless of the viewer's background and philosophy of life."

Conclusion

Our work as anatomists and scientists consists of making society aware of the great importance of body donation for research and teaching, making it possible to improve the training of future health professionals, increasing scientific research, oriented to a better development of medicine and health of the population.

The Medicine final aim is the person who suffers from illness, pain, and death. The practical activity in Anatomy constitutes for the health science student a determining opportunity to verify the structure of the human body and its relationship with illness and death. This learning method highlights the privilege of being the recipient of a disinterested will for the student to train as a doctor, a precious gift that materializes with the death of the donor. The apprentice thus sensitized will be able to assess the trust that his patient will later place in him so that he can be examined and trained as a doctor.

A solid anatomical knowledge, forged in practical work with real preparations, is decisive for the student to achieve a good performance in the clinical examination of their patients and in the practice of procedures efficiently and safely. Anatomical research and studies allow clinicians to transform ideas into innovations and advances in various branches of medicine.

These advances further highlight this generous gesture of men and women who donate themselves, a posthumous offering that will benefit future generations, contributing to the development of medical science.

References

1. von Hagens G. Impregnation of soft biological specimens with thermosetting resins and elastomers. Anat Rec. 1979;194(2):247–55. https://doi.org/10.1002/ar.1091940206.
2. von Hagens G. Heidelberg plastination folder. Collection of technical leaflets of plastination. Heidelberg: Biodur Products GmbH; 1986.
3. von Hagens G, Tiedemann K, Kriz W. The current potential of plastination. Anat Embryol. 1987;175(4):411–21. https://doi.org/10.1007/BF00309677.
4. Ottone NE. Lecture: "Donation of Human Bodies for Science, Teaching and Research in Chile" – In: Round Table "Body donation programs for teaching and research in the Region of the Americas. Its importance and legal, bioethical and administrative implications" – XX Pan American Congress of Anatomy, December 2, 2022. Quito, Ecuador.
5. Ottone NE. Extension Activity: "Seminar of Discussion and Dissemination on Donation of Human Bodies to Universities for Teaching and Research in Morphology". University of the Border. Extension Project without Financing - EXS18-0074. Directorate of Links with the Environment, Universidad de La Frontera, Temuco, Chile. July 6, 2018.
6. Ottone NE. Lecture: "The self-donation of corpses. Experience in Mexico and Chile". I International Course on Techniques for the Preparation and Conservation of Biological Material for the Teaching of Anatomy. Peruvian Society of Morphological Sciences, Federico Villarreal National University, "Hipólito Unanue" School of Medicine. Lima Peru. November 27, 2015.
7. International Federation of Associations of Anatomists (IFAA). Recommendations of good practice for the donation and study of human bodies and tissues for anatomical examination.

International Federation of Associations of Anatomists; 2023. https://ifaa.net/wp-content/uploads/2017/09/IFAA-guidelines-220811.pdf.

8. Bin P, Conti A, Buccelli C, Addeo G, Capasso E, Piras M. Plastination: ethical and medico-legal considerations. Open Med. 2016;11(1):584–6. https://doi.org/10.1515/med-2016-0095.

9. Jones DG. Anatomical investigations and their ethical dilemmas. Clin Anat. 2007;20(3):338–43. https://doi.org/10.1002/ca.20435.

10. Burr DB. Congress considering plastination import ban. Am Assoc Anat News. 2008;17:2–5.

11. Jones DG, Whitaker MI. Engaging with plastination and the Body Worlds phenomenon: a cultural and intellectual challenge for anatomists. Clin Anat. 2009;22(6):770–6. https://doi.org/10.1002/ca.20824.

12. Universum. Body Worlds Vital. Ciudad de México, Universidad Nacional Autónoma de México, 2014. https://www.universum.unam.mx/bodyworlds/en/vital/.

13. Khalifa University. Khalifa University - Body Worlds at Khalifa University for the first time in the Middle East. YouTube Channel, Khalifa University, 2020. https://www.youtube.com/watch?v=AqAG15ogWUU.

14. Body Museum. Web Site. Abu Dhabi, Khalifa University, 2023. https://bodymuseum.ku.ac.ae/.

15. Iwanaga J, Singh V, Takeda S, Ogeng'o J, Kim HJ, Moryś J, Ravi KS, Ribatti D, Trainor PA, Sañudo JR, Apaydin N, Sharma A, Smith HF, Walocha JA, Hegazy AMS, Duparc F, Paulsen F, Del Sol M, Adds P, Louryan S, Fazan VPS, Boddeti RK, Tubbs RS. Standardized statement for the ethical use of human cadaveric tissues in anatomy research papers: recommendations from Anatomical Journal Editors-in-Chief. Clin Anat. 2022;35(4):526–8. https://doi.org/10.1002/ca.23849. Epub 2022 Mar 11. PMID: 35218594.

16. Berriz BD. Plastination as a form of artistic expression or science. Infomed. 2017.

17. Bianucci R, Soldini M, Di Vella G, Verzé L, Day J. The body worlds exhibits and juvenile understandings of death: do we educate children to science or to voyeurism? Clin Ter. 2015;166(4):e264.

18. Kruse MHL. Anatomia: a ordem do corpo. Rev Bras Enferm. 2004;57(1):79–84.

19. von Hagens Plastination. Ethical Questions. Web Site, von Hagens Plastination, 2023. https://www.vonhagens-plastination.com/pages/medical-teaching-specimens/von-hagens-plastination.php/ethical-questions.

If you have any concerns about our products,
you can contact us on
ProductSafety@springernature.com

In case Publisher is established outside the EU,
the EU authorized representative is:
Springer Nature Customer Service Center GmbH
Europaplatz 3, 69115 Heidelberg, Germany

Printed by Libri Plureos GmbH
in Hamburg, Germany